W0037387

Ergebnisse der Mathematik und ihrer Grenzgebiete

Band 49

Herausgegeben von
P. R. Halmos · P. J. Hilton · R. Remmert · B. Szőkefalvi-Nagy

Unter Mitwirkung von
L. V. Ahlfors · R. Baer · F. L. Bauer · R. Courant · A. Dold
J. L. Doob · S. Eilenberg · M. Kneser · G. H. Müller · M. M. Postnikov
F. K. Schmidt · B. Segre · E. Sperner

Geschäftsführender Herausgeber: P. J. Hilton

C. A. Hayes · C. Y. Pauc

Derivation and Martingales

Springer-Verlag Berlin · Heidelberg · New York 1970

Professor Charles A. Hayes
University of California, Davis, California 95616

Professor Christian Y. Pauc
University of Nantes, F-44 Nantes

ISBN-13: 978-3-642-86182-6 e-ISBN-13: 978-3-642-86180-2
DOI: 10.1007/ 978-3-642-86180-2
 Library of Congress Catalog Card Number 78-88183.
Softcover reprint of the hardcover 1st edition 1970

Table of Contents

Introduction 1

Part I

Pointwise Derivation

Chapter I: Derivation Bases 5
1. Setting and general notation 5
2. DE POSSEL's derivation basis 6
3. Examples of bases 7
4. Pretopological notions 9
5. Comparison lemmas 12

Chapter II: Derivation Theorems for σ-additive Set Functions under Assumptions of the Vitali Type 13
1. The individual Vitali assumption 14
2. The individual full derivation theorem for Radon or μ-finite μ-integrals 19
3. The individual full derivation theorem for Radon measures 22
4. Class derivation theorems 23
5. Relation to YOUNOVITCH's derivation theorem 27
6. The strong Vitali property 27
7. Half-regular and regular branches of a derivation basis . . 29

Chapter III: The Converse Problem I: Covering Properties Deduced from Derivation Properties of σ-additive Set Functions 30
1. DE POSSEL's equivalence theorem 30
2. A necessary and sufficient condition for a weak derivation basis to derive a μ-finite $\mathscr{M}$-measure (Radon measure) ψ . 31
3. YOUNOVITCH's equivalence theorem 35
4. A converse theorem for bases deriving the $\mu^{(q)}$-functions, $q \geq 1$ 35

Chapter IV: Halo Assumptions in Derivation Theory. Converse Problem II 41
1. A. P. MORSE's halo properties 41
2. Abstract version of the strong Vitali theorem modelled after BANACH 42

3. Abstract version of the strong Vitali theorem modelled after CARATHÉODORY 51
4. Weak halo evanescence condition 59
5. Further criteria for the validity of the Density Theorem involving the weak halo 62
6. An individual derivability condition of BUSEMANN-FELLER type 65
7. The weak halo property in general bases 68
8. Product invariance of a weak halo property 73

Chapter V: The Interval Basis. The Theorem of JESSEN-MARCINKIEWICZ-ZYGMUND 78
1. The interval basis as a weak derivation basis 78
2. Theorem of JESSEN-MARCINKIEWICZ-ZYGMUND 83
3. Properties of the halo function as consequences of derivation properties 94
4. SAKS' counterexample 98
5. The parallelepipedon basis 104
6. SAKS' "rarity" theorem 107

Chapter VI: A. P. MORSE's Blankets 110
1. Nets 110
2. Hives 111
3. Fundamental covering theorems 112
4. Star blankets 114

Bibliography 120

Part II

Martingales and Cell Functions

Chapter I: Theory without an Intervening Measure 125
1. Additive functions 125
2. σ-additive functions 126
3. Premartingales, semi-martingales, and martingales 128
4. Ordered space of martingales of basis $(\mathscr{B}_\tau)$ 130
5. Integrals of premartingales 131
6. Martingales and additive functions 135
7. σ-additive martingales 136
8. Induced martingales 137
9. Premartingales and cell functions 138
10. Integrals of cell functions 141
11. Convergence theorems for martingales of bounded variation when $\mathscr{B}$ is a measure algebra 144

Chapter II: Theory in a Measure Space without Vitali Conditions 148
1. Preliminaries . . . 148
2. Absolutely continuous and singular premartingales . . . 150
3. Stochastic processes . . . 151
4. Stochastic convergence . . . 153
5. Mean convergence of order 1 . . . 161
6. Convergence in Orlicz spaces . . . 163
7. Cell functions . . . 164

Chapter III: Theory in a Measure Space with Vitali Conditions . 167
1. Preliminaries and definitions . . . 167
2. Vitali conditions . . . 168
3. Order convergence of martingales . . . 169
4. Necessity of the Vitali conditions . . . 170
5. Order convergence of submartingales . . . 170
6. Order convergence of cell functions . . . 170

Chapter IV: Applications . . . 172
1. Pointwise setting . . . 172
2. Specifically pointwise concepts and results. Convergence almost everywhere . . . 175
3. Martingales in the classical sense . . . 177
4. Product spaces . . . 177
5. The Radon-Nikodym integrand defined as a derivate . . . 177
6. Representation of the spaces L_x as spaces of cell functions 178
7. Pointwise derivation of cell functions . . . 178
8. Examples of concrete cell bases . . . 182
9. Stochastic bases on a group . . . 182

Bibliography . . . 182

Complements . . . 187
1°. Derivation of vector-valued integrals . . . 187
2°. Functional derivatives . . . 190
3°. Topologies generated by measures . . . 192
4°. Vitali's theorem for invariant measures . . . 194
5°. Global derivatives in locally compact topological groups . 195
6°. Submartingales with decreasing stochastic bases . . . 196
7°. Vector-valued martingales and derivation . . . 197
8°. A theorem of WARD for cell functions. A martingale convergence theorem of WARD's type . . . 198
9°. Derivation of measures . . . 198

Index . . . 201

Introduction

In Part I of this report the pointwise derivation of scalar set functions is investigated, first along the lines of R. DE POSSEL (abstract derivation basis) and A. P. MORSE (blankets); later certain concrete situations (e.g., the interval basis) are studied. The principal tool is a Vitali property, whose precise form depends on the derivation property studied.

The "halo" (defined at the beginning of Part I, Ch. IV) properties can serve to establish a Vitali property, or sometimes produce directly a derivation property. The main results established are the theorem of JESSEN-MARCINKIEWICZ-ZYGMUND (Part I, Ch. V) and the theorem of A. P. MORSE on the universal derivability of star blankets (Ch. VI).

In Part II, points are at first discarded; the setting is somatic. It opens by treating an increasing stochastic basis with directed index sets (Th. I. 3) on which premartingales, semimartingales and martingales are defined. Convergence theorems, due largely to K. KRICKEBERG, are obtained using various types of convergence: stochastic, in the mean, in L_p-spaces, in ORLICZ spaces, and according to the order relation. We may mention in particular Th. II. 4.7 on the stochastic convergence of a submartingale of bounded variation. To each theorem for martingales and semi-martingales there corresponds a theorem in the atomic case in the theory of cell (abstract interval) functions. The derivates concerned are global. Finally, in Ch. IV, points are re-introduced and results on pointwise convergence and on point derivates are deduced from results obtained in Chs. I, II, and III, under supplementary assumptions. Proofs are given for the main theorems, and generally for new results. In other cases, reference is given to the Bibliography.

The "Complements" consists of sketches of topics related to those of Parts I and II.

The writers have endeavored to follow rather faithfully certain conventions of notation, terminology, etc. Only in the Complements do we deviate deliberately from these conventions. There we use the original notation and meanings of the authors of the papers in question. We may mention some of our conventions for the benefit of the reader. We reserve the bold-face letters **N** and **R** for the set of natural integers (i.e., 1, 2, ...) and the set of real numbers, respectively. Axioms are shown in bold-face type in parentheses. Ordinary sets are denoted by italicized capital letters, sets of sets by script letters, and more complex systems, such as derivation bases, by Gothic letters. Sequences, whether MOORE-SMITH or FRÉCHET, are often shown in parentheses with a subscript index denoting a typical member of the directed set.

When we speak of an *increasing* or *decreasing* sequence, function, etc., we permit equality unless qualified by *strictly*.

By a *semi-integrable* function we mean one such that either its positive or its negative part has a finite integral; if both are finite, we say it is *integrable*.

We sometimes use δ to denote interchangeably a metric and also the diameters of sets taken with respect to the metric δ. We believe that the context makes clear the intended meaning in each case.

We use the term "iff" for "if and only if" since it has become an accepted word in the English language of mathematics.

We denote open and closed intervals on the real line by]a, b[and [a, b] respectively; half-open intervals are denoted in the obvious way using this type of notation. Intervals in n-dimensional space are denoted analogously.

We use bars over and under certain letters to denote measure-theoretic ideas at certain times, and to denote pretopological concepts at other times (cf. I. 1 and I. 4 of Part I). However, these notations occur only rarely in the latter sense and, in any case, we believe that the reader will have no difficulty to determine which situation is intended.

Parts I and II are essentially each self-contained. Each has its own Bibliography. External references are shown by italic numbers in parentheses corresponding to the appropriate Bibliography. In referring to a theorem, lemma, etc., we show only its decimal number if it occurs in the same chapter in which the reference arises; otherwise, its decimal number is preceded by the Roman numeral of the chapter in which it appears. Cross-references between Parts I and II carry the additional identifying Roman numeral of the Part in question.

The letter L is used to denote LEBESGUE or ORLICZ space as BANACH spaces, provided with a norm, whereas the corresponding semi-normed spaces are denoted by $\mathfrak{L}$ with a subscript or a superscript.

We wish to express our gratitude to Professor F. K. SCHMIDT who, in 1953, invited us to write a report on Derivation Theory to appear in the collection "Ergebnisse der Mathematik und ihrer Grenzgebiete", and maintained his offer despite a long delay in drafting and an expansion of the topics to "Derivation and Martingales". We appreciate very much his encouragement. We are also grateful for the assistance given us by Professor P. J. HILTON in technical matters related to the printing of the book.

PART I

Pointwise Derivation

Chapter I

Derivation Bases

1. Setting and general notation. Throughout this book, R denotes a set of points that is our universe, $\mathscr{S}$ denotes the Boolean σ-algebra (for the definition of Boolean σ-algebras and related terms see [*11*, 19–26] and [*10*, 1–27]) of all subsets of R, and $\emptyset$ is the empty subset of R. By *family of sets* we understand an indexed set of sets. For any two sets X and Y belonging to $\mathscr{S}$, $X \supset Y$ stands for the ordinary inclusion of Y in X, allowing the equality $X = Y$. We employ the lattice theoretical symbols $\bigcup$, $\bigcap$, $\cup$, $\cap$, and the algebraic symbols $+$, $-$, and $\cdot$ in STONE's sense; however, we generally use the latter only when STONE's and HAUSDORFF's (set-theoretical) meanings coincide.

$\mathscr{M}$ denotes henceforth a Boolean σ-algebra of subsets of R with R as unit, μ a σ-finite measure defined on $\mathscr{M}$, and $\mu^{\star}$ the completion of μ in $\mathscr{S}$, defined on $\mathscr{M}^{\star}$. Also, $\bar{\mu}$ and $\underline{\mu}$ denote the outer and inner measures derived from μ. We denote the family of μ-nullsets by $\mathscr{N}$; it is a σ-ideal in $\mathscr{M}$ (regarded as a Boolean σ-ring). $\mathscr{N}^{\star}$ denotes the family of $\mu^{\star}$-nullsets which is a σ-ideal in $\mathscr{S}$ (regarded as a Boolean σ-ring). $\mathscr{N}^{\star}$ is also the family of $\bar{\mu}$-nullsets.

We use the notation $X \supset Y \pmod{\mathscr{N}}$ to denote that $(Y - X \cdot Y) \in \mathscr{N}$; $X = Y \pmod{\mathscr{N}}$ means that STONE's difference $X - Y = (X - X \cdot Y) + (Y - X \cdot Y) \in \mathscr{N}$. Analogously, we define $X \supset Y \pmod{\mathscr{N}^{\star}}$ and $X = Y \pmod{\mathscr{N}^{\star}}$.

For any set $S \in \mathscr{S}$, a *μ-cover* (cf. [*10*, 68] for the definition of this expression) $\bar{S}$ of S is any $\mathscr{M}$-set for which $\bar{S} \supset S$ and $\bar{\mu}(S \cdot M) = \mu(\bar{S} \cdot M)$ for any $M \in \mathscr{M}$. Similarly, a *μ-kernel* $\underline{S}$ of S is any $\mathscr{M}$-set such that $\underline{S} \subset S$ and $\underline{\mu}(S \cdot M) = \mu(\underline{S} \cdot M)$ for any $M \in \mathscr{M}$.

Two sets S' and S'' are said to be *$\mu^{\star}$-entangled* iff they have positive outer measure and a common μ-cover.

We shall denote the set of those points $x \in R$ possessing a given property P by either $[P]$ or $E[x : P(x)]$, depending on the situation. In particular, if f and g are extended real-valued functions defined on a subset of R and α is a real number, then $[f > \alpha]$ denotes the set of points x such that $f(x) > \alpha$; similarly $[f < \alpha]$, $[f = \alpha]$, $[f > g]$, etc., are defined. If $A \subset R$ we sometimes write $f = g\ [A]$ to mean that f and g agree on A; similarly $f < g\ [A]$ means $f < g$ on A, etc.

2. De Possel's derivation basis [*39*, *28*]. We define a *prebasis* in the following manner. At the outset, we do not assume that a measure is given; we require only the Boolean σ-algebra $\mathcal{M}$. We assume that to each point x of a (fixed) subset E of R there correspond MOORE-SMITH sequences (families filtering to the right) of non-empty $\mathcal{M}$-sets, called *constituents*, which are said to *converge* to x, and are denoted generically by $(M_\iota(x))$, where ι is a typical member of the directed set and $<$ or $>$ denotes the ordering relation. Additionally, we assume DE POSSEL'S heredity (or FRÉCHET'S convergence) axiom, namely: each cofinal subsequence of an x-converging sequence itself converges to x. The family of all the sequences $(M_\iota(x))$ is our prebasis $\mathfrak{B}$. Thus the elements of $\mathfrak{B}$ are converging sequences together with their convergence points. This definition does not exclude the possibility that two or more points correspond to the same sequence. We denote by $\mathcal{D}$ the set of all sets occurring in the sequences $(M_\iota(x))$ for all $x \in E$. An x-converging sequence is called *x-contracting* iff there exists an index $\iota^\star$ such that $x \in M_\iota$ whenever $\iota > \iota^\star$.

If $\mathcal{M}$ is provided with a (non-negative) measure μ and if the sets of $\mathcal{D}$ are of finite measure, then $\mathfrak{B}$ is a *derivation basis.* If λ is a numerical function defined on the $\mathcal{D}$-sets, then we adopt the convention that $\lambda(M)/\mu(M)$ equals $+\infty, 0$, or $-\infty$ if $\mu(M) = 0$ and $\lambda(M) > 0$, $\lambda(M) = 0$, or $\lambda(M) < 0$, respectively.

We define, for each $x \in E$,

$$D^\star\lambda(x) = \sup\left[\limsup \frac{\lambda(M_\iota(x))}{\mu(M_\iota(x))}\right],$$

where the expression in brackets denotes the limit superior for any one x-converging sequence $(M_\iota(x))$ and the supremum is taken among all sequences converging to x. Evidently $-\infty \leqslant D^\star\lambda(x) \leqslant +\infty$; the infinite values are not excluded.

In exactly similar fashion we define

$$D_\star\lambda(x) = \inf\left[\liminf \frac{\lambda(M_\iota(x))}{\mu(M_\iota(x))}\right].$$

We call $D^\star\lambda(x)$ and $D_\star\lambda(x)$ the *upper* and *lower* $\mathfrak{B}$*-derivates at* x, respectively. If $D^\star\lambda(x) = D_\star\lambda(x)$ (finite or infinite), we say that the $\mathfrak{B}$*-derivative* $D\lambda(x) = D^\star\lambda(x) = D_\star\lambda(x)$ *exists at* x, or that λ is $\mathfrak{B}$*-derivable at* x. In case the sequences $(M_\iota(x))$ are subsequences of one universal sequence [*29*, *30*] then we can drop the prefixes "sup" and "inf" in the expressions for $D^\star\lambda(x)$ and $D_\star\lambda(x)$. In a derivation basis, the x-converging sequences are also called *deriving sequences at* x. In the sequel, we shall always assume that the constituents have not only finite, but also positive, μ measure.

If λ is a vector function defined on $\mathcal{D}$ into a topological vector space, then $D\lambda(x)$ is said to *exist at* x or λ to be *derivable at* x iff, for any x-converging sequence $(M_\iota(x))$, $\lim_\iota \lambda(M_\iota(x))/\mu(M_i(x))$ exists and equals $D\lambda(x)$.

We state another property which is needed sometimes; when it is required, it will be specifically included as part of the hypotheses.

Axiom (E): The converging sequences are contracting sequences.

3. Examples of bases.

3.1. Linear bases. $E = R = \mathbf{R}$ (the set of real numbers), μ is linear Borel measure, and the MOORE-SMITH sequences are ordinary sequences.

For each $x \in \mathbf{R}$, the x-convergent sequences are sequences of closed intervals $I_i = [\alpha_i, \beta_i]$, $\beta_i > \alpha_i$, satisfying one of the following requirements:

$(\mathfrak{B}_{1'}^1)$ x is the center of $[\alpha_i, \beta_i]$, $\lim_i \alpha_i = \lim_i \beta_i = x$;

$(\mathfrak{B}_{1''}^1)$ α_i or β_i coincide with x, $\lim_i \alpha_i = \lim_i \beta_i = x$;

$(\mathfrak{B}_2^1)$ $\alpha_i \leqslant x \leqslant \beta_i$, $\lim_i \alpha_i = \lim_i \beta_i = x$;

$(\mathfrak{B}_3^1)$ $\lim_i \alpha_i = \lim_i \beta_i = x$.

An important basis whose constituents are open intervals $]\alpha_i, \beta_i[$ is given by

$(\mathfrak{B}_4^1)$ $\alpha_i < x < \beta_i$, $\lim_i \alpha_i = \lim_i \beta_i = x$.

3.2. Plane bases. $E = R = \mathbf{R}^2$ (the cartesian plane), μ is plane Borel measure, and the MOORE-SMITH sequences are ordinary sequences. For each x in $\mathbf{R}^2$, the x-convergent sequences are defined as follows:

$(\mathfrak{B}_{1'}^2)$ Sequences of closed squares Q_i with sides parallel to the axes, with x as center and $\lim_i \mu(Q_i) = 0$;

$(\mathfrak{B}_{1''}^2)$ Sequences of closed circular discs K with x as center and radii tending to zero;

$(\mathfrak{B}_2^2)$ Sequences of closed intervals (I_i) (i.e., rectangles whose sides are parallel to the axes) containing x, and sides whose lengths tend to zero.

$(\mathfrak{B}_3^2)$ Sequences of open intervals (J_i) containing x, and sides whose lengths tend to zero.

3.3. A. P. Morse's blankets [*28*, *29*]. The bases in (3.1) and (3.2) are special instances of the A. P. MORSE bases, which we now define.

R is a metric space, $\delta(X)$ denotes the diameter of any subset X of R. To each point x of a given subset E of R there corresponds a set $\mathfrak{F}(x)$ of subsets of R such that $\inf \delta(M + \{x\}) = 0$, where $M \in \mathfrak{F}(x)$ and the infimum is taken over all sets in $\mathfrak{F}(x)$. $\mathfrak{F}$ is called a *blanket*. The x-convergent sequences are the (cofinal) subsequences of $\mathfrak{F}(x)$, which is a universal sequence. A blanket can thus be envisaged as a prebasis. In all blankets studied by A. P. MORSE the sets of $\mathscr{D} = \bigcup_{x \in E} \mathfrak{F}(x)$ are bounded and Borelian. R is provided with a Carathéodory measure function (outer measure) φ finite on bounded sets, satisfying the conditions (C 1), (C 2), and (C 3) of [*44*, 43]. Thus $\mathfrak{F}$ becomes a derivation basis if we take for μ the measure induced by φ, which takes finite values on the sets of $\mathscr{D}$.

An essential feature of A. P. MORSE's blankets is the presence of a metric in R and the definition of the convergence process by means of this metric.

3.4. Generally, any prebasis in which the process of convergence is defined by means of a metric or a topology will be called *metrical* or *topological*, respectively. In the course of the x-convergence process, the constituents are included in *metrically* or *topologically small* neighborhoods of x, respectively.

An example of the counterpart to such bases is the *strip basis* $\mathfrak{S}$. For this basis, $E = R$ is the unit square $[0,0;1,1]$, μ is plane Borel measure, and the x-converging sequences are ordinary sequences (M_i) of intervals $[\alpha_i,0;\beta_i,1]$ with $\alpha_i < \beta_i, \alpha_i \leqslant$ (abscissa of x) $\leqslant \beta_i$, $\lim_i \alpha_i = \lim_i \beta_i =$ abscissa of x. The x-contracting intervals become small in measure but not metrically small.

3.5. D-bases [*38*]. The bases $\mathfrak{B}_2^1$, $\mathfrak{B}_4^1$, $\mathfrak{B}_2^2$, $\mathfrak{B}_3^2$, and $\mathfrak{S}$ all belong to the class of D-bases, defined as follows: R is the unit of a Boolean σ-algebra $\mathscr{M}$, provided with a σ-finite measure μ, $\mathscr{U}$ is a set of $\mathscr{M}$-sets of finite, positive μ-measure, δ is a positive finite numerical function defined on $\mathscr{U}$. An x-converging (or x-contracting) sequence is any ordinary sequence of $\mathscr{U}$-sets (M_i) such that $x \in M_i$, $i = 1,2,\ldots$, and $\lim_i \delta(M_i) = 0$. Such a basis we term a D-*basis* [*8*, *18*], and we denote it by $[\mathscr{U},\delta]$. We regard the function δ as a *numerical index of contraction.* In $\mathfrak{B}_2^1$, $\mathfrak{B}_4^1$, $\mathfrak{B}_2^2$, and $\mathfrak{B}_3^2$ we can take for δ the diameter, and in $\mathfrak{S}$ the μ-measure, of a set. Once δ is chosen, a D-basis is defined only by the set of its constituents. Therefore, every subset $\mathscr{T}$ of $\mathscr{U}$ determines a D-subbasis $[\mathscr{T},\delta]$, the general D-basis subbasis of $[\mathscr{U},\delta]$. Its domain $D[\mathscr{T},\delta]$ (abbreviated $D(\mathscr{T})$) is no longer an arbitrary subset of $D(\mathscr{U})$, as is the case with a general subbasis. For instance, if $\mathfrak{B}$ is an $S^{(1)}$-basis (cf. II. 4.5 for definition), then the domain of any D-subbasis of $\mathfrak{B}$ is a $\mu^\star$-measurable set. A $\mathfrak{B}$-fine covering $\mathscr{V}(\text{mod}\,\mathscr{N}^\star)$ of X is characterized as a subfamily of $\mathscr{U}$ with $D(\mathscr{V}) \supset X(mod\,\mathscr{N}^\star)$ (cf. I. 4.1 for definitions).

3.6. Busemann-Feller bases [6]. Here R is euclidean n-space, μ is n-dimensional Borel measure, δ is set diameter, $\mathcal{U}$ is a set of bounded open sets, and the x-converging sequences are defined as in (3.5). Such a D-basis is called a Busemann-Feller basis when, with any set M in $\mathcal{U}$, $\mathcal{U}$ contains all sets homothetical to M.

3.7. Dieudonné bases. [9]. Here $I = [0,1]$; letting $I_n = I$ for $n = 1,2,\ldots$, we define $P = \prod_{n \in N} I_n = I^N$, where N is the set of positive integers. Also, if J and J' denote sets complementary in $\mathbf{N}$, i.e., $J \cup J' = N$, $J \cap J' = \emptyset$, then P may be identified with the product $I^J \times I^{J'}$. Hence, for $x = (x_n)$ in P, if x_J and $x_{J'}$ denote the projections of x, we have $x = (x_J, x_{J'})$, $V_{n,J}$ is the product of the cube of center x_J and sides of length $1/n$ with the set $I^{J'}$. The universal deriving sequence is $(V_{n,J})$, where $<$ is so defined that $(n_1,J_1) < (n_2,J_2)$ iff $n_1 \leqslant n_2$ and $J_1 \subset J_2$.

3.8. Ultrafilter bases [2]. In this situation, $\mathcal{A}$ denotes a Boolean algebra with or without unit, $\mathcal{U}$ is an ultrafilter of $\mathcal{A}$, u denotes the ultrafilter $\mathcal{U}$ regarded as a point. To each a in $\mathcal{A}$ there corresponds the set $P(a)$ consisting of those points u such that $a \in \mathcal{U}$. The family of sets $P(a)$ is the pointwise image of $\mathcal{A}$. The u-converging sequences are obtained by means of the universal sequence which is the P-image of the ultrafilter.

4. Pretopological notions [25, 38]. We shall define, by means of a prebasis, certain concepts, several of them resembling those introduced at the beginning of elementary topology. The pretopology will be a sort of substitute for the missing topology.

4.1. Concepts not depending on $\mathcal{N}$ or $\mathcal{N}^\star$.

Any subset $\mathfrak{B}^\star$ of $\mathfrak{B}$ that contains all the subsequences of any of its sequences and associates with them the corresponding convergence points is called a *subbasis of* $\mathfrak{B}$.

The *spread* of a subbasis $\mathfrak{B}^\star$ of $\mathfrak{B}$ is the set of $\mathfrak{B}$-constituents occurring in the $\mathfrak{B}^\star$-sequences and is denoted by $\mathcal{D}^\star = \mathcal{D}(\mathfrak{B}^\star)$; the *domain* of $\mathfrak{B}^\star$ is the set of those points that are the convergence points of at least one $\mathfrak{B}^\star$-sequence and is denoted by $D(\mathfrak{B}^\star)$.

If $X \subset R$, then a subbasis $\mathfrak{B}^\star$ of $\mathfrak{B}$ is said to be a *$\mathfrak{B}$-fine covering of* X iff $D(\mathfrak{B}^\star) \supset X$. A $\mathfrak{B}$-fine covering $\mathcal{V}$ of X may also be defined equivalently as a set of constituents containing, for each x in X, the sets of at least one sequence $(M_\iota(x))$.

The importance of this last notion for the theory of derivation results from the fact that if $X \subset [D^\star\lambda > \alpha]$, then the set of these constituents satisfying the inequality $\lambda(M) > \alpha\mu(M)$ is a $\mathfrak{B}$-fine covering of X; the same is true if $D^\star\lambda > \alpha$ is replaced by $D_\star\lambda < \alpha$ and $\lambda(M) > \alpha\mu(M)$ by $\lambda(M) < \alpha\mu(M)$.

The second or alternative definition of a $\mathfrak{B}$-fine covering is sometimes more than satisfied in the following sense: for each $x \in X$, every x-convergent sequence admits of a subsequence whose members all belong to $\mathscr{V}$. When this conditions holds, we say that $\mathscr{V}$ is a *full $\mathfrak{B}$-fine covering of X*. It is not difficult to confirm that the requirement just expressed is equivalent to the condition that for each $x \in X$ and every x-converging sequence (M_ι) there exists an index ι' such that $M_\iota \in \mathscr{V}$ whenever $\iota > \iota'$.

It is easily seen that the intersection of two full $\mathfrak{B}$-fine coverings of X is again a full $\mathfrak{B}$-fine covering of X, and that the intersection of a $\mathfrak{B}$-fine covering of X and a full $\mathfrak{B}$-fine covering of X is a $\mathfrak{B}$-fine covering of X.

Using the same notation as above, it is clear that the set of those constituents satisfying $\lambda(M) > \alpha\mu(M)$ is a full $\mathfrak{B}$-fine covering of any set $X \subset [D_\star \lambda > \alpha]$. The same is true if $\lambda(M) > \alpha\mu(M)$ is replaced by $\lambda(M) < \alpha\mu(M)$ and $D_\star \lambda > \alpha$ by $D^\star \lambda < \alpha$.

A point x is termed *totally interior* (with respect to $\mathfrak{B}$) to a subset X of R if $x \in E$ and for each x-converging sequence (M_ι), there exists an index ι' such that $M_\iota \subset X$ whenever $\iota > \iota'$. The set of those points that are totally interior to X is called the *T-interior* of X, and is denoted by $I(X)$. The set $I(X)$ need not be a subset of X. In the case of a blanket, $I(X)$ is the set $F \odot X$, in A. P. Morse's notation.

A point x is called *partially interior* (with respect to $\mathfrak{B}$) to X iff at least one x-converging sequence exists, the constituents of which are included in X. The set of those points that are partially interior to X is called the *P-interior of* X and is denoted by $\underline{X}$.

As an example, consider $\mathfrak{B} = \mathfrak{B}_2^2$; let X be a closed interval. Then $I(X)$ is the topological interior of X and $\underline{X} = X$. If Y is a closed disc, then $I(Y)$ is the topological interior of Y and Y is equal to Y minus the four vertices (i.e., points at which the circumference of the disc has a horizontal or vertical tangent).

We define the *T-hull* of $X \subset R$ as the set of those points x such that there exists at least one x-converging sequence, the constituents of which all have a non-empty intersection with X. We denote this set by $F(X)$.

The set of those points $x \in E$, for which every x-converging sequence possesses at least one subsequence, all of whose constituents have a non-empty intersection with X, is called the *P-hull* of X and is denoted by $\bar{X}$.

$I(X)$, $\underline{X}$, $F(X)$, and $\bar{X}$ are all subsets of E. The following relations evidently hold: $I(X) \subset \underline{X}$; $F(\bar{X}) \supset \bar{X}$; $F(\emptyset) = \emptyset$; $I(X \cdot Y) = I(X) \cdot I(Y)$; $F(X \cup Y) = F(X) \cup F(Y)$; $\underline{X \cdot Y} \subset \underline{X} \cdot \underline{Y}$; $\overline{X \cup Y} = \bar{X} \cup \bar{Y}$; $E - F(X) = I(R - X)$; $E - I(X) = F(R - X)$; $E - \bar{X} = \underline{R - X}$; $E - \underline{X} = \overline{R - X}$. In case Axiom (**E**) holds, then $I(X) \subset X$, $\underline{X} \subset X$, $F(X) \supset X$, and $\bar{X} \supset X$.

The operator F satisfies the KURATOWSKI hull axioms $(\mathbf{H}_0)$, $(\mathbf{H}_3)$, and $(\mathbf{H}_4)$ (cf. [*32*, 41–42]) for a topology; the operator $^{-}$ satisfies $\mathbf{H}_0$ and $\dot{\mathbf{U}}_4$; both satisfy $\dot{\mathbf{U}}_1$ if Axiom (**E**) holds. Each defines a pretopology that we call *T*- and *P-pretopology*, respectively [*38*, 72–74].

The cleavage between *T*- and *P*-concepts is due to the absence of an intersection axiom for the constituents of an x-convergent sequence. In the following, we shall have to do with the *T*-concepts.

A set $X \subset R$ is termed:

an *external* D-*open* set iff $E \cdot X \subset I(X)$;

an *external* D-*closed* set iff $E \cdot X \supset F(X)$;

an *internal* D-*open* set iff it is the intersection with E of an external D-open set;

an *internal* D-*closed* set iff it is the intersection with E of an external D-closed set.

Sometimes we call these concepts *strict* because they involve neither $\mathcal{N}$- nor $\mathcal{N}^{\star}$-sets.

4.2. Concepts depending on $\mathcal{N}$ or $\mathcal{N}^{\star}$.

We define a $\mathfrak{B}$-*fine covering* of $X \subset R$ (mod $\mathcal{N}$) or (mod $\mathcal{N}^{\star}$) as the spread of a subbasis $\mathfrak{B}^{\star}$ with $D(\mathfrak{B}^{\star}) \supset X$ (mod $\mathcal{N}$) or (mod $\mathcal{N}^{\star}$), respectively.

We obtain the definition of a *full* $\mathfrak{B}$-*fine covering of* X (mod $\mathcal{N}$) or (mod $\mathcal{N}^{\star}$) by replacing, in either of the two definitions of a full $\mathfrak{B}$-fine covering of X, the expression "for each $x \in X$" by "for μ-almost all x" or "for $\mu^{\star}$-almost all x", respectively.

An *external* D-*open set* (mod $\mathcal{N}$) or (mod $\mathcal{N}^{\star}$) is defined as a set X such that $E \cdot X \subset I(X)$ (mod $\mathcal{N}$) or (mod $\mathcal{N}^{\star}$), respectively; an *internal* D-*open set* (mod $\mathcal{N}$) or (mod $\mathcal{N}^{\star}$) is the intersection of an external D-open set (mod $\mathcal{N}$) or (mod $\mathcal{N}^{\star}$), respectively, with E. We use G as a generic notation for the external D-open sets (mod $\mathcal{N}^{\star}$), O for the internal D-open sets (mod $\mathcal{N}^{\star}$); $\mathcal{G}$ denotes the family of the sets G, and $\mathcal{O}$ denotes the family of the sets O.

An *external* D-*closed set* (mod $\mathcal{N}$) or (mod $\mathcal{N}^{\star}$) is defined as a set X such that $E \cdot X \supset F(X)$ (mod $\mathcal{N}$) or (mod $\mathcal{N}^{\star}$), respectively; an *internal* D-*closed set* (mod $\mathcal{N}$) or (mod $\mathcal{N}^{\star}$) is the intersection of an external D-closed set (mod $\mathcal{N}$) or (mod $\mathcal{N}^{\star}$), respectively, with E. We use A as the generic name for the external D-closed sets (mod $\mathcal{N}^{\star}$), C for the internal D-closed sets (mod $\mathcal{N}^{\star}$); $\mathcal{A}$ denotes the family of the sets A, and $\mathcal{C}$ denotes the family of the sets C.

The prefix "D" refers to DENJOY [*8*], who introduced the internal D-open sets (mod $\mathcal{N}^{\star}$) and the internal D-closed sets (mod $\mathcal{N}^{\star}$) under the names of "ensembles-enveloppes" and "ensembles-noyaux", respectively, for his special bases (cf. 3.5), and used them as approximation sets.

If $\mathscr{V}$ is a $\mathfrak{B}$-fine covering of X (strictly or mod $\mathscr{N}^{\star}$), and if the external D-open set G includes X (strictly or mod $\mathscr{N}^{\star}$), then the family $\mathscr{V}_G$ of those constituents in G is still a $\mathfrak{B}$-fine covering of X (strictly or mod $\mathscr{N}^{\star}$). This fact is called the *G-pruning principle.*

5. Comparison lemmas.

For $S \subset R$, we denote by $S \cdot \mathscr{M}$ the family of sets of the form $S \cdot M$, where $M \in \mathscr{M}$, and by μ_S the restriction of $\bar{\mu}$ to $S \cdot \mathscr{M}$. Thus, for $M \in \mathscr{M}$,

$$\mu_S(S \cdot M) = \bar{\mu}(S \cdot M) = \mu(\bar{S} \cdot M).$$

For $X \subset S$, we have $\bar{\mu}_S(X) = \bar{\mu}(X)$. A real-valued function h defined on S is said to be *μ_S-measurable* if the Lebesgue sets $[h < \alpha]$ $(-\infty < \alpha < \infty)$ belong to $S \cdot \mathscr{M}$.

5.1. Lemma: *We suppose that*

(A 1) f and g are real-valued functions defined on P and Q, respectively, where $Q \subset P \subset R$.

(A′ 2) Whenever A and B are $\mu^{\star}$-entangled sets of finite outer measure for which $A \cup B \subset Q$, then there exist no two numbers α and β such that $\alpha < \beta$, $A \subset [f < \alpha]$, and $B \subset [g > \beta]$.

Then $f \geqslant g$ (mod $\mathscr{N}^{\star}$) on Q; that is, $Q \cdot [f < g] \in \mathscr{N}^{\star}$.

Proof. We assume the assertion to be false; that is, $\bar{\mu}(Q \cdot [f<g]) > 0$. There exist two (rational) numbers α and β such that $\bar{\mu}(Q \cdot [f < \alpha < \beta < g]) > 0$. We take for A and B two equal subsets of $Q \cdot [f < \alpha < \beta < g]$, of positive finite outer measure. Clearly $A \subset Q \cdot [f < \alpha]$, $B \subset Q \cdot [\beta < g]$, $A = B$, and $\bar{\mu}(A) = \bar{\mu}(B) > 0$. This contradicts (A′ 2).

5.2. Lemma: *We assume that (A 1) holds and, in addition, (A″ 2): Whenever A and B are any two $\mu^{\star}$-entangled sets of finite outer measure for which $A \cup B \subset Q$, then there exist no two numbers α and β such that $\alpha < \beta$, $A \subset [f > \beta]$, and $B \subset [g < \alpha]$.*

Then $f \leqslant g$ (mod $\mathscr{N}^{\star}$) on Q; that is, $Q \cdot [f > g] \in \mathscr{N}^{\star}$.

Proof. Replace f and g in Lemma 5.1 by $-f$ and $-g$, respectively.

The formulation of the following lemma, which may seem unnecessarily sophisticated for numerical functions, is intended for the more general case where f and g take their values in a separable Banach space.

5.3. Lemma: *We assume (A 1) holds and also (A 2):*

(A 2) There exist no two $\mu^{\star}$-entangled sets A and B of finite outer measure with $A \cup B \subset Q$, such that the convex closures of $f(A)$ and $g(B)$ have positive distance apart.

Then $f = g$ (mod $\mathscr{N}^{\star}$) on Q; that is $Q \cdot [f \neq g] \in \mathscr{N}^{\star}$. Also the restriction $f|Q$ of f to Q, and g, are both $\mu_Q^{\star}$-measurable.

Proof. It is readily seen that (A′ 2) and (A″ 2) together are equivalent to (A 2); application of Lemmas 5.1 and 5.2 completes the proof of the first part of the lemma.

We attend to the second part. Since μ is σ-finite, $R = \bigcup_n R_n$, where $R_n \in \mathcal{M}$ and $\mu(R_n) < \infty$ for $n = 1, 2, \ldots$ Hence $Q = \bigcup_n Q_n$, where $Q_n = Q \cdot R_n$ for $n = 1, 2, \ldots$ Since the $\mu^\star_{Q_n}$-measurability of $g|Q_n$ (restriction of g to Q_n), for all n, implies the $\mu^\star_Q$-measurability of $g|Q = g$, we may limit our proof to the case where $\bar{\mu}(Q) < \infty$. We assume that g is not $\mu^\star_Q$-measurable; there exists a rational number δ such that $D = [g \leqslant \delta]$ is not $\mu^\star_Q$-measurable. We denote by $\bar{D}$ and $\underline{D}$ a $\mu^\star_Q$-cover and a $\mu^\star_Q$-kernel of D, respectively. We let $D' = D - \underline{D}$, $D'' = \bar{D} - D$. The $\mu^\star_Q$-non-measurability of D implies that $\mu^\star_Q(D') = \bar{\mu}(D')$ and $\mu^\star_Q(D'') = \bar{\mu}(D'')$ are both positive. Thus, for a suitable $\beta > \delta$, the set $S = D'' \cdot [g > \beta]$ is of positive outer measure.

The difference set

$$D^0 = \bar{D} - \underline{D} = D' + D'' \in Q \cdot \mathcal{M}^\star;$$

hence there exists a $\mu^\star_Q$-cover $\bar{S}$ of S which is included in D, so that $\bar{S} = \bar{S} \cdot D' + \bar{S} \cdot D''$. Since $f = g(\text{mod}\,\mathcal{N}^\star)$ in Q, then $D = [f \leqslant \delta] \cdot Q(\text{mod}\,\mathcal{N}^\star)$; defining $A = \bar{S} \cdot D'[f \leqslant \delta]$, then $A = S \cdot D'(\text{mod}\,\mathcal{N}^\star)$. Due to the definition of D'', we have $\mu^\star_Q(D'') = 0$; thus $\bar{S} \cdot D''$ includes no $\mu^\star_Q$-measurable set of positive $\mu^\star_Q$-measure. Since $\bar{S} = A + \bar{S} \cdot D''(\text{mod}\,\mathcal{N}^\star)$ and $A \subset \bar{S}$, it follows that $\bar{S}$ is a $\mu^\star_Q$-cover for A. Let $B = S$. Then A and B are $\mu^\star_Q$-entangled, hence $\mu^\star$-entangled. If α denotes a (rational) number between δ and β, we have $A \subset [f < \alpha]$, $B \subset [g > \beta]$ contradicting (A' 2), implied by (A 2).

5.4. Corollary. If $P = Q = R$ (mod $\mathcal{N}^\star$), then (A 1) and (A 2) imply $f = g(\text{mod}\,\mathcal{N}^\star)$ and the $\mu^\star$-measurability of f and g.

Remark. Lemmas 5.1 and 5.2 will be used when f is a Radon-Nikodym $\mu^\star$-integrand and g a derivate. They are analogous to De Possel's lemma [*39*, 394]. Lemma 5.3 can be used when f and g are the extreme derivates. If we know somehow that both f and g are $\mu^\star$-measurable, we can formulate (A' 2) and (A'' 2) considering only $\mu^\star$-measurable sets A and B. The entanglement condition then means $A = B\,(\text{mod}\,\mathcal{N}^\star)$ and $\mu^\star(A) = \mu^\star(B) > 0$.

Chapter II

Derivation Theorems for σ-additive Set Functions under Assumptions of the Vitali Type

The classical Vitali theorem on the real line **R** asserts that if $\mathcal{V}$ is a closed interval covering of a set X (mod $\mathcal{N}^\star$) such that almost all points of X belong to intervals of $\mathscr{V}$ of arbitrarily small length, then for any $\varepsilon > 0$ there exists an enumerable (countable) disjoint subfamily $\{V_n\}$ of $\mathscr{V}$ covering X (mod $\mathscr{N}^\star$) and satisfying $\bar{\mu}(S - S \cdot X) < \varepsilon$, where $S = \bigcup_n V_n$ and μ denotes Borel measure on $\mathbf{R}$.

The Vitali theorem, sometimes called the "strong Vitali theorem", is the main tool to prove that a function of bounded variation has a derivative almost everywhere (i.e., mod$\mathscr{N}^\star$).

In what follows, we shall make variations in the strong Vitali property and obtain corresponding derivation theorems.

1. The individual Vitali assumption.

1.1. Preliminary definitions. By *$\mathscr{M}$-function* we shall mean a real-valued function defined on $\mathscr{M}$; by *$\mathscr{M}$-measure,* a non-negative σ-additive $\mathscr{M}$-function; and by *signed $\mathscr{M}$-measure,* a σ-additive $\mathscr{M}$-function of variable sign.

An $\mathscr{M}$-function is said to be *μ-finite* if it is finite on the $\mathscr{M}$-sets of finite μ-measure. Thus a *μ-finite μ-integral* is a μ-integral of the form $\psi(M) = \int_M f d\mu$ finite on the $\mathscr{M}$-sets of finite μ-measure.

We say that the property (G_σ) holds iff R is the union of enumerably many non-decreasing $\mathscr{G}$-sets G_n^0 such that $G_n^0 \in \mathscr{M}$ and $\mu(G_n^0) < \infty$, $n = 1,2,\ldots$.

If such a sequence exists, then a set X is said to be *bounded* iff it is included in one of the sets G_n^0. Thus our notion of boundedness depends on the special sequence of $\mathscr{G}$-sets occurring in the formulation of (G_σ).

When (G_σ) holds, we adopt the following definitions: A *Radon μ-integral* is any (indefinite) μ-integral $\psi(M) = \int_M f d\mu$, bounded in the sets G_n^0; that is, there exists for $n = 1,2,\ldots$, a number $\beta(n)$ such that if $M \in \mathscr{M}$ and $M \subset G_n^0$, then $|\psi(M)| \leqslant \beta(n)$. A *Radon measure* is an $\mathscr{M}$-measure bounded in the sets G_n^0; a *signed Radon measure* is a σ-additive $\mathscr{M}$-function (of variable sign) bounded in the sets G_n^0. A Radon measure in the classical sense is a Radon measure in $\mathbf{R}^m$ in the sense just defined, where $\mathscr{M}$ is the set of the Borel subsets of $\mathbf{R}^m$ and the reference sequence G_1^0, G_2^0, ... consists of countably many concentric open balls whose radii tend to infinity. A *σ-bounded function* is any extended real-valued function defined on R and bounded on each set G_n^0.

We state some useful classical decomposition theorems. Each μ-finite signed $\mathscr{M}$-measure is the sum of a μ-finite integral and a μ-finite singular part; each signed Radon measure is the sum of a Radon μ-integral and a μ-singular part. Also, each signed Radon measure ψ is the difference of two Radon measures ψ^+ and ψ^-; the sum $\tau = \psi^+ + \psi^-$ is the *total variation* of ψ. If (G_σ) is not assumed, then the term "Radon" may often be replaced by "μ-finite" in these theorems.

Henceforth, when any concept involving boundedness is considered, it will be tacitly understood that (G_σ) is presupposed.

Remarks. In the formulation of Lemmas I. 5.1, I. 5.2, and I. 5.3, the expression "of finite outer measure" may be replaced by "bounded" when (G_σ) holds.

In the subsequent sections, we state "full derivation theorems" for functions ψ of the types just described, namely, theorems asserting the existence almost everywhere (i.e., mod $\mathcal{N}^\star$) on E of the $\mathfrak{B}$-derivative $D\psi$, and its coincidence on E with a Radon-Nikodym μ- or $\mu^\star$-integrand. We avoid the use of such terms as "*R-N* derivative" [*11*, 133] and "pseudo-derivée" [*37*; *39*, 396], reserving "derivate" and "derivative" for functions defined by means of a convergence process, either pointwise, as usual, or globally, as in Part II. IV. 3, under "*L*-derivée". In fact, we shall see in Part II that any *R-N* integrand can be interpreted as a global derivative. In the (G_σ) case, the existence $\mu^\star$-almost everywhere of $D\psi$, and its coincidence with a Radon-Nikodym integrand is proved as follows: We denote by $\mathfrak{B}_n$ the set of those deriving sequences $(M_\iota(x))$ of $\mathfrak{B}$ whose constituents are included in G_n^0 under preservation of the convergence points (i.e., $\mathfrak{B}_n$ is the G_n^0-pruned basis). For any function λ defined on the spread $\mathcal{D}$ of $\mathfrak{B}$, the $\mathfrak{B}$-derivates coincide with the $\mathfrak{B}_n$-derivates on $E \cdot G_n^0$ $\mu^\star$-almost everywhere. To establish the existence of $D\psi$ and its coincidence $\mu^\star$-almost everywhere with the Radon-Nikodym integrand, it suffices to prove it for $\mathfrak{B}_n$, since $E = E \cap \bigcup_{n=1}^{\infty} G_n^0$.

1.2. Definitions. By M-*family* we mean an enumerable family of sets each with an associated multiplicity [*47*, 277]. Equivalently, an M-family may be defined as any sequence of sets, the multiplicity of a set coinciding with the number of appearances in the sequence. In the latter formulation, the order of the appearances of any set is disregarded. Certain advantages arise from the use of M-families instead of ordinary families in the work to follow. For instance, the *frequency* (defined a few lines farther on) is additive; thus, if $\mathscr{E}$ and $\mathscr{F}$ are M-families and $\mathscr{G}$ is the M-family obtained by uniting them, then $\varphi_{\mathscr{E}} + \varphi_{\mathscr{F}} = \varphi_{\mathscr{G}}$. On the other hand, it is only subadditive for ordinary families. Also, any μ-measurable function on R, taking only positive integral values, may be regarded as the frequency function of a measurable M-family covering R. Awkward limitations occur if we restrict ourselves to families without repetition. In natural fashion, we may define the limit of a sequence $\mathscr{E}_1, \mathscr{E}_2, \ldots, \mathscr{E}_n, \ldots$ of M-families as the M-family $\mathscr{E}$, if it exists, such that $\lim_n \varphi_{\mathscr{E}_n} = \varphi_{\mathscr{E}}$. So defined, $\mathscr{E}$ has an *overlap* (defined just below) which is conveniently represented by use of the Lebesgue convergence theorem.

If $\mathscr{E}$ is an M-family, then $\sigma\mathscr{E}$ will denote the union of the sets occurring in $\mathscr{E}$. By *$\mathscr{E}$-frequency $\varphi_{\mathscr{E}}(x)$ at the point x* we shall mean the number of $\mathscr{E}$-sets (possibly ∞) to which x belongs; by *$\mathscr{E}$-excess function* we shall mean the function $\in_{\mathscr{E}}$ defined on $\sigma\mathscr{E}$ by $\in_{\mathscr{E}}(x) = \varphi_{\mathscr{E}}(x)-1$. We define $\theta\,\mathscr{E} = [\varphi_{\mathscr{E}} > 1] = [\in_{\mathscr{E}} \geqslant 1]$, and call $\theta\mathscr{E}$ the *$\mathscr{E}$-overlap set.*

Henceforth, we assume that the $\mathscr{E}$-sets belong to $\mathscr{M}$. Then $\varphi_{\mathscr{E}}$ and $\in_{\mathscr{E}}$ are μ-measurable. If ψ is any $\mathscr{M}$-measure, then we define the ψ-overlap of $\mathscr{E}$ by

$$\omega(\mathscr{E},\psi) = \int_{\sigma\mathscr{E}} \in_{\mathscr{E}} d\mu \,.$$

In case $\psi(\sigma\mathscr{E})$ is finite, we note that

$$\omega(\mathscr{E},\psi) = \sum_{M\in\mathscr{E}} \psi(M) - \psi(\sigma\mathscr{E}) \,.$$

In the particular case $\psi = \mu$, the foregoing equations define the μ-overlap of $\mathscr{E}$, which is of somewhat special importance [*17*, 193].

If $X \subset R$, M is a μ-cover for X, and ψ is any $\mathscr{M}$-measure, then the *ψ-overflow of $\mathscr{E}$ with respect to X and M* is defined as $\psi(\sigma\mathscr{E} - M\cdot\sigma\mathscr{E})$. If ψ is μ-absolutely continuous, then this overflow does not depend on the particular μ-cover M, but is the same for each set $\bar{X}$, so that the terminal expression "and M" may be dropped. In particular, if $\psi = \mu$, then $\mu(\sigma\mathscr{E} - \bar{X}\cdot\sigma\mathscr{E})$ is the μ-overflow of $\mathscr{E}$ with respect to X.

If $X \subset R$, then we define the *μ-deficiency of covering of X by $\mathscr{E}$* as $\bar{\mu}(X - X\cdot\sigma\mathscr{E})$, and we denote this by the notation $\gamma(\mathscr{E},X,\mu)$. $\mathscr{E}$ is said to be an *ε-covering in measure of X* iff $\gamma(\mathscr{E},X,\mu) < \varepsilon$; it is said to be an *0-covering in measure of X* if $X \subset \sigma\mathscr{E} (\text{mod}\,\mathscr{N}^\star)$.

1.3. Definitions. If ψ denotes a (non-negative) $\mathscr{M}$-measure, then we say that the basis $\mathfrak{B}$ possesses the *Vitali ψ-property* iff for any set $X \subset E$ of finite outer measure, any $\mathfrak{B}$-fine covering $\mathscr{V}$ of X, any μ-cover M of X, and any $\varepsilon > 0$, there exists an (enumerable) M-family $\mathscr{E}$ of $\mathscr{V}$-sets such that, putting $S = \sigma\mathscr{E}$, we have

(V 1) $X - X\cdot S \in \mathscr{N}^\star$ ($\mathscr{E}$ is an 0-covering of X);

(V 2) $\psi(S - S\cdot M) < \varepsilon$ (the ψ-overflow of $\mathscr{E}$ with respect to X and M is less than ε);

(V 3) $\omega(\mathscr{E},\psi) < \varepsilon$ (the ψ-overlap is less than ε). (K. O. Househam has suggested the term *ψ-redundancy of $\mathscr{E}$ with respect to X and M* for the sum of the ψ-overflow and the ψ-overlap.)

In case only (V 1) and (V 3) hold, we say that $\mathfrak{B}$ possesses the *reduced Vitali ψ-property.*

Remarks. If $\mathfrak{B}$ possesses the Vitali property corresponding to ψ, then it evidently possesses the Vitali property corresponding to all $\psi' \leqslant \psi$; that is, the Vitali property has a hereditary character. In particular, if ψ is a Radon or a μ-finite $\mathscr{M}$-measure, then $\mathfrak{B}$ possesses the Vitali property corresponding to the μ-absolutely continuous part of ψ. Some equivalent formulations of the Vitali ψ-property are possible. The requirement (V 1) may be replaced by an ε-covering condition; simultaneously, "enumerable" may be replaced by "finite". That such an ε-covering version implies the original version can be shown by an exhaustion process. The requirement that X be of finite outer measure may be dropped. In the (G_σ) case, the phrase "of finite outer measure" may be replaced by "bounded".

1.4. Definition. We define the *upper μ-approximation property of the $\mathcal{M}$-sets by the $\mathcal{G}$-sets* (abbreviated (UG)) as follows: Corresponding to any $\mathcal{M}$-set M of finite μ-measure and any $\eta > 0$, there exists [*38*, 83] a $\mathcal{G}$-set $G \in \mathcal{M}$ for which $M \subset G$ and $\mu(G-M) < \eta$.

We note that (UG) implies (G_σ). (UG) is not altered if the condition "of finite measure" is waived.

1.5. Proposition. If (UG) holds, ψ is either a non-negative μ-finite or Radon μ-integral and $\mathfrak{B}$ possesses the reduced Vitali ψ-property, then $\mathfrak{B}$ enjoys the Vitali ψ-property.

Proof. We take an arbitrary subset X of E of finite outer measure (in particular, bounded if ψ is a Radon measure). We take any $\mathfrak{B}$-fine covering $\mathcal{V}$ of X and any positive number ε. We use (UG) to find a $\mathcal{G}$-set $G' \in \mathcal{M}$, with $G' \supset X$ and $\mu(G'-\bar{X}) < 1$. Since ψ is μ-absolutely continuous and $\psi(X)$ is finite, there exists $\eta = \eta(X,\psi,\varepsilon) > 0$ such that $|\psi(\bar{X})-\psi(M)| < \varepsilon$ whenever $M \in \mathcal{M}$, $M \subset G'$ and $\mu(M-\bar{X}) < \eta$, where $M-\bar{X}$ denotes Stone's difference. Invoking (UG) and the fact that the intersection of two D-open sets is again D-open, we find a $\mathcal{G}$-set $G \in \mathcal{M}$ satisfying $G' \supset G \supset \bar{X}$ and $\mu(G-\bar{X}) < \eta$. We apply the Vitali ψ-property to the G-pruned family $\mathcal{V}_G$ to obtain an M-family $\mathcal{E}$ satisfying (V 1) and (V 3). Since the $\mathcal{E}$-sets lie in G, we have $\mu(S-S\cdot\bar{X}) \leqslant \mu(G-G\cdot\bar{X}) < \eta$. Thus $\psi(S) < \psi(\bar{X}) + \varepsilon$, whence $\psi(S-S\cdot\bar{X}) < \varepsilon$, and (V 2) holds, as required.

1.6. Definition. We say that *Haupt's adaptation property* holds *iff* there exists a σd-family $\mathcal{G}^0$ of $\mathcal{G}$-sets that is a Borel generator for $\mathcal{M}$ (that is, $\mathcal{M}$ is the smallest $\sigma\delta$-family including $\mathcal{G}^0$), [*14*, 173].

Remark. It is possible, by a relatively simple purely set-theoretic argument, to show that $R \in \mathcal{G}^0$. However, from the assumed σ-finiteness of μ, one can see the validity of this assertion immediately as a consequence of Prop. 1.7. by taking $M = R$. Hence, if we let $\mathcal{A}^0$ denote the set of complements in R of the members of $\mathcal{G}^0$, then $\mathcal{A}^0$ is a subset of $\mathcal{A}$ all of whose members belong to $\mathcal{M}$.

1.7. Proposition. Haupt's adaptation property implies the following (which includes (UG)). For any σ-finite $\mathcal{M}$-measure (in particular, any

Radon measure) ψ, any $\mathcal{M}$-set M, and any $\varepsilon > 0$, there exists a $\mathcal{G}^0$-set G such that $M \subset G$ and $\psi(G-M) < \varepsilon$.

Proof. We let ψ denote any σ-finite $\mathcal{M}$-measure (which may, in particular, be a Radon measure). We let $\mathcal{H}$ denote the family of those sets $H \in \mathcal{M}$ such that for each $\varepsilon > 0$ there exists a set $G^0 \in \mathcal{G}^0$ for which $H \subset G^0$ and $\psi(G^0 - H) < \varepsilon$. Since evidently $\mathcal{G}^0 \subset \mathcal{M}$, then we have $\mathcal{G}^0 \subset \mathcal{H} \subset \mathcal{M}$.

Next, we take any sequence $H_1, H_2, \ldots, H_n, \ldots$, in $\mathcal{H}$ and suppose $\varepsilon > 0$. For each $n = 1,2,\ldots$ there exists a set $G_n^0 \in \mathcal{G}^0$ satisfying $H_n \subset G_n^0$ and $\psi(G_n^0 - H_n) < \varepsilon/2^n$. Now we let $G^0 = \bigcup_{n=1}^{\infty} G_n^0$, $H = \bigcup_{n=1}^{\infty} H_n$, and note that $G \in \mathcal{G}^0$, $H \subset G^0$, and $\psi(G^0 - H) \leqslant \sum_{n=1}^{\infty} \psi(G_n^0 - H) < \varepsilon$. Thus $H \in \mathcal{H}$ and $\mathcal{H}$ is a σ-family.

Again we consider a sequence of sets $H_1, H_2, \ldots, H_n, \ldots$ in $\mathcal{H}$ and let $H' = \bigcap_{n=1}^{\infty} H_n$. We assume first that one of the sets H_n, say H_1, is of finite ψ-measure. To each set H_n, $n = 1,2,\ldots$, there corresponds a set $G_n^0 \in \mathcal{G}^0$ with $H_n \subset G_n^0$ and $\psi(G_n^0 - H_n) < \varepsilon/2^{n+1}$. We let $G' = \bigcap_{n=1}^{\infty} G_n^0$. Clearly, for each positive integer k we have

$$H' \subset G' \subset \bigcap_{n=1}^{k} G_n^0, \qquad \lim_m \psi\left(\left(\bigcap_{n=1}^{m} G_n^0\right) - H'\right) = \psi(G' - H').$$

Thus, for a suitably large positive integer N, we have

$$\psi\left(\left(\bigcap_{n=1}^{N} G_n^0\right) - H\right) < \psi(G' - H') + \varepsilon/2 < \sum_{n=1}^{\infty} \psi(G_n^0 - H_n) + \varepsilon/2 < \varepsilon.$$

Since $\mathcal{G}$ is a $\sigma\,d$-family, it follows that $\bigcap_{n=1}^{N} G_n^0 \in \mathcal{G}^0$, and so $H' \in \mathcal{H}$.

In case all the sets H_n are of infinite ψ-measure, we may utilize the σ-finiteness of ψ to determine an expanding sequence of $\mathcal{M}$-sets M_k, $k = 1, 2, \ldots$, each of finite ψ-measure, with $R = \bigcup_{n=1}^{\infty} M_n$. Then, by the argument just given, we can assert that $H' \cdot M_k \in \mathcal{H}$. Since $H' = \bigcup_{k=1}^{\infty} (H' \cdot M_k)$ and $\mathcal{H}$ is a σ-family, then $H' \in \mathcal{H}$. Accordingly, $\mathcal{H}$ is a $\sigma\delta$-family including $\mathcal{G}^0$, whence $\mathcal{H} = \mathcal{M}$. From the arbitrary nature of ψ, the truth of the proposition follows.

Remarks. The property described in Prop. 1.7. is called the *universal upper approximation property for* $\mathcal{G}^0$*-sets*. It holds [*31*, 244–245] in the special case where R is a metric space, ψ is a classical finite Radon measure, and $\mathcal{G}^0$ is the family of the open sets.

Correspondingly, the family $\mathscr{A}^0$ possesses a universal lower approximation property.

Finally, it may be observed that when Haupt's adaptation property holds, ψ is a μ-finite $\mathscr{M}$-measure iff it is a Radon measure.

1.8. Proposition. If ψ is a μ-finite $\mathscr{M}$-measure, Haupt's adaptation property and the reduced Vitali ψ-property both hold, then the Vitali ψ-property holds.

Proof. This follows closely the proof of Prop. 1.5., except that we take a μ-cover M of X, and use Prop. 1.7. directly to find a $\mathscr{G}^0$-set $G \supset M$ with $\psi(G-M) < \varepsilon$. As before, we find an M-family $\mathscr{E}$ satisfying (V 1) and (V 3) with members lying in G. Thus $\psi(S-S\cdot M) \leqslant \psi(G-M) < \varepsilon$, and (V 2) holds.

2. The individual full derivation theorem for Radon or μ-finite μ-integrals.

2.1. Proposition. If ψ is a non-negative μ-finite (or Radon) μ-integral $\int f\,d\mu$ and $\mathfrak{B}$ possesses the Vitali μ-property, then $D_\star\psi \geqslant f \pmod{\mathscr{N}^\star}$ on E.

Proof. According to the Remarks following Def. 1.1., we need treat only the case where ψ is μ-finite. We shall obtain a contradiction from the assumed existence of two $\mu^\star$-entangled subsets A and B of E of finite outer measure, and two numbers α, β such that $\alpha < \beta$, $A \subset [f > \beta]$, and $B \subset [g < \alpha]$, where $g = D_\star \psi$. Since $[f > \beta] \in \mathscr{M}$, then $A' = \bar{A}\cdot[f > \beta]$ is a μ-cover for A; since $\mu(\bar{A}) > 0$, we have

$$\psi(A') = \int_{A'} f\,d\mu > \beta\mu(A') > 0. \tag{2.1.1}$$

On the other hand, the family $\mathscr{V}$ of the constituents satisfying

$$\psi(V) < \alpha\mu(V) \tag{2.1.2}$$

is a $\mathfrak{B}$-fine covering of $B \subset [g < \alpha]$. Thus, by virtue of the Vitali μ-property, for any positive integer n, there exists an M-family $\mathscr{E}_n$ of $\mathscr{V}$-sets V_{ni} such that if $S_n = \sigma\mathscr{E}_n$, then

$$B - B\cdot S \in \mathscr{N}^\star; \quad \mu(S_n - S_n\cdot B) < 2^{-n}; \quad \omega(\mathscr{E}_n;\mu) < 2^{-n}. \tag{2.1.3}$$

Using (2.1.3) and (2.1.2) we obtain

$$\psi(\bar{B}) \leqslant \psi(S_n) \leqslant \sum_i \psi(V_{ni}) < \alpha\sum_i \mu(V_{ni}) < \alpha(\mu(S_n) + 2^{-n}), \tag{2.1.4}$$

and $\lim_n \mu(S_n) = \mu(\bar{B})$. Combining, we obtain

$$\psi(\bar{B}) \leqslant \alpha\mu(\bar{B}), \tag{2.1.5}$$

which, since $\psi(A') = \psi(\bar{B})$ and $\alpha < \beta$, is a contradiction of (2.1.1). From Lemma I. 5.2 follows the assertion $f \leqslant g \pmod{\mathscr{N}^\star}$.

Remarks. If an ε-covering version of the Vitali μ-property is used in place of the 0-covering version, then the first statement in (2.1.3) has to be replaced by $\mu(\bar{B}-\bar{B}\cdot S_n) < \eta_n$; and, because of the μ-absolute continuity of ψ, η_n can be so chosen that $\psi(\bar{B}) \leqslant \psi(S_n) + 2^{-n}$; (2.1.4) has to be altered accordingly.

An example of a blanket possessing the Vitali μ-property and a function $g \in \mathfrak{L}^p$ for each p, $1 \leqslant p < \infty$ such that its integral ψ has $D\psi = \infty$ everywhere, is known [*19*, 293]. The domain is the open unit square.

2.2. Proposition. If ψ is a non-negative μ-finite (or Radon) μ-integral $\int f\,d\mu$ and $\mathfrak{B}$ possesses the Vitali ψ-property, then $D^\star\psi \leqslant f\,(\mathrm{mod}\,\mathcal{N}^\star)$ on E.

Proof. As in the preceding proposition, we may and do assume that ψ is μ-finite. We assume that A and B are two $\mu^\star$-entangled sets of finite outer measure, α and β two numbers such that $\alpha < \beta$, $A \subset [f < \alpha]$, and $B \subset [g > \beta]$, where $g = D^\star\psi$. Since $[f < \alpha] \in \mathcal{M}$, $A' = A\cdot[f < \alpha]$ is a μ-cover for A. Since $\mu(\bar{A}) > 0$, we obtain

$$\psi(A') = \int_{A'} f\,d\mu < \alpha\mu(A'). \tag{2.2.1}$$

The family $\mathscr{V}$ of the constituents satisfying

$$\psi(V) > \beta\mu(V)] \tag{2.2.2}$$

is a $\mathfrak{B}$-fine covering of $B \subset [g > \beta]$. We use the Vitali ψ-property to determine, for each positive integer n, an M-family $\mathscr{E}_n$ of $\mathscr{V}$-sets V_{ni} such that if $S_n = \sigma\mathscr{E}_n$, then

$$B - B\cdot S_n \in \mathcal{N}^\star\colon \quad \psi(S_n - S_n\cdot\bar{B}) < 2^{-n}; \quad \omega(\mathscr{E}_n, \psi) < 2^{-n}. \tag{2.2.3}$$

The ψ-overlap condition yields

$$\psi(S_n) > \sum_i \psi(V_{ni}) - 2^{-n};$$

hence, using (2.2.2),

$$\psi(S_n) > \beta\sum_i \mu(V_{ni}) - 2^{-n} \geqslant \beta\mu(S_n) - 2^{-n}.$$

This last and (2.2.3) together yield, for $n = 1, 2, \ldots$

$$\psi(\bar{B}) + 2^{-n} > \beta\mu(\bar{B}) - 2^{-n}; \tag{2.2.4}$$

hence $\psi(\bar{B}) \geqslant \beta\mu(\bar{B})$, which contradicts (2.2.1), since $\alpha < \beta$ and $\psi(A') = \psi(\bar{B})$. Thus Lemma I. 5.1 applies and $D^\star\psi \leqslant f\,(\mathrm{mod}\,\mathcal{N}^\star)$ on E.

Remark. With the ε-covering version of the Vitali ψ-property, we must replace the first statement in (2.2.3) by $\mu(\bar{B}-\bar{B}\cdot S_n) < 2^{-n}$, hence $\mu(S_n) > \mu(\bar{B}) - 2^{-n}$; and in (2.2.4), we have to replace $\mu(\bar{B})$ by $\mu(\bar{B}) - 2^{-n}$.

2.3. Theorem. *If ψ is a non-negative μ-finite (or Radon) μ-integral and $\mathfrak{B}$ possesses the Vitali μ-property and the Vitali ψ-property, then the*

$\mathfrak{B}$-derivative $D\psi$ exists μ-almost everywhere on E and is equal (mod $\mathcal{N}^\star$) *to $f|E$, where f is any Radon-Nikodym $\mu^\star$-integrand of ψ.*

Proof. This is an immediate consequence of Prop. 2.1 and 2.2.

2.4. Definition. An $\mathcal{M}$-function ψ is said to be *majorized* or *dominated* by the $\mathcal{M}$-function ψ^0 if $|\psi(M)| \leqslant \psi^0(M)$ for each $M \in \mathcal{M}$.

We note that a signed μ-finite $\mathcal{M}$-measure (resp., Radon measure) dominated by a μ-integral is itself a μ-integral; also a finitely additive $\mathcal{M}$-function dominated by a μ-finite $\mathcal{M}$-measure (resp. Radon measure) is a signed μ-finite $\mathcal{M}$-measure (resp., signed Radon measure).

2.5. Theorem. *If ψ^0 is a non-negative μ-finite (or Radon) μ-integral that possesses the Vitali μ-property and the Vitali ψ^0-property, then for any μ-finite signed $\mathcal{M}$-measure (or Radon measure) ψ dominated by ψ^0, the $\mathfrak{B}$-derivative $D\psi$ exists $\mu^\star$-almost everywhere on E and is equal to the E-restriction of a Radon-Nikodym integrand of ψ.*

Proof. This follows immediately upon decomposing ψ into ψ^+ and ψ^- and using the hereditary character of the Vitali ψ-property.

2.6. Theorem. *If ψ^0 is a non-negative Radon μ-integral, $\mathfrak{B}$ possesses the Vitali μ-property and the Vitali ψ^0-property, ψ is a signed Radon measure, and there corresponds to each set G_n^0 a positive finite number $\kappa(n)$ such that $|\psi(M)| \leqslant \kappa(n)\cdot\psi^0(M)$ for each $\mathcal{M}$-set $M \subset G_n^0$ (this is the ψ^0-Lipschitz condition), then $D\psi$ exists $\mu^\star$-almost everywhere on E and is equal to the E-restriction of a Radon-Nikodym integrand of ψ.*

Proof. Apply Th. 2.5 to each set G_n^0 used as an autonomous domain of $\mathfrak{B}$-derivation, with $\kappa(n)\,\psi^0$ as majorant.

Remark. If we know that the extreme derivates are $\mu^\star$-measurable then, from the Remarks under Cor. I. 5.4, it follows that Ths. 2.3, 2.5, and 2.6 remain valid if, in the definition of the Vitali property, X is taken from $\mathcal{M}$.

2.7. Definition. The special case of the Vitali property, wherein $\psi = \mu$, is the so-called *weak Vitali property*, and a basis $\mathfrak{B}$ possessing it is called a *weak* derivation basis.

Remarks. By Th. 2.6, such a basis derives (in De Possel's sense), the uniformly μ-Lipschitzian integrals; explicitly, if ψ is a σ-additive $\mathcal{M}$-function for which $|\psi(M)| \leqslant \kappa\mu(M)$, where κ is a constant, then $D\psi$ exists almost everywhere on E and is equal to the E-restriction of a Radon-Nikodym integrand of ψ. In De Possel's version, a weak derivation basis derives the μ-integral of any essentially bounded μ-measurable function, and in the Radon case, the integrals of functions that are μ-measurable and essentially bounded on each set G_n^0. Five equivalent properties defining these bases in the case $E = R$ (mod $\mathcal{N}^\star$) are given in [*39*, 403–405].

2.8. Proposition. If each $\mathfrak{B}$-fine covering of any subset X of E admits an enumerable subfamily covering X (mod $\mathcal{N}^\star$), then for any $\mathcal{G}$-set G, we have $\overline{E \cdot G} \subset G$ (mod $\mathcal{N}^\star$) or, equivalently $E \cdot G \subset \underset{\sim}{G}$ (mod $\mathcal{N}^\star$). If, in addition, $E = R$, then the $\mathcal{G}$-sets are $\mu^\star$-measurable.

Proof. The family $\mathcal{W}$ of the $\mathfrak{B}$-constituents included in G is a $\mathfrak{B}$-fine covering of $E \cdot G$; thus there exists an M-family $\mathcal{E} \subset \mathcal{W}$ with $E \cdot G \subset \sigma\mathcal{E}$ (mod $\mathcal{N}^\star$), whence $\overline{E \cdot G} \subset \sigma\mathcal{E}$ (mod $\mathcal{N}^\star$) and $\overline{E \cdot G} \subset G$ (mod $\mathcal{N}^\star$). Since $\sigma\mathcal{E} \subset G$, it is apparent that $E \cdot G \subset \underset{\sim}{G}$ (mod $\mathcal{N}^\star$). If $E = R$, then $\bar{G} \subset G$ (mod $\mathcal{N}^\star$), hence $G = \bar{G}$ (mod $\mathcal{N}^\star$).

3. The individual full derivation theorem for Radon measures.

3.1. Lemma: *If ψ is a Radon measure (or a μ-finite $\mathcal{M}$-measure), $\mathfrak{B}$ possesses the Vitali ψ-property, $Q \subset E$, $\bar{\mu}(Q) < \infty$, $0 < \eta < \infty$, and there exists a $\mathfrak{B}$-fine covering $\mathcal{V}$ of Q such that for all $\mathcal{V}$-sets V,*

$$\psi(V) \geqslant \eta\,\mu(V) \tag{3.1.1}$$

then $\psi(M) \geqslant \eta\,\bar{\mu}(Q)$ whenever $M \in \mathcal{M}$ and $Q \subset M$.

Proof. The Remarks following Defs. 1.1 permit us to consider only the case of a μ-finite $\mathcal{M}$-measure. We may evidently also assume $\mu(M) < \infty$. We take an arbitrary positive number ε, let T denote a μ-cover of Q for which $Q \subset T \subset M$, and invoke the Vitali ψ-property to obtain an M-family $\mathcal{E}$ of sets V_i, $i = 1, 2, \ldots$, for which, putting $\sigma\mathcal{E} = S$, we have

$$Q - Q \cdot S \in \mathcal{N}^\star; \quad \omega(\mathcal{E}, \psi) < \varepsilon; \quad \psi(S - S \cdot T) < \varepsilon. \tag{3.1.2}$$

From (3.1.1) and the first two conditions of (3.1.2) we obtain

$$\psi(S) > \sum_i \psi(V_i) - \varepsilon \geqslant \eta \sum_i \mu(V_i) - \varepsilon \geqslant \eta\,\mu(S) - \varepsilon \geqslant \eta\,\bar{\mu}(Q) - \varepsilon;$$

This result combined with the last inequality of (3.1.2), yields

$$\psi(M) \geqslant \psi(T) \geqslant \psi(T \cdot S) > \psi(S) - \varepsilon \geqslant \eta\,\bar{\mu}(Q) - 2\varepsilon$$

which, since ε is arbitrary, yields the desired relation.

3.2. Theorem. *If ψ is a Radon measure (or a μ-finite $\mathcal{M}$-measure), and $\mathfrak{B}$ possesses the Vitali μ- and ψ-properties, then ψ has μ-almost everywhere a $\mathfrak{B}$-derivative that is equal on E to a Radon-Nikodym integrand of ψ.*

Proof. We decompose ψ into the μ-absolutely continuous part ψ_r and the μ-singular part ψ_s, denoting by N_0 an $\mathcal{N}$-set on which ψ_s is concentrated; that is $\psi_s(R - N_0) = 0$. $\mathfrak{B}$ possesses the Vitali μ- and ψ-properties; hence, in accordance with the Remarks under Defs. 1.3, $\mathfrak{B}$ has also the ψ_r-property. Because of Th. 2.3, we need to prove only that $D^\star \psi_s = 0$ (mod $\mathcal{N}^\star$).

To this end, we let $A_n = [D^\star \psi_s > n^{-1}] \cdot (R - N_0)$ for $n = 1, 2, \ldots$. The family of $\mathfrak{B}$-constituents V for which $\psi_s(V) \geqslant n^{-1}\mu(V)$ is a $\mathfrak{B}$-fine covering of A_n. In accordance with Lemma 3.1, $\psi_s(M) \geqslant n^{-1}\bar{\mu}(A_n)$ for any set $M \in \mathcal{M}$ with $A_n \subset M$; in particular, this holds for $M = R - N_0$, thus $0 \geqslant n^{-1}\mu(A_n)$, so that $\bar{\mu}(A_n) = 0$ and A_n is an $\mathcal{N}^\star$-set. Now, since

$$[D^\star \psi_s > 0] \cdot (R - N_0) = \bigcup_n [D^\star \psi > n^{-1}] \cdot (R - N_0) = \bigcup_n A_n,$$

and

$$[D^\star \psi > 0] = [D^\star \psi_S > 0] \cdot (R - N_0) \cup [D^\star \psi_S > 0] \cdot N_0,$$

it follows that $D^\star \psi = 0 \pmod{\mathcal{N}^\star}$.

The following results are immediate consequences of Ths. 2.5 and 2.6.

3.3. Theorem. *If ψ^0 is a Radon measure (resp., μ-finite $\mathcal{M}$-measure) and $\mathfrak{B}$ possesses the Vitali μ- and ψ^0-properties, then $\mathfrak{B}$ derives any signed Radon (resp., μ-finite) $\mathcal{M}$-measure dominated by ψ^0.*

3.4. Theorem. *If ψ^0 is a Radon measure, $\mathfrak{B}$ possesses the Vitali μ- and ψ^0-properties, and ψ is such a signed Radon measure that there corresponds to any set G_n^0 a positive (finite) number $\kappa(n)$ such that $|\psi(M)| \leqslant \kappa(n)\psi^0(M)$ for any $\mathcal{M}$-set $M \subset G_n^0$, then $\mathfrak{B}$ derives ψ.*

The remark following Th. 2.6 applies here also.

4. Class derivation theorems.

4.1. Definition. If $\mathfrak{B}$ possesses the Vitali ψ-property for *every* non-negative Radon (resp., μ-finite) μ-integral ψ, then we say that $\mathfrak{B}$ has the *Vitali property for non-negative Radon* (resp., *μ-finite*) *μ-integrals.*

4.2. Theorem. *If $\mathfrak{B}$ has the Vitali property for non-negative Radon (resp., μ-finite) μ-integrals, then $\mathfrak{B}$ derives every Radon (resp., μ-finite) μ-integral.*

Proof. This follows from Th. 2.3.

4.3. Definition. If $\mathfrak{B}$ possesses the Vitali ψ-property for every Radon (resp., μ-finite) $\mathcal{M}$-measure ψ, then we say that $\mathfrak{B}$ has the *Vitali property for Radon* (resp., *μ-finite*) *$\mathcal{M}$-measures.*

4.4. Theorem. *If $\mathfrak{B}$ has the Vitali property for Radon (resp., μ-finite) $\mathcal{M}$-measures, then $\mathfrak{B}$ derives every Radon (resp., μ-finite) $\mathcal{M}$-measure.*

Proof. This follows from Th. 3.2.

Remark. In case $\mathfrak{B}$ is a blanket, the Vitali property for Radon measures is the "pseudo-strength" of [*19*]. The existence of the derivative of any classical Radon measure is established in [*30*] for star blankets.

4.5. Definitions. We shall introduce a chain of properties between the Vitali μ-property and the Vitali property for non-negative μ-integrals, under the assumption that (G_σ) holds. We let p and q denote two numbers, both greater than 1, for which $p^{-1} + q^{-1} = 1$. By *$\mu^{(q)}$-functions* we shall mean those Radon μ-integrals ψ of the form $\psi(M) = \int_M f\,d\mu$ for bounded

$M \in \mathcal{M}$, where f is such a function that for any given positive integer n, $\int\limits_{G_n^0} |f|^q d\mu < \infty$. By $\mathfrak{L}^q$*-functions* we shall mean those functions f that are integrands of $\mu^{(q)}$-functions. We shall say that $\mathfrak{B}$ is an S^p*-basis* iff for each subset $X \subset E$ of finite outer measure, each $\mathfrak{B}$-fine covering $\mathscr{V}$ of X, and each $\varepsilon > 0$, there exists an M-family $\mathscr{E}$ of $\mathscr{V}$-sets for which, putting $S = \sigma\mathscr{E}$,

(I) $\mathscr{E}$ is an 0-covering of X;

(II) the μ-overflow of $\mathscr{E}$ with respect to X is less than ε;

(III) the $\mathfrak{L}^p$-semi-norm of $\mathscr{E}$, i. e., $\left(\int\limits_S (\in_{\mathscr{E}})^p d\mu\right)^{1/p}$, is less than ε.

We shall refer to the $\mathfrak{L}^p$-semi-norm of $\mathscr{E}$ as the $\mathfrak{L}^p$*-overlap of* $\mathscr{E}$, and shall denote it by $\omega^{(p)}(\mathscr{E})$.

Statements (I), (II), and (III) are meaningful for $p = 1$. Accordingly, we define an S^1*-basis* as one having these properties, with $p = 1$. We define as $\mu^{(\infty)}$*-functions* all integrals of μ-measurable functions that are essentially bounded on each set G_n^0.

Remarks. Comparison with Defs. 1.3 shows that for any $p \geqslant 1$, conditions (I) and (II) are the same as (V 1) and (V 2), while (III) is at least as strong as (V 3) for $\psi = \mu$; hence, every S^p-basis, for $p \geqslant 1$, possesses the Vitali μ-property, and, in accordance with the Remarks following Def. 2.7, derives the $\mu^{(\infty)}$-functions. The following is an extension of this result.

4.6. Theorem. *If* $p > 1$ *and* $\mathfrak{B}$ *is an* S^p*-basis, then* $\mathfrak{B}$ *derives the* $\mu^{(q)}$*-functions.*

Proof. We let $\mathfrak{B}$ denote any S^p-basis. From the property (G_σ), it follows that we may restrict our proof to the case where the domain E of $\mathfrak{B}$ lies in one set G_N^0; that is, E may be assumed to be bounded. Furthermore, it follows from the remarks just above, and from Th. 2.5, that we need prove only that for each non-negative $\mu^{(q)}$-function ψ, defined by $\psi(M) = \int\limits_M f\,d\mu$ for bounded $M \in \mathcal{M}$, $\mathfrak{B}$ possesses the Vitali ψ-property. Thus we may and do assume $f \geqslant 0$.

Accordingly, we let X denote any subset of E (necessarily of finite outer measure), $\mathscr{V}$ any $\mathfrak{B}$-fine covering of X, and ε any positive number. We put $G = G_N^0$ and define ε' as any positive number such that

$$\varepsilon'\left(\int\limits_G f^q d\mu\right)^{1/q} < \varepsilon. \tag{4.6.1}$$

From the μ-absolute continuity of ψ on the $\mathcal{M}$-subsets of G, it follows that there exists a positive number η for which

$$|\psi(M') - \psi(M'')| < \varepsilon \tag{4.6.2}$$

whenever $\mu(M'-M'') < \eta$, where $M' \in \mathcal{M}$, $M'' \in \mathcal{M}$, $M' \subset G$, $M'' \subset G$, and $M'-M''$ denotes STONE's difference. We may and do assume that $\eta < \varepsilon'$.

We may assume $\mathscr{V}$ to be G-pruned. We invoke the S^p-properties of $\mathfrak{B}$ to determine an M-family $\mathscr{E}$ of $\mathscr{V}$-sets for which, putting $S = \sigma\mathscr{E} \subset G$, we have

$$X - X \cdot S \in \mathcal{N}^\star; \quad \mu(S - S \cdot \bar{X}) < \eta; \quad \Big(\int_S (\in_{\mathscr{E}})^p d\mu\Big)^{1/p} < \eta. \qquad (4.6.3)$$

Evidently $\mathscr{E}$ satisfies (V 1) of Defs. 1.3. From the first relation of (4.6.3) and Prop. 2.8, we see that $S \subset G \pmod{\mathcal{N}^\star}$ and $\bar{X} \subset G \pmod{\mathcal{N}^\star}$; hence, because of (4.6.2),

$$|\psi(S) - \psi(\bar{X} \cdot S)| < \varepsilon;$$

therefore, $\psi(S - S \cdot \bar{X}) < \varepsilon$, and ($V$ 2) holds.

Using Hölder's inequality (4.6.1) and the last relation in (4.6.3), we have

$$\int_{\sigma\mathscr{E}} \in_{\mathscr{E}} d\psi = \int_S \in_{\mathscr{E}} f\, d\mu \leqslant \Big(\int_S (\in_{\mathscr{E}})^p d\mu\Big)^{1/p} \Big(\int_S f^q d\mu\Big)^{1/q} < \eta \Big(\int_G f^q d\mu\Big)^{1/q} < \varepsilon.$$

Hence (V 3) holds, and the proof is complete.

Remark. In [*20*, 378], there is given an example of an S^p-basis ($p > 1$) and a function which is a $\mu^{(q')}$-function for each q', $q' < p/(p-1)$, whose derivative is infinite everywhere. In this example, as in many counter-examples known tu us in the theory of derivation, a derivate is infinite on a set of positive measure. In this connection, it is interesting to observe that ZYGMUND's proof [*49*] depends upon the summability of the derivates, which prevents a "flight to infinity" on a set of positive measure.

4.7. Theorem. *In the definition of an S^p-basis, the* 0-*covering condition may be replaced by an ε-covering condition; simultaneously $\mathscr{E}$ may be required to be finite.*

Proof. Since we are merely relaxing the initial definition of an S^p-basis, we have to prove only that any S^p-basis under the ε-covering definition is an S^p-basis under the 0-covering definition. We thus assume that for any subset $X \subset E$ of finite outer measure, any $\mathfrak{B}$-fine covering $\mathscr{V}$ of X and any $\varepsilon > 0$, there exists a finite family $\mathscr{F}$ of $\mathscr{V}$-constituents such that

$$\mu(\bar{X} - \bar{X} \cdot \sigma\mathscr{F}) < \varepsilon; \quad \mu(\sigma\mathscr{F} - \bar{X} \cdot \sigma\mathscr{F}) < \varepsilon; \quad \Big(\int_{\sigma\mathscr{F}} (\in_{\mathscr{F}})^p d\mu\Big)^{1/p} < \varepsilon. \qquad (4.7.1)$$

We take a subset X of E of finite outer measure, a $\mathfrak{B}$-fine covering $\mathscr{V}$ of X, and a positive number ε which we may assume to be less than 1. We choose a sequence of positive numbers $\eta_1, \eta_2, \ldots, \eta_n, \ldots$ whose sum is less than ε^p.

We shall determine inductively a sequence of finite families $\mathscr{F}_1, \mathscr{F}_2, \ldots, \mathscr{F}_n, \ldots$ of $\mathscr{V}$-constituents such that, for $n = 1, 2, \ldots,$

(a) $\mathscr{F}_1 \subset \mathscr{F}_2 \subset \cdots \subset \mathscr{F}_n \subset \cdots$;

(b) $\mu(\bar{X} - \bar{X} \cdot \sigma\mathscr{F}_n) < \eta_n$;

(c) $\mu(\sigma\mathscr{F}_n - \bar{X} \cdot \sigma\mathscr{F}_n) < \sum_{i=1}^{n} \eta_i = \zeta_n$;

(d) $\int_{\sigma\mathscr{F}_n} (\in_{\mathscr{F}_n})^p \, d\mu < \zeta_n$.

The existence of a family $\mathscr{F}_1$, satisfying conditions (b), (c), and (d) for $n = 1$, follows from our hypotheses as expressed in (4.7.1). We now assume the existence of a nested sequence of families $\mathscr{F}_1, \mathscr{F}_2, \ldots, \mathscr{F}_n$ satisfying (a), (b), (c), and (d), and proceed to find $\mathscr{F}_{n+1}$ also satisfying them.

We put $\sigma\mathscr{F}_n = S$, $X - X \cdot S = Y$; then $\bar{Y} = \bar{X} - \bar{X} \cdot S$ is a μ-cover for Y. From (d) and the fact that $\mu(S) < \infty$, it follows that

$$\int_S (\varphi_{\mathscr{F}_n})^p \, d\mu < \infty ;$$

thus we may find a positive number $\gamma = \gamma(\eta_{n+1})$ such that

$$\int_M (\varphi_{\mathscr{F}_n})^p \, d\mu < \eta_{n+1}/2^p, \tag{4.7.2}$$

whenever M is an $\mathscr{M}$-set, $M \subset S$, and $\mu(M) < \gamma$. We may and do assume that $\gamma < \eta_{n+1}/2 < (\eta_{n+1})^{1/p}/2$.

Again recalling (4.7.1), we find a finite subfamily $\mathscr{H}$ of $\mathscr{V}$ for which, putting $\sigma\mathscr{H} = T$,

$$\mu(\bar{Y} - \bar{Y} \cdot T) < \gamma ; \quad \mu(T - \bar{Y} \cdot T) < \gamma ; \quad \int_T (\in_{\mathscr{H}})^p \, d\mu < \gamma^p . \tag{4.7.3}$$

Noting that $S \cdot T \subset T - \bar{Y} \cdot T$, using (4.7.2), and the second relation of (4.7.3), we obtain

$$\int_{S \cdot T} (\varphi_{\mathscr{F}_n})^p \, d\mu < \eta_{n+1}/2^p . \tag{4.7.4}$$

We define $\mathscr{F}_{n+1} = \mathscr{F}_n \cup \mathscr{H}$ and let $\sigma\mathscr{F}_{n+1} = U$.

We observe that $\bar{X} - \bar{X} \cdot U = \bar{Y} - \bar{Y} \cdot T$ and $(U - \bar{X} \cdot U) \subset (S - \bar{X} \cdot S) \cup (T - \bar{Y} \cdot T)$, whence from (4.7.3) and (c) we obtain

$$\mu(\bar{X} - \bar{X} \cdot U) < \gamma < \eta_{n+1} ; \; \mu(U - \bar{X} \cdot U) < \zeta_n + \gamma < \zeta_n + \eta_{n+1}$$
$$= \sum_{i=1}^{n} \eta_i = \zeta_{n+1},$$

which establishes (b) and (c) as applied to $\mathscr{F}_{n+1}$.

Next,

$$\int_U (\in_{\mathscr{F}_{n+1}})^p \, d\mu = \int_{U-T} (\in_{\mathscr{F}_{n+1}})^p \, d\mu + \int_{U \cdot T} (\in_{\mathscr{F}_{n+1}})^p \, d\mu . \tag{4.7.5}$$

Since $(U - T) \subset S$ and $\in_{\mathscr{F}_{n+1}} = \in_{\mathscr{F}_n}$ on $U - T$, then, by (d),

$$\int_{U-T} (\in_{\mathscr{F}_{n+1}})^p \, d\mu \leqslant \int_S (\in_{\mathscr{F}_n})^p \, d\mu < \zeta_n . \tag{4.7.6}$$

Using (4.7.3), (4.7.4), and Minkowski's inequality, we also obtain

$$\begin{aligned}\int_T (\in_{\mathscr{F}_{n+1}})^p \, d\mu &= \int_T (\varphi_{\mathscr{F}_n} + \varphi_{\mathscr{H}} - 1)^p \, d\mu \\ &\leqslant \left[\left(\int_T (\varphi_{\mathscr{F}_n})^p \, d\mu)^{1/p} + \int_T (\varphi_{\mathscr{H}} - 1)^p \, d\mu \right)^{1/p} \right]^p \\ &= \left[\left(\int_{T \cdot S} (\varphi_{\mathscr{F}_n})^p \, d\mu \right)^{1/p} + \left(\int_T (\in_{\mathscr{H}})^p \, d\mu^{\;1/p} \right)^p \right] < \eta_{n+1} \, .\end{aligned}$$

Putting this last inequality and (4.7.6) into (4.7.5), we establish (d) as applied to $\mathscr{F}_{n+1}$.

Finally, we define $\mathscr{E} = \bigcup_{n=1}^{\infty} \mathscr{F}_n$. It is clear from (a), (b), (c), and (d) that $\mathscr{E}$ is an M-family of $\mathscr{V}$-constituents satisfying conditions (I), (II), and (III) of Def. 4.5, which completes our proof.

Remarks. No change in the definition of an S^p-basis results if X is required to be a bounded subset of E. For then any subset $Y \subset E$ of finite outer measure may be decomposed into a countable sum of bounded disjoint subsets of E, for each of which the S^p-property holds, and by the method of the preceding theorem a countable family $\mathscr{F}$ of $\mathscr{V}$-sets may be found which is an 0-covering of Y, with $\mathfrak{L}^p$-overlap and (μ, Y)-overflow of $\mathscr{F}$ both as small as desired.

If, in the proof of Th. 4.6, we adopt the ε-covering version of the definition of S^p-bases, then the first relation in (4.6.3) must be replaced by $\mu(\bar{X} - \bar{X} \cdot S) < \varepsilon$. The Vitali ψ-property is established in the ε-covering version.

5. Relation to Younovitch's derivation theorem [*48*; *25*, 1.7]. We mention this work in passing for the interested reader. YOUNOVITCH operates under hypotheses that are in some ways more restrictive, in others less so, than ours. Unfortunately, neither proofs nor even sketches of them appear in the abstract [*48*] of his work. If one can prove under his assumptions that the (extreme) derivates are μ-measurable (see Remarks after Th. 2.6), or that any $\mathfrak{B}$-fine covering of a set X is a $\mathfrak{B}$-fine covering of any measure cover for X, then Younovitch's theorem follows from Th. 4.2.

6. The strong Vitali property.

6.1. Definitions. We shall say that the basis $\mathfrak{B}$ has the *strong Vitali property* (abbreviated (S.V.)) or is a *strong Vitali basis* iff for each $\varepsilon > 0$, each set $X \subset E$ of finite outer measure, and each $\mathfrak{B}$-fine covering $\mathscr{V}$ of X, there exists an (enumerable) M-family $\mathscr{E}$ of $\mathscr{V}$-constituents such that, putting $S = \sigma\mathscr{E}$,

(S.V. 1) $(X - X \cdot S) \in \mathscr{N}^\star$;

(S.V. 2) $\mu(S - S \cdot \bar{X}) < \varepsilon$;

(S.V. 3 strict) the $\mathscr{E}$-constituents are pairwise disjoint.

If, in (S.V. 3 strict), we replace strict disjunction by 0-disjunction, i.e., disjunction mod $\mathcal{N}^\star$, we obtain the *strong Vitali property* mod $\mathcal{N}^\star$. If we discard (S.V. 2), we have the *reduced strong Vitali property* (abbreviated (R.S.V.). Recalling Defs. 4.5, we find it convenient to designate as an S^∞-basis, any basis having the property (S.V. mod $\mathcal{N}^\star$).

The straightforward proofs of the following are omitted.

6.2. Proposition. (S.V. mod $\mathcal{N}^\star$) implies the Vitali property for μ-finite μ-integrals.

6.3. Proposition. (R.S.V.) and Haupt's adaptation property together imply the Vitali property for Radon measures.

Remarks. The strong Vitali properties lack the flexibility of the Vitali properties of II. 1. In their formulation, one cannot replace the 0-covering condition by an ε-covering one, nor in the (G_σ) case, replace the phrase "of finite measure" by "bounded". However, such alterations are permissible if the constituents are $\mathcal{A}$-sets.

6.4. An example [*19*, 291]. We show here an example of a blanket (I. 3.3) that has the Vitali property for Radon measures but not the strong Vitali property. Here, $R = \mathbf{R}^2$ (the cartesian euclidean plane), $E = S =]0,0; 1,1[$ (the open unit square), μ is plane Borel measure, n is an arbitrary positive integer, and t is an arbitrary point of S. We let $I_{n,t}$, $\dot{I}_{n,t}$, and $\ddot{I}_{n,t}$ denote closed squares centered at t, with sides parallel to the axes and of length 2^{-n}, $3 \cdot 2^{-n}$, and $9 \cdot 2^{-n}$, respectively. For each point $(x,y) = t$ we let K_t denote the set of points in $\mathbf{R}^2$ of the form $(x+r, y+s)$, where r and s are arbitrary rational numbers. Finally, we define $J_{n,t} = I_{n,t} \cup K_t \cdot \ddot{I}_{n,t}$. Thus $J_{n,t}$ consists of a "solid core" surrounded by a "weightless cloud" $K_t \cdot (\ddot{I}_{n,t} - I_{n,t})$.

To each set B of the form $B = J_{n,t}$, there correspond uniquely the sets $B' = I_{n,t}$ and $B'' = \dot{I}_{n,t}$, and conversely. Whenever we speak of corresponding sets in the work at hand it is to be understood in this sense. It is easy to see that

$$\mu(B) = \mu(B') = (1/9)\,\mu(B''); \qquad (6.4.1)$$

$$\text{if} \quad B_1 \cdot B_2 = \emptyset, \quad \text{then} \quad B_1' \cdot B_2' = B_1'' \cdot B_2'' = \emptyset. \qquad (6.4.2)$$

We define $\mathfrak{T}$ as that blanket with domain S for which the universal deriving sequences at t are $(J_{n,t})$. We assert that $\mathfrak{T}$ has not the strong Vitali property.

To show this, we take a subblanket of $\mathfrak{T}$, say $\mathfrak{T}_0$, with domain S, each member of whose spread is included in S, and pick an arbitrary countable disjoint subfamily $\mathcal{V}$ of this spread. We denote by $\mathcal{V}''$ the family of the sets B'' corresponding to the sets B in $\mathscr{V}$. We note that $S'' = \cup \mathscr{V}'' \subset S$, use (6.4.1) and (6.4.2), and obtain

$$\mu(S - \cup \mathscr{V}) \geqslant \mu(S) - \sum_{B \in \mathscr{V}} \mu(B) = 1 - 1/9 \sum_{B \in \mathscr{V}} \mu(B'')$$
$$= 1 - (1/9)\mu(S'') \geqslant 8/9\,.$$

Hence $\mathfrak{T}$ has not the strong Vitali property.

We further declare that $\mathfrak{T}$ possesses the Vitali property for Radon measures ψ. We first select an arbitrary ψ, let M denote the set of points $u \in \mathbf{R}^2$ for which $\psi(\{u\}) > 0$, and define N as the set of points of the form $(x + r, y + s)$, where r and s are arbitrary rational numbers and $(x, y) \in M$. We choose an arbitrary subblanket of $\mathfrak{T}$, say $\mathfrak{T}_1$, with domain $S_1 \subset (S - S \cdot N)$. Since N is countable, making $\mu(N) = 0$, the Vitali property for Radon measures will be established if we can find in the spread of $\mathfrak{T}$ a countable family $\mathscr{V}$ for which

$$\mu(S - S \cdot \cup \mathscr{V}) = 0\,, \qquad \int_{\cup \mathscr{V}} \epsilon_{\mathscr{V}}\, d\psi = 0\,.$$

We so define the blanket $\mathfrak{U}_1$ that for $t \in S_1$, the universal sequence associated with t by $\mathfrak{U}_1$ is

$$\mathfrak{U}_1(t) = \bigcup_{B \in \mathfrak{T}_1(t)} \{B'\}$$

and use the strong Vitali property for $\mathfrak{U}_1$ (6.1) to find a countable disjoint subfamily $\mathscr{H}$ of the spread of $\mathfrak{U}_1$ for which

$$\mu(S_1 - S_1 \cdot \cup \mathscr{H}) = 0\,.$$

There exists a corresponding countable family (not necessarily disjoint)

$$\mathscr{V} = \bigcup_{B' \in \mathscr{H}} \{B\}$$

in the spread of $\mathfrak{T}_1$. Clearly $\cup \mathscr{H} \subset \cup \mathscr{V}$ and $\mu(S_1 - S_1 \cdot \cup \mathscr{V}) = 0$. From the disjointedness of $\mathscr{H}$ we know that no two members of $\mathscr{V}$ have "core" points in common, so that the set of those points $u \in \cup \mathscr{V}$ for which $\epsilon_{\mathscr{V}}(u) > 0$ must lie in the countable set Q which embraces all "cloud" points of members of $\mathscr{V}$. However, the centers of members of $\mathscr{H}$ cannot be points of N, so that the "cloud" points of members of $\mathscr{V}$ cannot be points of M; thus $\psi(\{u\}) = 0$ for each $u \in Q$. Therefore,

$$\int_{\cup \mathscr{V}} \epsilon_{\mathscr{V}}\, d\psi \leqslant \int_{Q} \epsilon_{\mathscr{V}}\, d\psi = 0$$

and $\mathfrak{T}$ enjoys the Vitali property for Radon measures.

7. Half-regular and regular branches of a derivation basis [*15*, 9.7; *25*, 3.3].

We merely mention this topic for the interested reader. It concerns a generalization of the idea of "regular" sequences converging to a point in the classical situation [cf. *44*, 106]. Some of the consequences of this concept are proved in [*25*, 3.3].

Chapter III

The Converse Problem I: Covering Properties Deduced from Derivation Properties of σ-additive Set Functions

1. De Possel's equivalence theorem [*38*, 403–405].

1.1. Definition. A derivation basis $\mathfrak{B}$ possesses the *density property* iff it derives the μ-integrals of the characteristic functions of all μ-measurable sets; that is, iff for any $\mathcal{M}$-set M, the density at x, defined as the limit of $\mu(M_\iota(x)\cdot M)/\mu(M_\iota(x))$, exists and equals $c_M(x)$ (characteristic function of M at x) $\mu^\star$-almost everywhere on E.

1.2. Theorem. *The density property and the Vitali μ-property are equivalent.*

Proof. As we noted in the Remarks following Def. II. 2.7, the Vitali μ-property implies the density property. We have to prove the converse.

We assume that the density property holds and let X denote a subset of E of finite positive outer measure, $\mathcal{V}$ a $\mathfrak{B}$-fine covering of X, and ε a positive number. We select α, $0<\alpha<1$, so that

$$0<(\alpha^{-1}-1)\mu(\bar{X})<\varepsilon. \tag{1.2.1}$$

If Y is any subset of X such that $\bar{\mu}(Y)>0$, then we define $\mathcal{V}(Y,\alpha)$ as that family of $\mathcal{V}$-sets V for which

$$\mu(\bar{Y}\cdot V)>\alpha\mu(V), \tag{1.2.2}$$

and μ_Y as the supremum of the numbers $\mu(V)$, for $V\in\mathcal{V}(Y,\alpha)$.

If $\bar{\mu}(Y)>0$, then it follows from the density property that the density of Y equals 1 for at least one point $y\in Y$. Hence there exists at least one y-converging sequence in $\mathfrak{B}$ whose constituents belong to $\mathcal{V}$ and satisfy (1.2.2). The family $\mathcal{V}(Y,\alpha)$ is thus non-vacuous, hence $\mu_Y>0$. In case $Y\subset X$ and $\bar{\mu}(Y)=0$, we put $\mu_Y=0$.

We fix a number κ, $0<\kappa<1$. By the definition of μ_X, there exists a $\mathcal{V}$-set V_1 such that

$$\mu(V_1)>\kappa\mu_X,\ \mu(\bar{X}\cdot V_1)>\alpha\mu(V_1).$$

We let $X_1=X$, $X_2=X_1-V_1\cdot X_1$. From this point we proceed inductively, assuming that sets $V_i\in\mathcal{V}$ have been defined for $i=1,2,\ldots,n$, satisfying the relations

$$\mu(\bar{X}_i \cdot V_i) > \alpha\mu(V_i), \mu(V_i) > \kappa\mu_{X_i}, \tag{1.2.3}$$

where

$$X_{i+1} = X_1 - X_1 \cdot \left(\bigcup_{j=1}^{i} V_j\right).$$

In case $\bar{\mu}(X_{n+1}) = 0$, we stop the process; in case $\bar{\mu}(X_{n+1}) > 0$, we define a new $\mathscr{V}$-constituent V_{n+1} such that

$$\mu(V_{n+1}) > \kappa\mu_{X_{n+1}}, \quad \mu(\bar{X}_{n+1} \cdot V_{n+1}) > \alpha\mu(V_{n+1}).$$

The process just described leads to the construction of an M-family $\mathscr{E}$ defined by a finite or infinite sequence of sets (V_i) taken from $\mathscr{V}$, satisfying (1.2.3) for $i = 1, 2, \ldots$. Since also the sets $(V_i \cdot X_i)$ are disjoint (mod $\mathscr{N}^\star$), we have

$$\mu(\bar{X}) \geqslant \mu\left(\bar{X} \cdot \left(\bigcup_i V_i\right)\right) \geqslant \mu\left(\bigcup_i (\bar{X}_i \cdot V_i)\right) = \sum_i \mu(\bar{X}_i \cdot V_i) > \alpha \sum_i \mu(V_i);$$

consequently,

$$\sum_i \mu(V_i) < \alpha^{-1}\mu\left(\bar{X} \cdot \left(\bigcup_i V_i\right)\right) < \infty. \tag{1.2.4}$$

Putting $S = \sigma\mathscr{E}$ and combining (1.2.1) with (1.2.4), we obtain

$$\left(\sum_i \mu(V_i) - \mu(S)\right) + \mu(S - S \cdot \bar{X}) = \sum_i \mu(V_i) - \mu(S\,\bar{X}) < (\alpha^{-1} - 1)\mu(S \cdot \bar{X}) < \varepsilon. \tag{1.2.5}$$

Hence conditions (V 2) and (V 3) of Defs. II. 2.7, with $\psi = \mu$, hold.

To show that (V 1) holds for our family $\mathscr{E}$, we note that if the sequence of the sets (V_i) is finite, then for some positive integer N we have $\mathscr{E}$: $(V_1, V_2, \ldots, V_N)$, and $\bar{\mu}(X_{N+1}) = 0$. Thus $\bar{\mu}(X - X \cdot \sigma\mathscr{E}) = 0$, as required by ($V$ 1). If the sequence is infinite, then from (1.2.3) and (1.2.4) we see that

$$\kappa \sum_i \mu_{X_i} < \sum_i \mu(V_i) < \alpha^{-1}\mu(\bar{X}) < \infty.$$

Hence $\lim_i \mu_{X_i} = 0$. We let $X_\infty = X - X \cdot \left(\bigcup_i V_i\right)$. Since $X_\infty \subset X_n$, and hence $\mathscr{V}(X_\infty, \alpha) \subset \mathscr{V}(X_n, \alpha)$, for $n = 1, 2, \ldots$, then $\mu_{X_\infty} = 0$. This means that $\sigma\mathscr{E} = \bigcup_i V_i \supset X$ (mod $\mathscr{N}^\star$) as required, and the proof is complete.

2. A necessary and sufficient condition for a weak derivation basis to derive a μ-finite $\mathscr{M}$-measure (Radon measure) ψ. We assume that $\mathfrak{B}$ is a weak derivation basis; that is, $\mathfrak{B}$ possesses the Vitali μ-property.

We let f denote a μ-measurable non-negative function, finite μ-almost everywhere in R. By f_n we shall mean that function for which $f_n(x) = f(x)$ if $f(x) < n$, and $f_n(x) = 0$ if $f(x) \geqslant n$, $n = 1, 2, \ldots$. We further define $r_n(x) = f(x) - f_n(x)$; and, for $M \in \mathscr{M}$, we define ψ, ψ_n, and ρ_n so that for each value $n = 1, 2, \ldots$,

$$\psi(M) = \int_M f\,d\mu; \quad \psi_n(M) = \int_M f_n\,d\mu; \quad \rho_n(M) = \int_M r_n\,d\mu.$$

Since f_n is a μ-measurable bounded function, ψ_n is $\mathfrak{B}$-derivable $\mu^\star$-almost everywhere in E; that is, $D\psi_n$ exists $\mu^\star$-almost everywhere in E and equals f_n (mod $\mathcal{N}^\star$). We have

$$D^\star\psi = D\psi_n + D^\star\rho_n \ (\mathrm{mod}\ \mathcal{N}^\star)$$

hence

$$D^\star\psi = f_n + D^\star\rho_n (\mathrm{mod}\, \mathcal{N}^\star)$$

on E. In accordance with the definition of f_n and the finiteness of f we have $\lim_n f_n = f$ $\mu^\star$-almost everywhere on E. This leads to the following result.

2.1. Lemma. *A necessary and sufficient condition for a weak derivation basis $\mathfrak{B}$ to derive ψ is*

$$\bar{\mu}\left(\left[\lim_n D^\star\rho_n > 0\right]\right) = 0\,.$$

If $\bar{\mu}(E)$ is finite, then the condition $\lim_n \bar{\mu}([D^\star\rho_n > \varepsilon]) = 0$ for each positive ε implies $\bar{\mu}\left(\left[\lim_n D^\star\rho_n > 0\right]\right) = 0.$

2.2. Corollary. If $\mathfrak{B}$ derives the non-negative μ-integral ψ, then it derives any $\mathcal{M}$-measure ψ' for which $\psi' \leqslant \psi$.

This result can be extended to any μ-finite or Radon $\mathcal{M}$-measure ψ. In fact, for $M \in \mathcal{M}$,

$$\psi(M) = \psi_s(M) + \int_M f\,d\mu\,,$$

where f represents a (μ-measurable) Radon-Nikodym integrand of ψ. We suppose that ψ' is any $\mathcal{M}$-measure, $\psi' \leqslant \psi$. Then

$$\psi'(M) = \psi'_s(M) + \int_M f'\,d\mu$$

is the corresponding decomposition for ψ'.

If N_0 denotes an $\mathcal{N}$-set for which $\psi_s(R-N_0) = 0$, then ψ is μ-absolutely continuous on the $\mathcal{M}$-subsets of $R-N_0$; consequently, so is ψ'. Therefore $\psi'_s(R-N_0) = 0$,

$$\psi'_r(M) = \psi'(M\cdot(R-N_0)) \leqslant \psi(M\cdot(R-N_0)) = \psi_r(M)\,,$$
$$\psi'_s(M) = \psi'(M\cdot N_0) \leqslant \psi(M\cdot N_0) = \psi_s(M)\,.$$

Thus ψ'_r and ψ'_s are dominated by ψ_r and ψ_s, respectively. The assumption that $\mathfrak{B}$ derives ψ means that $D\psi = f\,(\mathrm{mod}\,\mathcal{N}^\star)$ on E, hence $D\psi_r = f$ (mod $\mathcal{N}^\star$) on E, and $D\psi_s = 0\,(\mathrm{mod}\,\mathcal{N}^\star)$ on E. Since $D^\star\psi'_s \leqslant D^\star\psi_s$, then $D^\star\psi'_s$ exists and equals zero (mod $\mathcal{N}^\star$) on E. Thus, we have the following general result.

2.3. Theorem. *If a weak derivation basis $\mathfrak{B}$ derives the μ-finite (resp., Radon) $\mathcal{M}$-measure ψ, then $\mathfrak{B}$ derives any μ-finite (resp., Radon) $\mathcal{M}$-measure dominated by ψ.*

2.4. Corollary. If a weak derivation basis derives the total variation τ of a signed μ-finite (Radon) $\mathcal{M}$-measure ψ, then it derives ψ itself.

As a special case, if the weak basis $\mathfrak{B}$ derives the integral $\int_M |f| \, d\mu$, where f is a σ-bounded measurable function, then $\mathfrak{B}$ derives $\int_M f \, d\mu$.

Remarks. If we wish, as does DE POSSEL, to "anchor" the sets V_n to points of X, we can extract V_n from an x-converging sequence, whose constituents belong to $\mathscr{V}(X_n, \alpha)$, and such that $x_n \in X_n$.

In any Euclidean space, the interval basis $\mathfrak{J}$ possesses the density property (V. 1; see also [*44*, 129]), therefore, by Theorem 1.2, it is a weak derivation basis. There exists [*33*] an example of an integrable function f in the plane whose indefinite integral is $\mathfrak{J}$-derivable (*strongly* derivable), although the integral of $|f|$ is not.

2.5. Lemma. *If the weak derivation basis $\mathfrak{B}$ derives the μ-finite (Radon) $\mathcal{M}$-measure ψ, $M \in \mathcal{M}$, and $\tau = \psi + \mu$, then the τ-density*

$$\lim_{\iota} (\tau (M \cdot M_\iota (x))/\tau (M_\iota (x)))$$

exists for $\mu^\star$-almost all x in E and equals $c_M(X)$ (the value of the characteristic function of M at x).

Proof. We let f denote a Radon-Nikodym integrand of ψ. Then, for $M' \in \mathcal{M}$,

$$\begin{aligned} \tau_M(M') &= \tau(M' \cdot M) = \psi(M' \cdot M) + \mu(M' \cdot M) \\ &= \int_{M' \cdot M} f \, d\mu + \psi_s(M' \cdot M) + \mu(M' \cdot M) \\ &= \int_{M' \cdot M} (f + 1) \, d\mu + \psi_s(M' \cdot M) \\ &= \int_{M'} c_M (f + 1) \, d\mu + \psi_s(M' \cdot M), \end{aligned}$$

where ψ_s is the μ-singular part of ψ. Since $\mathfrak{B}$ derives τ,

$$\lim_{\iota} (\tau (M_\iota (x))/\mu (M_\iota (x)))$$

exists and equals $f(x) + 1$ for $\mu^\star$-almost all $x \in E$. But $\mathfrak{B}$ also derives τ_M, so that $\lim_{\iota} (\tau (M_\iota(x) \cdot M)/\mu (M_\iota(x)))$ exists, and, by above, is equal to $c_M(x) \cdot (f(x) + 1)$ for $\mu^\star$-almost all $x \in E$. Hence, by division, $\lim_{\iota} (\tau (M_\iota(x) \cdot M))/\tau (M_\iota(x))$ exists and equals $c_M(x)$ for $\mu^\star$-almost all $x \in E$.

2.6. Lemma. *If $\mathfrak{B}$, ψ, and τ are as in the preceding lemma, X is a subset of E of finite outer measure, M_1 is a measure-cover of X, $\mathscr{V}$ is a $\mathfrak{B}$-fine covering of X, ε is a positive number, and $0 < \alpha < 1$, then there exists a finite or infinite sequence of $\mathscr{V}$-sets (V_n) for which*

$$\bigcup_n V_n \supset X\,(\operatorname{mod}\mathscr{N}^\star), \qquad \sum_n \tau(V_n) < \tau\Big(M_1\cdot\Big(\bigcup_n V_n\Big)\Big)/\alpha\,. \tag{2.6.1}$$

Proof. For any $\mathscr{M}$-set M such that $\bar{\mu}(M\cdot E) > 0$, we define $\mathscr{V}(\tau, M, \alpha)$ as the family of $\mathscr{V}$-sets V for which

$$\tau(M\cdot V) > \alpha\tau(V) \tag{2.6.2}$$

and $\mu(\tau, M)$ as the supremum of the numbers $\mu(V)$ for $V \in \mathscr{V}(\tau, M, \alpha)$. From Lemma 2.5, it follows that there is at least one point $x \in M\cdot E$ at which the τ-density of M equals 1, hence $\mathscr{V}(\tau, M, \alpha)$ is non-vacuous and $\mu(\tau, M) > 0$. In case $M \in \mathscr{M}$ and $\bar{\mu}(M\cdot E) = 0$, we define $\mu(\tau, M) = 0$.

From this point on, the proof follows closely that of Th. 1.2, with τ replacing μ and the measure covers having to be specially selected, since τ need not be μ-absolutely continuous. By a process similar to that of Th. 1.2, for fixed κ, $0 < \kappa < 1$, we determine inductively a finite or infinite sequence $V_1, V_2, \ldots, V_n \ldots$ of $\mathscr{V}$-sets with properties as follows. We put $X_1 = X$ and, for any positive integer $n \geqslant 1$,

$$X_{n+1} = X_1 - X_1\cdot\Big(\bigcup_{i=1}^{n} V_i\Big);$$

M_{n+1} denotes a measure cover of X_{n+1} contained in $M_n - M_n\cdot\Big(\bigcup_{i=1}^{n} V_i\Big)$. If $\bar{\mu}(X_{n+1}) > 0$, then V_{n+1} is so chosen from $\mathscr{V}$ that

$$\tau(M_{n+1}\cdot V_{n+1}) > \alpha\tau(V_{n+1}), \qquad \mu(V_{n+1}) > \kappa\mu(\tau, M_{n+1}). \tag{2.6.3}$$

If $\bar{\mu}(X_{n+1}) = 0$, this process stops.

Our choice of the sets M_n ensures that the sets $M_n\cdot V_n$ are strictly disjoint; hence, using (2.6.3), we have

$$\begin{aligned}\tau(M_1) &\geqslant \tau\Big(M_1\cdot\Big(\bigcup_n V_n\Big)\Big) \geqslant \tau\Big(\bigcup_n (M_n\cdot V_n)\Big)\\ &= \sum_n \tau(M_n\cdot V_n) > \alpha\sum_n \tau(V_n),\end{aligned} \tag{2.6.4}$$

which is the second relation of (2.6.1).

If the sequence (V_n) is finite, then $\bar{\mu}(X_N) = 0$ holds for some positive integer N, and the first relation of (2.6.1) clearly holds. If this sequence is infinite, we let $X_\infty = X - X\cdot \bigcup_n V_n$; we may, and do, choose a μ-cover M_∞ of X_∞, included in $\bigcap_n M_n$. From (2.6.3) and (2.6.4) we have $\lim_n \mu(\tau, M_n) = 0$. Since $M_\infty \subset M_n$, we have $\mathscr{V}(\tau, M_\infty, \alpha) \subset \mathscr{V}(\tau, M_n, \alpha)$ and

$\mu(\tau, M_\infty) \leqslant \mu(\tau, M_n)$ for $n = 1, 2, \ldots$; thus $\mu(\tau, M_\infty) = 0$, and $\bar{\mu}(M_\infty \cdot E) = 0$. But $X_\infty \subset M_\infty \cdot E$; hence $\bar{\mu}(X_\infty) = 0$, and the first condition of (2.6.1) holds.

2.7. Theorem. *If a weak derivation basis $\mathfrak{B}$ derives the μ-finite (Radon) $\mathcal{M}$-measure ψ, then $\mathfrak{B}$ possesses the Vitali ψ-property.*

Proof. Taking X, M_1, $\mathcal{V}$, and ε as in the statement of Lemma 2.6, we select α so that

$$0 < (\alpha^{-1} - 1)\tau(M_1) < \varepsilon, \tag{2.7.1}$$

and choose an M-family $\mathscr{E}$ in accordance with Lemma 2.6, satisfying (2.6.1). For $S = \sigma\mathscr{E}$, the (τ, M_1)-redundancy of covering is given by

$$\left(\sum_n \tau(V_n) - \tau(S)\right) + \tau(S - S \cdot M_1) = \sum_n \tau(V_n) - \tau(S \cdot M_1),$$

which, by (2.6.1) and (2.7.1), is less than ε. Thus the τ-overlap of $\mathscr{E}$ and (τ, M_1)-overflow of $\mathscr{E}$ are each less than ε, so that the Vitali τ-property holds. Since $\psi \leqslant \tau$, the Vitali ψ-property also holds.

Remark. If desired, the sets V_n may be "anchored" to points of X_n, as in the DE POSSEL theorem.

Combining Th. 2.7 and Th. II. 3.2, we obtain the following criterion of derivability of an individual $\mathcal{M}$-measure.

2.8. Theorem. *A necessary and sufficient condition for a weak derivation basis $\mathfrak{B}$ to derive the μ-finite (Radon) $\mathcal{M}$-measure ψ is the validity of the Vitali ψ-property.*

2.9. Theorem. *The Vitali property for μ-finite (resp., Radon) μ-integrals is equivalent to the $\mathfrak{B}$-derivability of every μ-finite (resp., Radon) μ-integral; the Vitali property for μ-finite (resp., Radon) $\mathcal{M}$-measures is equivalent to the $\mathfrak{B}$-derivability of every μ-finite (resp., Radon) $\mathcal{M}$-measure.*

Proof. This follows from Ths. 1.2, 2.7 and II. 3.2.

3. Younovitch's equivalence theorem [*48*]. YOUNOVITCH (cf. II. 5.) formulates a Vitali μ-property and asserts its equivalence with the density property. That is, YOUNOVITCH asserts the truth of Th. 1.2 under his assumptions with a weakened Vitali μ-property in which X is required to belong to $\mathcal{M}$. He also formulates a criterion for the derivation of μ-integrals, which are necessarily finite since he assumes $\mu(R) < \infty$.

4. A converse theorem for bases deriving the $\mu^{(q)}$ functions, $q \geqslant 1$. In what follows, we assume that $\mathfrak{B}$ is a general derivation basis and that R has the property (G_σ).

For any M-family $\mathscr{E}$ of $\mathfrak{B}$-constituents and any positive numbers r and α, we denote by $\mathscr{E}(\alpha, r)$ the family of those sets $V \in \mathscr{E}$ for which

$$\int_V (\in_{\mathscr{E}})^r \, d\mu > \alpha \mu(V),$$

and we further let $S(\alpha, r, \mathscr{E})$ denote the union of the sets of $\mathscr{E}(\alpha, r)$. Clearly, if $r' > r''$, then $\mathscr{E}(\alpha, r') \supset \mathscr{E}(\alpha, r'')$ and $S(\alpha, r', \mathscr{E}) \supset S(\alpha, r'', \mathscr{E})$.

4.1. Lemma. *If $\mathscr{H}$ represents the* M*-family of those sets $V \in \mathscr{E}$ such that*

$$\int_V (\in_{\mathscr{E}})^r d\mu \leqslant \alpha \mu(V),$$

that is, if $\mathscr{H} = \mathscr{E} - \mathscr{E}(\alpha, r)$, then

$$(\omega^{(r+1)}(\mathscr{H}))^{r+1} \leqslant \alpha \sum_{V \in \mathscr{H}} \mu(V),$$

where $\omega^{(r+1)}(\mathscr{H})$ denotes the $\mathfrak{L}^{r+1}(\mu)$-overlap of $\mathscr{H}$.

Proof.

$$(\omega^{(r+1)}(\mathscr{H}))^{r+1} = \int_{\sigma\mathscr{H}} (\in_{\mathscr{H}})^{r+1} d\mu = \int_{\sigma\mathscr{H}} (\varphi_{\mathscr{H}} - 1)^{r+1} d\mu \leqslant \int_{\sigma\mathscr{H}} (\varphi_{\mathscr{E}} - 1)^r \varphi_{\mathscr{H}} d\mu$$

$$= \sum_{V \in \mathscr{H}} \int_V (\in_{\mathscr{H}})^r d\mu \leqslant \alpha \sum_{V \in \mathscr{H}} \mu(V).$$

In the preceding considerations, if $r = 0$, we shall interpret $(\varphi_{\mathscr{E}} - 1)^r$ as that function defined on $\sigma\mathscr{E}$, taking the value zero if $\in_{\mathscr{E}}(x) = 0$, or the value one if $\in_{\mathscr{E}}(x) \geqslant 1$; thus, it is the restriction to $\sigma\mathscr{E}$ of the characteristic function $c_{\theta\mathscr{E}}$ of the $\mathscr{E}$-overlap set $\theta\mathscr{E}$ (see Defs. II. 1.2). $\mathscr{E}(\alpha, 0)$ is the family of those sets $V \in \mathscr{E}$ for which $\mu(V \cdot \theta\mathscr{E}) > \alpha\mu(V)$.

The above lemma remains true when $r = 0$, since

$$\omega^{(1)}(\mathscr{H}) = \omega(\mathscr{H}, \mu) = \int_{\sigma\mathscr{H}} (\varphi_{\mathscr{H}} - 1) d\mu \leqslant \int_{\sigma\mathscr{H}} c_{\theta\mathscr{H}} \cdot \varphi_{\mathscr{H}} d\mu$$

$$\leqslant \sum_{V \in \mathscr{H}} \int_V c_{\theta\mathscr{E}} d\mu \leqslant \alpha \sum_{V \in \mathscr{H}} \mu(V).$$

4.2. Definition. We say that the basis $\mathfrak{B}$ has the property (H_p), for $p > 1$, iff for any bounded set $X \subset E$, any $\mathfrak{B}$-fine covering $\mathscr{V}$ of X, any $z^\star > \bar{\mu}(X)$, and any two positive numbers $\varepsilon^\star$ and $\alpha^\star$, there exists a finite M-family $\mathscr{F} \subset \mathscr{V}$ such that

$$\bar{\mu}(X - X \cdot \sigma\mathscr{F}) < \varepsilon^\star; \quad \sum_{V \in \mathscr{F}} \mu(V) < z^\star; \quad \mu(S(\alpha^\star, p-1, \mathscr{F})) < \varepsilon^\star. \qquad (4.2.1)$$

Remarks. Without the third condition, we have the Vitali μ-property in the ε-version, and for bounded subsets of E. As was noted earlier, this is equivalent to the original definition of the Vitali μ-property.

We observe that if $p' > p''$, then $(H_{p'})$ implies $(H_{p''})$.

4.3. Lemma. *If $\mathfrak{B}$ is an S^z basis, $z \geqslant 1$, and if $\mathfrak{B}$ does not possess the property $(H_{p'})$, where $p' > 1$, then there exists a bounded set $X_0 \subset E$, a bounded $\mathscr{G}$-set $G_0 \supset X_0$, a $\mathfrak{B}$-fine covering $\mathscr{V}_0$ of X_0, and positive numbers ε_0, α_0 such that for every M-family $\mathscr{F} \subset \mathscr{V}_0$ satisfying the relations*

$$\bar{\mu}(X_0 - X_0 \cdot \sigma\mathscr{F}) < \varepsilon_0, \quad \mu(\sigma\mathscr{F} - \sigma\mathscr{F} \cdot \bar{X}_0) < \varepsilon_0, \quad (\omega^{(z)}(\mathscr{F})) < \varepsilon_0, \quad \sigma\mathscr{F} \subset G_0, \tag{4.3.1}$$

we have

$$\mu(S(\alpha_0, p'-1, \mathscr{F})) > 2\varepsilon_0.$$

Proof. If G is a bounded $\mathscr{G}$-set, $X \subset E \cdot G$, $\mathscr{V}$ is a $\mathfrak{B}$-fine covering of X, α and ε are both positive numbers, then we call $(X, G, \mathscr{V}, \alpha, \varepsilon)$ an *admissible quintuple*. Since $\mathfrak{B}$ is an S^z-basis, then for any such quintuple there exist M-families $\mathscr{F} \subset \mathscr{V}$ for which

$$\bar{\mu}(X - X \cdot \sigma\mathscr{F}) < \varepsilon, \quad \mu(\sigma\mathscr{F} - \bar{X} \cdot \sigma\mathscr{F}) < \varepsilon, \quad (\omega^{(z)}(\mathscr{F}))^z < \varepsilon, \quad \text{and} \quad \sigma\mathscr{F} \subset G.$$

For any such family $\mathscr{F}$ we therefore have

$$\sum_{V \in \mathscr{F}} \mu(V) = \mu(\sigma\mathscr{F}) + \omega(\mathscr{F}, \mu) \leqslant \mu(\sigma\mathscr{F}) + (\omega^{(z)}(\mathscr{F}))^z < \bar{\mu}(X) + \varepsilon + \varepsilon = \bar{\mu}(X) + 2\varepsilon.$$

For any fixed admissible quintuple, we let η denote the infimum, among all such families $\mathscr{F}$, of the numbers $\mu(S(\alpha, p'-1, \mathscr{F}))$. It follows that if, for each admissible quintuple, the corresponding η were zero, then $\mathfrak{B}$ would have the property $(H_{p'})$, contrary to hypothesis. Thus, for some admissible quintuple $(X_0, G_0, \mathscr{V}_0, \alpha_0, \varepsilon_0)$, the corresponding η_0 is a positive number, and for each finite M-family $\mathscr{F}$ of $\mathscr{V}$-sets satisfying the relations (4.3.1), we have

$$\mu(S(\alpha_0, p'-1, \mathscr{F})) \geqslant \eta_0 > 0. \tag{4.3.2}$$

Now, if $\mathscr{F}$ is any M-family that satisfies the relations obtained from (4.3.1) merely by replacing ε_0 by a smaller positive number, then $\mathscr{F}$ necessarily satisfies the unchanged relations (4.3.1). Thus, we may assume that ε_0 has been chosen so small that $0 < \varepsilon_0 < \frac{1}{2}\eta_0$ which, in the light of (4.3.2), completes the proof.

Henceforth, any quintuple $(X_0, G_0, \mathscr{V}_0, \alpha_0, \varepsilon_0)$ satisfying the conditions of Lemma 4.3 will be called a *privileged quintuple*.

4.4. Lemma. *If the basis $\mathfrak{B}$ possesses the property (H_p), where $p > 1$, then $\mathfrak{B}$ is an S^p-basis.*

Proof. By virtue of the Remarks following Th. II. 4.7, it is sufficient to show that the S^p-properties hold when X is any bounded subset of E. Thus we take a bounded set $X \subset E$, a $\mathfrak{B}$-fine covering $\mathscr{V}$ of X, choose any positive number ε, and select $z^\star$ so that $\bar{\mu}(X) < z^\star < \bar{\mu}(X) + \frac{1}{2}\varepsilon$. We let $\varepsilon^\star = \frac{1}{4}\varepsilon$ and select any positive number $\alpha^\star$ for which $\alpha^\star z^\star < \varepsilon^p$.

We invoke the property (H_p) to find a finite M-family $\mathscr{H}$ of $\mathscr{V}$-constituents satisfying (4.2.1). We define $\mathscr{F}$ as the family of those $\mathscr{H}$-sets V for which

$$\int_V (\in_{\mathscr{H}})^{p-1} d\mu \leqslant \alpha^\star \mu(V).$$

If $Y = X \cdot \sigma\mathscr{H}$, then $\mathscr{F}$ covers $Y - Y \cdot S(\alpha^\star, p-1, \mathscr{H})$. Hence by (4.2.1) and the definition of $\varepsilon^\star$, we have

$$\bar{\mu}(X - X \cdot \sigma\mathscr{F}) \leqslant \bar{\mu}(X - X \cdot \sigma\mathscr{H}) + \mu(S(\alpha^\star, p-1, \mathscr{H})) < 2\varepsilon^\star = \tfrac{1}{2}\varepsilon\,;$$

that is to say, $\mathscr{F}$ is an $\frac{\varepsilon}{2}$-covering of X.

Using Lemma 4.1 with $r = p-1 > 0$, and taking account of the second relation in (4.2.1) and the choice of $\alpha^\star$, we obtain

$$(\omega^{(p)}(\mathscr{F}))^p \leqslant \alpha^\star \sum_{V \in \mathscr{F}} \mu(V) \leqslant \alpha^\star \sum_{V \in \mathscr{H}} \mu(V) < \alpha^\star z^\star < \varepsilon^p\,.$$

Finally, from conditions (4.2.1) we have

$$\begin{aligned}\mu(\sigma\mathscr{F} - \bar{X} \cdot \sigma\mathscr{F}) &\leqslant \mu(\sigma\mathscr{H} - \bar{X} \cdot \sigma\mathscr{H}) = \mu(\bar{X} - \bar{X} \cdot \sigma\mathscr{H}) + \mu(\sigma\mathscr{H}) - \mu(\bar{X}) \\ &\leqslant \mu(\bar{X} - \bar{X} \cdot \sigma\mathscr{H}) + \left(\sum_{V \in \mathscr{H}} \mu(V) - \mu(\bar{X})\right) < \varepsilon\,,\end{aligned}$$

which completes the proof that $\mathfrak{B}$ is an S^p-basis.

4.5. Hypothesis and notations. We assume $\mathfrak{B}$ to be a D-basis $[\mathscr{U}, \delta]$ (I. 3.5). For $\mathscr{V} \subset \mathscr{U}$ and $\eta > 0$, we denote by $\mathscr{V}_\eta$ the subfamily of $\mathscr{V}$ consisting of those sets $V \in \mathscr{V}$ for which $\delta(V) < \eta$. For a family (or an M-family) $\mathscr{F}$ of $\mathscr{U}$-sets U, we define the *δ-fineness* or *δ-norm* $\nu(\mathscr{F})$ as $\sup \delta(U)$, for $U \in \mathscr{F}$.

4.6. Lemma. *If $\mathfrak{B}$ is a D-basis $[\mathscr{U}, \delta]$, $p' > z \geqslant 1$, and $\mathfrak{B}$ is an S^z-basis but not an $S^{p'}$-basis, then there exists a bounded $\mathscr{G}$-set G_0 and a set X_0, with $G_0 \cdot E \supset X_0$, a $\mathfrak{B}$-fine covering $\mathscr{V}_0$ of X_0, positive numbers α_0 and ε_0, and a sequence $\mathscr{F}_1, \mathscr{F}_2, \ldots, \mathscr{F}_n, \ldots$ of finite M-families of $\mathscr{V}_0$-sets for which*

$$\lim_n \nu(\mathscr{F}_n) = 0, \qquad S_n = \sigma\mathscr{F}_n \quad \textit{for} \quad n = 1, 2, \ldots\,; \tag{4.6.1}$$

and

$$\begin{aligned}&\bar{\mu}(X_0 - X_0 \cdot S_n) < \varepsilon_0, \ (\omega^{(z)}(\mathscr{F}_n))^z < \varepsilon_0 / 2^{n+1}\,, \\ &\mu(S(\alpha_0, p'-1, \mathscr{F}_n)) > 2\varepsilon_0\,, \quad \textit{for} \quad n = 1, 2, \ldots\,.\end{aligned} \tag{4.6.2}$$

Proof. By Lemma 4.4, $\mathfrak{B}$ does not possess the property $(H_{p'})$, and we may apply Lemma 4.3 to find a privileged quintuple $(X_0, G_0, \mathscr{V}_0, \alpha_0, \varepsilon_0)$. Since $\mathfrak{B}$ is an S^z-basis, we define $\mathscr{F}_n$ as a finite M-family of $(V_0)_{1/n}$-sets included in G_0, satisfying the first two relations of (4.6.2). The last relation in (4.6.2) holds due to our choice of a privileged quintuple, and (4.6.1) is clearly valid.

4.7. Lemma. *We let $\mathfrak{B}$ denote an S^1-basis which is also a D-basis $[\mathscr{U}, \delta]$. We define p_0 as the supremum of numbers p such that $\mathfrak{B}$ is an S^p-basis, and assume $p_0 < \infty$. We define q_0 so that $p_0^{-1} + q_0^{-1} = 1$ if $p_0 > 1$; otherwise $q_0 = \infty$ if $p_0 = 1$. Then, for any number q, $1 < q < q_0$,* *there exists a $\mu^{(q)}$-function ψ_0, a positive number α_0, and a subset C_0 of E of positive outer measure such that*

$$\psi_0(M) = \int_M f_0 \, d\mu \quad \text{for} \quad M \in \mathcal{M}, \quad \text{and} \quad \int_R |f_0|^q \, d\mu < \infty; \tag{4.7.1}$$

$$f_0(x) = 0 \quad \text{for each} \quad x \in C_0; \tag{4.7.2}$$

$$D^\star \psi_0 \geqslant \alpha_0 > 0 \quad \text{for each} \quad x \in C_0. \tag{4.7.3}$$

Proof. We define p so that $p^{-1} + q^{-1} = 1$. Our hypotheses on q ensure that $p_0 < p < \infty$. In case $p_0 > 1$, we clearly have

$$0 < q(p_0 - 1) < q_0(p_0 - 1) = p_0;$$

hence we can choose a number p' so that

$$0 < q(p' - 1) < p_0 < p' < p.$$

Even in the case $p_0 = 1$, this last inequality may be satisfied for a suitable choice of p'.

In either case, we so define q' that $(p')^{-1} + (q')^{-1} = 1$; clearly, then, $q < q' < q_0$. We let z denote the larger of the two numbers $q(p'-1)$ and 1.

From our assumptions, it follows that $\mathfrak{B}$ is an S^z-basis but not an $S^{p'}$-basis. Lemma 4.6 asserts the existence of a privileged quintuple $(X_0, G_0, \mathcal{V}_0, \alpha_0, \varepsilon_0)$ and a sequence $\mathcal{F}_1, \mathcal{F}_2, \ldots, \mathcal{F}_n, \ldots$ of finite M-families of $\mathcal{V}_0$-sets satisfying (4.6.1) and (4.6.2). We let

$$S_n = \sigma \mathcal{F}_n, \qquad O_n = \theta \mathcal{F}_n \quad (\mathcal{F}_n\text{-overlap set}), \qquad D = \bigcup_{n=1}^{\infty} O_n,$$
$$H_n = S(\alpha_0, p'-1, \mathcal{F}_n), \qquad Q_n = H_n - H_n \cdot D, \qquad C_0 = \limsup_n Q_n,$$
$$\in_n(x) = \in_{\mathcal{F}_n}(x) \quad \text{if} \quad x \in O_n, \qquad \in_n(x) = 0 \quad \text{if} \quad x \notin O_n.$$

We have

$$\mu(D) \leqslant \sum_{n=1}^{\infty} \mu(O_n) \leqslant \sum_{n=1}^{\infty} \omega^{(1)}(\mathcal{F}_n) \leqslant \sum_{n=1}^{\infty} (\omega^{(z)}(\mathcal{F}_n))^z < \varepsilon_0.$$

Since $\mu(H_n) > 2\varepsilon_0$ for $n = 1, 2, \ldots$, then $\mu(Q_n) > \varepsilon_0$ for each such n. Since $Q_n \subset G_0$ for $n = 1, 2, \ldots$, and $\bar{\mu}(G_0) < \infty$, then $\mu(C_0) \geqslant \varepsilon_0$. We define

$$f_0 = \sum_{n=1}^{\infty} (\in_n)^{p'-1}; \qquad \psi_0(M) = \int_M f_0 \, d\mu \quad \text{for} \quad M \in \mathcal{M}. \tag{4.7.4}$$

Evidently f_0 is non-negative and vanishes on $R - D$, hence in particular on $C_0 \subset (R - D)$. This confirms (4.7.2).

From the definition of f_0 and Minkowski's inequality, we have

$$\left(\int_R |f_0|^q d\mu\right)^{1/q} \leqslant \sum_{n=1}^{\infty}\left(\int_R (\epsilon_n)^{(p'-1)q} d\mu\right)^{1/q} \leqslant \sum_{n=1}^{\infty}\left(\int_R (\epsilon_n)^z d\mu\right)^{1/q}$$
$$= \sum_{n=1}^{\infty} (\omega^{(z)}(\mathscr{F}_n))^{z/q} \leqslant \sum_{n=1}^{\infty}\left(\frac{\varepsilon_0}{2^{n+1}}\right)^{1/q} = \varepsilon_0^{1/q} \sum_{n=1}^{\infty} \rho^{n+1},$$

where $\rho = 2^{-1/q}$. Since $q > 1$, then $\rho < 1$ and the sum of the geometric series is finite. Thus (4.7.1) holds.

Since $H_n = S(\alpha_0, p'-1, \mathscr{F}_n)$, it follows from the definition of this last expression that for each point $x \in H_n$, $n = 1,2,\ldots$, there exists a set $V \in \mathscr{F}_n$ for which

$$\int_V (\epsilon_n)^{p'-1} d\mu > \alpha_0 \mu(V), \quad x \in V. \tag{4.7.5}$$

Consequently, to each point $x \in C_0 \subset \limsup_n H_n$, there exists a sequence (n_j) of natural numbers such that

$$x \in V_{n_j}, \qquad V_{n_j} \in \mathscr{F}_{n_j}(\alpha_0, p'-1), \qquad j = 1,2,\ldots.$$

The sequence (V_{n_j}), $j = 1,2,\ldots$ is an x-contracting sequence of sets belonging to $\mathfrak{B}$; thus $x \in E$ and $C_0 \subset E$.

From (4.7.4) and (4.7.5) it follows that

$$\psi_0(V_{n_j})/\mu(V_{n_j}) \geqslant \left(\int_{V_{n_j}} (\epsilon_{n_j})^{p'-1} d\mu\right) \Big/ \mu(V_{n_j}) > \alpha_0,$$

from which we obtain (4.7.3).

4.8. Theorem. *If $\mathfrak{B}$ is a* D-*basis which derives the $\mu^{(q)}$-functions, where $1 < q < \infty$, and if p is defined so that $p^{-1} + q^{-1} = 1$, then $\mathfrak{B}$ is an $S^{p'}$-basis for each number p' such that $1 \leqslant p' < p$.*

Proof. Since $q > 1$, and $\mathfrak{B}$ derives the $\mu^{(q)}$-functions, then $\mathfrak{B}$ must derive the $\mu^{(\infty)}$-functions, that is, the integrals of μ-measurable functions that are bounded on each set G_n^0. By Th. 1.2, $\mathfrak{B}$ is an S^1-basis. Next, we define p_0 and q_0 as in Lemma 4.7. In case $p_0 = \infty$, it is clear that $\mathfrak{B}$ is an $S^{p'}$-basis and the theorem holds. In case $1 \leqslant p_0 < \infty$, Lemma 4.7 tells us that for each number q' such that $1 < q' < q_0$, there exists at least one $\mu^{(q')}$-function that $\mathfrak{B}$ fails to derive. Our hypotheses thus compel us to conclude that $q \geqslant q_0$, hence $p \leqslant p_0$, from which the statement of the theorem is seen to be true.

Remarks. Th. 4.8 is not a clear-cut converse theorem because it does not say that $\mathfrak{B}$ is an S^p-basis. We conjecture that a D-basis can be constructed that is an $S^{p'}$-basis for each $p' < p$, yet fails to be an S^p-basis.

Chapter IV

Halo Assumptions in Derivation Theory. Converse Problem II

In the classical proof by CARATHÉODORY [7, 299–307] of the Lebesgue derivation theorem for the cube basis, the preliminary Vitali theorem is deduced from a "halo property" of cubes, namely: If, for any cube V_0 (the *nucleus*), $H(V_0)$ (the *halo*) denotes the union of those cubes V which are not greater than V_0 and intersect V_0, then the *dilation*, that is, the ratio of the measure of the halo to the measure of V_0, is uniformly bounded for all V_0; in fact, it is equal to 3^n, where n denotes the dimension of the euclidean space. In the following, we shall consider various halo properties differing mainly by the *incidence* requirements; in the example just given, the nonvacuity of $V \cdot V_0$ was demanded. From these halo properties we shall deduce Vitali properties and thus, by virtue of Ch. II, derivation properties.

1. A. P. Morse's halo properties.

1.1. Definitions. We say that Δ is a (MORSE) *disentanglement function* iff Δ is a non-negative finite function defined on the spread $\mathcal{D}$ (the family of the constituents) of the basis $\mathfrak{B}$ (I. 4.1).

If α is a fixed number, $\alpha \geqslant 1$, Δ is a disentanglement function, V_0 (the *nucleus*) is a $\mathfrak{B}$-constituent, then the *Morse halo* $H(\Delta,\alpha,V_0)$ (abbreviated $H(V_0)$ when no confusion can arise with regard to Δ and α) is the union of those $\mathfrak{B}$-constituents V that intersect V_0 and satisfy the relation $\Delta(V) \leqslant \alpha\Delta(V_0)$. The *halo dilation* $\rho(\Delta,\alpha,V_0)$ (abbreviated $\rho(V_0)$) is defined as the ratio $\bar{\mu}(H(\Delta,\alpha,V_0))/\mu(V_0)$.

Remarks. (1) The term "halo" was first used by K. O. HOUSEHAM in his talks in Cape Town, 1950, on A. P. MORSE's derivation theory, to denote Morse's set $\Delta:\beta$ [*29*, 207]. We diverge from the conventional use of the term by permitting our halo to have points in common with the nucleus, or even to include its nucleus. However, all our halo conditions control the proper halo, that is, the part of the halo outside its nucleus, thus retaining the basic meaning of the term.

(2) The definition of the halo involves only the spread $\mathcal{D}$ and the disentanglement function Δ; it refers neither to the basis $\mathfrak{B}$ (the pretopology) nor to the measure μ. The halo dilation depends on μ but not on $\mathfrak{B}$.

(3) For a recent generalization of Morse's work, see [*1*].

1.2. Examples. $R = \mathbf{R}^2$, the cartesian Euclidean plane, μ is Borel measure in R^2.

(1) $\mathcal{D}$ is the set of closed squares I; $\Delta(I)$ is the diameter of I, $\alpha = 1$. In this case the halo of a square I_0 with center at x and sides of length

s_0 is a concentric closed square with sides of length $3\,s_0$ and $\rho(I_0) = 3^2 = 9$.

(2) $\mathscr{D}$ is the set of circular discs, Δ is the diameter function defined on $\mathscr{D}$, and $\alpha = 1$. It is clear that the halo of a disc D in $\mathscr{D}$ is a concentric disc of three times the diameter of D, and $\rho(D) = 9$.

(3) $\mathscr{D}$ is the set of closed intervals I in $\mathbf{R}^2$, $\Delta(I)$ is the Borel measure of I, and $\alpha \geqslant 1$. This time, $H(I)$ is a cross-like set that stretches to infinity in the directions of the axes, and $\rho(I) = \infty$.

(4) $\mathscr{D}$ is the set of closed intervals I in $\mathbf{R}^2$, $\Delta(I)$ is the diameter of I, and $\alpha \geqslant 1$. In this case $\rho(I)$ is finite; however, elementary geometric considerations show that $\rho(I)$ is not bounded on $\mathscr{D}$.

1.3. Definitions. As auxiliary notions in the proofs of §§ 2 and 3 that follow, we define:

(1) The *sharp* (R. S. V.) *property* for a basis $\mathfrak{B}$ representing a strengthening of the (R. S. V.) property (II. 6.1), under which finitely many arbitrarily prescribed disjoint $\mathscr{V}$-sets appear among the sets of the family $\mathscr{E}$.

(2) The *sharp* (S. V.) *property* for a basis $\mathfrak{B}$, which is a strengthened (S. V.) property, such that for any given $\varepsilon > 0$, finitely many arbitrarily prescribed disjoint $\mathscr{V}$-sets whose total (X, μ)-overflow is less than ε, occur in the family $\mathscr{E}$.

2. Abstract version of the strong Vitali theorem modelled after Banach.

2.1. Introductory considerations. We shall follow the procedure of A. P. Morse, but with an important difference. Morse assumes the space to be metric; pointwise contraction is defined with respect to this metric, and the (R. S. V.) property is established. In fact, the (S. V.) property follows from the latter in Morse's setting, since, for any $\varepsilon > 0$ and any set X such that $\bar{\mu}(X) < \infty$, we can associate an open set G for which $X \subset G$ and $\bar{\mu}(G) < \bar{\mu}(X) + \varepsilon$ (property $(U\,G)$, Def. II. 1.4). We then discard all $\mathscr{V}$-sets not included in G; the set of the remaining $\mathscr{V}$-sets still comprises a $\mathfrak{B}$-fine covering (mod $\mathscr{N}^\star$) of X (I. 4.2), and for every sequence of such sets the (X, μ)-overflow is clearly less than ε. These simple considerations explain why so many authors manage using only the (R. S. V.) property, although the stronger (S. V.) property is required in the full derivation theorem.

We now transfer Morse's theory to the general setting of Ch. I; the discarded metric topology is replaced by the pretopological considerations of Ch. I.

2.2. Theorem. *(Special* B-*Vitali theorem.) We assume that*

(1) $\mathfrak{B}$ *is a derivation basis with domain E whose spread consists of* $\mathscr{A}$*-sets (Def.* I. 4.2*), and for which Axiom* (**E**) *holds (cf.* I. 2*);*

(2) *there exists a disentanglement function* Δ, *bounded by* $\kappa < \infty$; *and* $\alpha > 1$, $\beta > 0$ *are such real numbers that* $\rho(V) < \beta$ *for each set* V *in the spread of* $\mathfrak{B}$.

(3) *if* $X \subset E$ *and* $\bar{\mu}(X) < \infty$, *then there exists* $G \in \mathscr{G}$ *with* $X \subset G$, $\bar{\mu}(G) < \infty$.

Then $\mathfrak{B}$ *has the sharp (R.S.V.) property.*

Proof. We take an arbitrary set $X \subset E$ with $\bar{\mu}(X) < \infty$, let G be a $\mathscr{G}$-set such that $X \subset G$, $\bar{\mu}(G) < \infty$, and let $\mathscr{V}$ be any $\mathfrak{B}$-fine covering of $X \pmod{\mathscr{N}^\star}$. Since G is an external D-open set (mod $\mathscr{N}^\star$), the G-pruned family $\mathscr{V}_G$ is a $\mathfrak{B}$-fine covering (mod $\mathscr{N}^\star$) of X, so there is no loss of generality in assuming that all members of $\mathscr{V}$ are subsets of G. We also let $W_1, W_2, \ldots, W_p$ be any finite disjoint subfamily of the spread of $\mathfrak{B}$.

We first show that $\mathfrak{B}$ has the (R.S.V.) property. To this end, we let $\mathscr{V}_1 = \mathscr{V}$ and set $\Delta_1 = \sup_{V \in \mathscr{V}_1} \Delta(V)$; evidently $0 \leqslant \Delta_1 \leqslant \kappa$ and there is a set $V_1 \in \mathscr{V}_1$ such that $\Delta(V_1) \geqslant \Delta_1/\alpha$. We let $\mathscr{V}_2$ denote the family of those members of $\mathscr{V}$ that do not intersect V_1. In case $\mathscr{V}_2 \neq \emptyset$, we let $\Delta_2 = \sup_{V \in \mathscr{V}_1} \Delta(V)$ and select a set $V_2 \in \mathscr{V}_2$ with $\Delta(V_2) \geqslant \Delta/\alpha$. In case $\mathscr{V}_2 = \emptyset$, the process stops.

We proceed thus inductively. If λ is any ordinal number such that to each ordinal number γ, $\gamma < \lambda$, there corresponds a set $V_\gamma \in \mathscr{V}$, each pair of such sets being disjoint, then we define $\mathscr{V}_\lambda$ as the family of those $\mathscr{V}$-sets that do not intersect V_γ for any $\gamma < \lambda$. If $\mathscr{V}_\lambda \neq \emptyset$, we let $\Delta_\lambda = \sup_{V \in \mathscr{V}_\lambda} \Delta(V)$ and choose a set $V_\lambda \in \mathscr{V}_\lambda$ satisfying $\Delta(V_\lambda) \geqslant \Delta_\lambda/\alpha$. In case $\mathscr{V}_\lambda = \emptyset$, the process terminates.

Thus we define a disjoint sequence (V_λ), possibly transfinite, such that each set $V \in \mathscr{V}$ intersects some set V_λ of the sequence. Since $\bar{\mu}(G) < \infty$ and each set of the sequence is included in G, it follows that the sequence is countable. Accordingly, we may rearrange its terms in the form of an ordinary sequence $V_1, V_2, \ldots, V_n, \ldots$. For each positive integer i we let

$$H_i = H(V_i), \quad S_i = V_1 + V_2 + \cdots + V_i, \quad \text{and} \quad \bar{\mu}(X - X \cdot S_i) = \varepsilon_i .$$

For any such i, we let $X_i' = X - X \cdot S_i$. Because $\mathscr{V} \subset \mathscr{A}$, then $S_i \in \mathscr{A}$ and, for $\mu^\star$-almost all $x \in X_i'$, there exists an x-converging sequence (M_ι) each of whose sets belong to $\mathscr{V}$, each contains x by virtue of Axiom (E), and each is disjoint from S_i.

Consider any set V in such a sequence (M_ι). There exists a certain ordinal number γ_0 such that V_{γ_0} is the first term of the (possibly transfinite) sequence (V_λ) that intersects V. Then $V \in \mathscr{V}_{\gamma_0}$, $\Delta(V) \leqslant \Delta_{\gamma_0} \leqslant \alpha \Delta(V_{\gamma_0})$, and so $V \subset H(V_{\gamma_0})$. Let i_0 be the positive integer for which $V_{i_0} = V_{\gamma_0}$. Since $V \cdot S_i = \emptyset$, we must have $i_0 > i$, and so $x \in V \subset H(V_{i_0}) \subset \bigcup_{j>i} H_j$. Accordingly, $X_i' \subset \bigcup_{j>i} H_j \pmod{\mathscr{N}^\star}$ for any positive integer i. Thus

$$\bar{\mu}(X_i') \leqslant \sum_{j>i} \bar{\mu}(H_j) < \beta \sum_{j>i} \mu(V_j) .$$

Because the sets of the sequence (V_j) belong to $\mathcal{M}$, are pairwise disjoint, are all included in G, and $\bar{\mu}(G) < \infty$, we have

$$\sum_j \mu(V_j) = \mu\Big(\bigcup_j V_j\Big) \leqslant \bar{\mu}(G) < \infty,$$

whence, from above, we obtain $\lim_i \varepsilon_i = 0$. Thus $X \subset \bigcup_j V_j \pmod{\mathcal{N}^\star}$ and $\mathfrak{B}$ has the (R.S.V.) property.

To show that $\mathfrak{B}$ has the sharp (R.S.V.) property, we take X, G, $\mathscr{V}$, and $W_1, W_2, \ldots, W_p$ as in the opening paragraph of our proof and let $X' = X - X \cdot \Big(\bigcup_{j=1}^{p} W_j\Big)$. The set $\bigcup_{j=1}^{p} W_j$ is then an $\mathscr{A}$-set and one can take a $\mathfrak{B}$-fine covering $\mathscr{V}'$ of X' by considering only those sets V in $\mathscr{V}$ such that $V \cap \Big(\bigcup_{j=1}^{p} W_j\Big) = \emptyset$.

Applying the results just obtained, we determine a disjoint countable sequence (V_i) of these sets, such that $X' \subset \bigcup_i V_i \pmod{\mathcal{N}^\star}$. The disjoint sequence $W_1, W_2, \ldots, W_p, V_1, V_2, \ldots, V_n, \ldots$ evidently covers $X \pmod{\mathcal{N}^\star}$, and so $\mathfrak{B}$ has the sharp (R.S.V.) property.

Remark. To show the essential role played by Axiom (**E**) in the foregoing theorem, we take $\mathfrak{B} = \mathfrak{B}_3^1$ (I. 3.1), let X denote a CANTOR discontinuum of positive Borel measure, and define $\mathscr{V}$ to be the system of closed intervals not intersecting X. $\mathscr{V}$ is then a $\mathfrak{B}$-fine covering of X, but no set in $\mathscr{V}$ covers a single point of X.

2.2.1. Definition. A basis $\mathfrak{B}$ satisfying all the hypotheses of Th. 2.2, is called a *special* MORSE *basis*. If it satisfies only the condition (2), then it is said to have the *special* MORSE *halo property*.

The (R.S.V.) and sharp (R.S.V.) properties can be formulated for families of $\mathcal{M}$-sets without reference to a basis or points. The following lemma asserts the σ-invariance of such a sharp (R.S.V.) property. It is proved by a simple adaptation of MORSE's theory to our setting.

2.3. Lemma. *We assume that*

(1) $X = \lim_i \cdot X_i \pmod{\mathcal{N}^\star}$, *where the sets* X_i, $i = 1, 2, \ldots$ *form a non-decreasing sequence, each of finite outer measure.*

(2) $\mathscr{V}$ *is a family of* $\mathcal{M}$-*sets possessing the sharp (R.S.V.) property with respect to each set* X_i, $i = 1, 2, \ldots$. *That is, given any finite disjoint sequence of sets* $W_1, W_2, \ldots W_p$ *in* $\mathscr{V}$, *and any set* X_i, *there exists a countable disjoint subfamily of* $\mathscr{V}$ *that includes* $W_1, W_2, \ldots, W_p$ *and covers* $X_i \pmod{\mathcal{N}^\star}$.

Then $\mathscr{V}$ *possesses the sharp (R.S.V.) property with respect to* X.

Proof. We take any finite disjoint subfamily of $\mathscr{V}$, namely $W_1, W_2, \ldots, W_p$, and any decreasing nullsequence of positive numbers $\varepsilon_1, \varepsilon_2, \ldots, \varepsilon_n, \ldots$ By virtue of (2), there exists a disjoint sequence of $\mathscr{V}$-sets $V_1^1, V_1^2, \ldots,$

$V_1^i, \ldots$ such that $V_1^i = W_i$ for $i = 1, 2, \ldots, p$ and $X_1 \subset \bigcup_i V_1^i \pmod{\mathcal{N}^\star}$. For any $i = 1, 2, \ldots$, we let $S_1^i = \bigcup_{1 \leqslant i' \leqslant i} V_1^{i'}$ and, since $\bar{\mu}(X_1) < \infty$, we may choose $p_1 \geqslant p$ so that $\mu(\bar{X}_1 - \bar{X}_1 \cdot S_1^{p_1}) < \varepsilon_1$.

We now apply the sharp (R.S.V.) property to $\mathcal{V}$ and $X_2 = X_1 - X_1 \cdot S_1^{p_1}$ with $V_1^1, V_1^2, \ldots, V_1^{p_1}$ as the prescribed disjoint $\mathcal{V}$-sets, and obtain a disjoint sequence $V_2^1, V_2^2, \ldots, V_2^i, \ldots$ of $\mathcal{V}$-sets with $V_2^i = V_1^i, i = 1, 2, \ldots, p_1$, and $X_2 \subset \bigcup_i V_2^i \pmod{\mathcal{N}^\star}$. For any $i = 1, 2, \ldots$, we let $S_2^i = \bigcup_{i' \leqslant i} V_2^{i'}$ and, since $\bar{\mu}(X_i) < \infty$, we may choose $p_2 \geqslant p_1$ so that $\mu(\bar{X}_2 - \bar{X}_2 \cdot S_2^{p_2}) < \varepsilon_2$. Proceeding thus inductively, we are led to the disjoint sequence of $\mathcal{V}$-sets $V_1^1, V_1^2, \ldots, V_1^{p_1}, V_2^{p_1+1}, V_2^{p_1+2}, \ldots, V_2^{p_2}, \ldots, V_{n+1}^{p_n+1}, V_{n+1}^{p_n+2}, \ldots, V_{n+1}^{p_{n+1}}, \ldots$ whose first p terms are $W_1, W_2, \ldots, W_p$. We let S denote the union of all sets in this sequence. If $\nu \geqslant n$, then evidently

$$\mu(\bar{X}_\nu - \bar{X}_\nu \cdot S_\nu^{p_\nu}) < \varepsilon_\nu \leqslant \varepsilon_n. \tag{2.3.1}$$

The sequence $(X_\nu - X_\nu \cdot S)$ is non-decreasing and $X - X \cdot S = \lim_\nu (X_\nu - X_\nu \cdot S) \pmod{\mathcal{N}^\star}$. Hence, using (2.3.1), we obtain $\mu(\bar{X} - \bar{X} \cdot S) \leqslant \varepsilon_n$ for each positive integer n. Accordingly, $\mu(\bar{X} - \bar{X} \cdot S) = 0$ and $X \subset S$ $\pmod{\mathcal{N}^\star}$, as required.

2.4. The general B-Vitali theorem.

2.4.1. Definition. We say that $\mathfrak{B}$ has the *generalized* MORSE *halo property*, or that $\mathfrak{B}$ is a *generalized* MORSE *basis* [29, 213, Def. 6.4] iff there exists $\alpha > 1$ and a disentanglement function Δ (Def. 1.1) for which

$$\sup \{\limsup_\iota [\Delta(M_\iota(x)) + \rho(\Delta, \alpha, M_\iota(x))]\} < \infty$$

holds for $\mu^\star$-almost all points $x \in E$. Here, as in I. 2.7, the limit superior is taken for an arbitrary sequence $(M_\iota(x))$ and then the supremum of these numbers for all x-converging sequences is found.

2.4.2. Theorem. (General B-Vitali theorem). *We assume that*

(1) $\mathfrak{B}$ *is a generalized* MORSE *basis whose constituents are* $\mathcal{A}$*-sets, and that Axiom* (**E**) *holds;*

(2) *Corresponding to each set* $X \subset E$ *with* $\bar{\mu}(X) < \infty$, *there exists a* $\mathcal{G}$*-set* G *with* $X \subset G$ *and* $\bar{\mu}(G) < \infty$.

Then $\mathfrak{B}$ *has the sharp (R.S.V.) property.*

Proof. We take any set $X \subset E$ with $\bar{\mu}(X) < \infty$, select $G \in \mathcal{G}$ with $X \subset G$ and $\bar{\mu}(G) < \infty$, choose any $\mathfrak{B}$-fine covering $\mathcal{V}$ of $X \pmod{\mathcal{N}^\star}$, and select $\alpha > 1$ and Δ in conformity with Def. 2.4.1. We also suppose given a finite disjoint set of $\mathcal{V}$-sets $W_1, W_2, \ldots, W_p$; we put $F = \bigcup_{i=1}^{p} W_i$ and $X' = X - X \cdot F$.

For each positive integer n, we let X_n denote the set of those points $x \in X$ for which

$$\sup \{\lim_{\iota} \sup [\Delta (M_\iota(x)) + \rho(\Delta, \alpha, M_\iota(x))]\} < n, \qquad (2.4.2.1)$$

and we set $X'_n = X_n - X_n \cdot F$.

Now (X_n) is an expanding sequence and $X = \lim_n X_n \pmod{\mathcal{N}^\star}$ because of our hypotheses. Thus $\mathscr{V}$ is *a fortiori* a $\mathfrak{B}$-fine covering of X_n $(\operatorname{mod} \mathcal{N}^\star)$, and so of X'_n $(\operatorname{mod} \mathcal{N}^\star)$, for each $n = 1, 2, \ldots$. Hence, because of (2.4.2.1), for each $x \in X'_n$ there exists at least one x-converging sequence $(M_\iota(x))$, each set V of which satisfies

$$\Delta(V) + \rho(\Delta, \alpha, V) < n, \qquad V \in \mathscr{V}. \qquad (2.4.2.2)$$

Thus, for each such n, we may define a subbasis $\mathfrak{B}_n$ of $\mathfrak{B}$ with spread $\mathscr{V}_n$ and domain X'_n, such that $\mathscr{V}_n \subset \mathscr{V}$ and each set $V \in \mathscr{V}_n$ satisfies (2.4.2.2). Because $G \in \mathscr{G}$, there is no loss of generality in assuming that all the members of $\mathscr{V}_n$ are included in G. Also, because F is an $\mathscr{A}$-set, we may assume that the sets of $\mathscr{V}_n$ do not intersect F.

It may now be seen that $\mathfrak{B}_n$ satisfies the hypotheses of Th. 2.2, whence there exists a countable disjoint subfamily $\mathscr{F}'$ of $\mathscr{V}_n$ covering X'_n $(\operatorname{mod} \mathcal{N}^\star)$. The family $\mathscr{F}$ consisting of $\mathscr{F}'$ together with the sets $W_1, W_2, \ldots, W_p$ is thus a countable disjoint subfamily of $\mathscr{V}$ covering X_n $(\operatorname{mod} \mathcal{N}^\star)$, and containing the given sets $W_1, W_2, \ldots, W_p$. Hence $\mathscr{V}$ has the sharp (R.S.V.) property of Lemma 2.3 with respect to X_n for $n = 1, 2, \ldots$, and, therefore, with respect to X. It now follows that $\mathfrak{B}$ has the sharp (R.S.V.) property, as we wished to prove.

Remarks on the general B-*Vitali theorem.* Since the conclusion of this theorem asserts the existence of a covering $(\operatorname{mod} \mathcal{N}^\star)$, we could assume $\mu = \mu^\star$ in the formulation of the theorem. This means ultimately a weakening of the assumption that the $\mathfrak{B}$-constituents are $\mathscr{A}$-sets. The assumption (2) remains effectively unchanged. With a view to the derivation of countably additive but not necessarily μ-absolutely continuous $\mathscr{M}$-functions (for instance, in MORSE's case, general Radon measures), we have not assumed the equality of μ and $\mu^\star$.

In the applications known to us, it is always true that $\lim_\iota \Delta(M_\iota(x)) = 0$ and $\lim_\iota \mu(M_\iota(x)) = 0$ for all x-converging sequences $(M_\iota(x))$, therefore, the hypothesis that $\mathfrak{B}$ is a generalized MORSE basis reduces to the condition that $\lim_\iota \sup \rho(\Delta, \alpha, M_\iota(x)) < \infty$ for all x-converging sequences, for $\mu^\star$-almost all $x \in E$, whenever $(M_\iota(x))$ is an x-converging sequence. In brief, the halos tend to vanish with respect to $\bar{\mu}$, and for this reason we have introduced the term "halo evanescence condition" to describe this situation (cf. Def. IV. 4.2).

The two following examples show the essential nature of the assumption of the closedness of the constituents.

2.4.3. Example. We let $R = \mathbf{R}^2$ (Euclidean plane), let μ denote plane Borel measure, and take for E the open unit square with principal vertices at $(0, 0)$ and $(1, 1)$. To avoid repetition, throughout this discussion t will denote an arbitrary point of E, and n will denote an arbitrary positive integer. We let T_n denote the set of points in R of the form $(r/2^n, s/2^n)$, where r and s are arbitrary integers. $\mathscr{K}_n$ denotes the family of closed squares whose four vertices are points of T_n, with sides of length 2^{-n}. Each point t lies in or on the boundary of at least one square in $\mathscr{K}_n$; we associate, with each such t, exactly one square $I_{n,t}$ in $\mathscr{K}_n$ such that $t \in I_{n,t}$. We define $I'_{n,t}$ as the square concentric with $I_{n,t}$, with sides parallel to the axes and three times as long as those of $I_{n,t}$.

At each point $z \in T_n$ we construct a square centered at z, with sides parallel to the axes and of length 2^{-2n}. We let $\mathscr{H}_n$ denote the family of all such squares, and we define

$$J_n = \bigcup_{m=n+1}^{\infty} (\cup \mathscr{H}_m) .$$

We further define $I''_{n,t} = I_{n,t} \cup I'_{n,t} \cdot J_n$. Finally, we define the basis $\mathfrak{B}$ with domain E by associating with each point $x \in E$ the sequence of sets $(I''_{n,t})$ and, of course, all its subsequences.

For each integer $m \geqslant n + 1$, there are not over $16 \cdot 2^{2m-2n}$ points of T_m lying on or in $I_{n,t}$, thus not over $16 \cdot 2^{2m-2n}$ members of $\mathscr{H}_m$, each of μ-measure 2^{-4m}, intersecting $I''_{n,t}$. Therefore,

$$\mu(I'_{n,t} \cdot J_n) \leqslant 16 \sum_{m=n+1}^{\infty} 2^{-2m} \cdot 2^{-2n} < 2^{-2n+3} \mu(I_{n,t});$$

since $n \geqslant 1$, we have

$$\mu(I''_{n,t}) \leqslant \mu(I_{n,t}) + \mu(I'_{n,t} \cdot J_n) < 3\,\mu(I_{n,t}) < (1/2)\,\mu(I'_{n,t}). \quad (2.4.3.1)$$

We consider a $\mathfrak{B}$-fine covering of E, say $\mathscr{V}$, whose members are all contained in the open set E. We let $\mathscr{E}$ be any countable subfamily of $\mathscr{V}$ whose μ-overlap is zero. Each set $B \in \mathscr{E}$ is a set $I''_{n,t}$, and we may associate with B the corresponding set $B' = I'_{n,t}$; we let $\mathscr{E}'$ denote the family of these corresponding sets. Due to our construction, it follows that $\mathscr{E}'$ has μ-overlap zero.

Using (2.4.3.1) and the fact that $\cup \mathscr{E}' \subset E$, we obtain

$$\mu(E - \cup \mathscr{E}') \geqslant \mu(E) - \sum_{B \in \mathscr{E}} \mu(B) > \mu(E) - (1/2) \sum_{B' \in \mathscr{E}'} \mu(B')$$
$$= 1 - (1/2)\,\mu(\cup \mathscr{E}') \geqslant 1 - 1/2 = 1/2 .$$

Thus, no countable subfamily of $\mathscr{V}$ whose μ-overlap is zero can cover μ-almost all of E. At the same time, if we define $\Delta(B) = \operatorname{diam} B$ for each $B \in \mathscr{V}$, then it is clear that MORSE's halo property (Def. 2.2.1) holds.

2.4.4. Example. In this case we let R and E both be the set of all real numbers and we take for μ linear Borel measure. We let V^0 denote a fixed open subset of the open interval $J = (-1, 1)$, containing the point $x = 0$, everywhere dense in J, with $\mu(V^0) = 2\theta^0$, where $0 < \theta^0 < 1$. For each $x \in R$ and each $\zeta > 0$, we define $V(x,\zeta)$ as the open set image of V^0 by the direct homothetical transformation that carries the interval $(x-\zeta, x+\zeta)$ onto J. The basis $\mathfrak{B}$ is defined so that to each $x \in E = R$ there is associated the family of sets $\{V(x,\zeta)\}$, $\zeta > 0$, with the convergence being defined with respect to the diameters tending to zero. We define $\Delta(V) = \mu(V)$ for $V(x,\zeta)$; thus $\Delta(V) = 2\zeta \cdot \theta^0$. From this it follows that $H(\Delta, \alpha, V(x,\zeta)) = (x-\zeta(2\alpha+1), x+\zeta(2\alpha+1))$; hence $\rho(\Delta,\alpha,V) = (2\alpha+1)/\theta^0$. Since obviously $\limsup_{\zeta \to 0} \Delta(V(x,\zeta)) = 0$ holds for each $x \in E$, then Morse's halo property (Def. 2.2.1) is valid.

Now we take a $\mathfrak{B}$-fine covering $\mathscr{V}$ of J such that the closure of each member of $\mathscr{V}$ is a subset of J, and consider any countable subfamily $\mathscr{E}$ of $\mathscr{V}$ whose μ-overlap is zero. If $B = V(x,\zeta) \in \mathscr{E}$, then the closure $\bar{B}$ of B is the closed interval $[x-\zeta, x+\zeta]$ and $\mu(B) = \theta^0 \mu(\bar{B})$. If $\bar{\mathscr{E}}$ denotes the family of the corresponding sets $\bar{B}$, it follows from the density of B in $\bar{B}$ that the μ-overlap of $\bar{\mathscr{E}}$ is zero. Thus, since $\bigcup \mathscr{E} \subset \bigcup \bar{\mathscr{E}} \subset J$, we have

$$\begin{aligned}\mu(J - \bigcup \mathscr{E}) &\geqslant \mu(J) - \sum_{B \in \mathscr{E}} \mu(B) = 2 - \theta^0 \sum_{B \in \mathscr{E}} \mu(\bar{B}) \\ &= 2 - \theta^0 \mu(\bigcup \bar{\mathscr{E}}) \geqslant 2 - 2\theta^0 > 0 .\end{aligned}$$

Accordingly, $\mathfrak{B}$ does not posses the (S.V.) property in spite of having Morse's halo property.

2.5. Theorem. *If $\mathfrak{B}$ is such a basis that each $\mathfrak{B}$-constituent of each x-converging sequence at any point $x \in E$ includes a measurable $\mathscr{A}$-set containing x, and if Haupt's adaptation property (*II. 1.6*) and the generalized Morse's halo property (Def.* 2.2.1*) both hold, then $\mathfrak{B}$ possesses the Vitali property for Radon measures.*

Proof. We let ψ denote an arbitrary Radon measure, X an arbitrary bounded subset of E, $\mathscr{V}$ any $\mathfrak{B}$-fine covering of X, and ε an arbitrary positive number. Since X is bounded, it is included in a set $G_N^0 \in \mathscr{G} \cap \mathscr{M}$ Because of Prop. II. 1.8, we have only to show the existence of an M-family $\mathscr{E}$ of $\mathscr{V}$-sets such that

$$\bar{\mu}(X - X \cdot \sigma\mathscr{E}) < \varepsilon, \ \omega(\mathscr{E},\psi) < \varepsilon .$$

For $n = 1, 2, \dots$ we denote by $\mathfrak{B}_n$ the basis comprising the set of those $\mathfrak{B}$-sequences $(M_\iota(x))$ such that $x \in X$, whose constituents V belong to $\mathscr{V}$, are included in G_N^0, and satisfy the relation

$$\Delta(V) + \rho(\Delta,\alpha,V) < n . \tag{2.5.1}$$

We choose a number η, $0 < \eta < 1$, and let X_n denote the domain of $\mathfrak{B}_n$. Evidently the sequence (X_n) is non-decreasing. Since G_N^0 is a $\mathscr{G}$-set, $\mathscr{V}$ is a $\mathfrak{B}$-fine covering of X, and the generalized Morse's halo property holds, it follows that $X = \lim_n X_n \,(\mathrm{mod}\, \mathscr{N}^\star)$, hence $\mu(\bar{X}) = \lim_n \mu(\bar{X}_n)$. Since $\mu(\bar{X})$ is finite, we may and do choose n_0 so that

$$\mu(\bar{X}) - \mu(\bar{X}_{n_0}) < \varepsilon . \tag{2.5.2}$$

We put $\psi^0 = \psi(G_N^0) < \infty$ and select δ to satisfy

$$0 < \delta < 1 , \quad 0 < (\delta^{-1} - 1)\psi^0 < \varepsilon . \tag{2.5.3}$$

Recalling the Remarks following Prop. II. 1.7, given any $\varepsilon' > 0$ and any $\mathfrak{B}_{n_0}$-constituent V, there exists an $\mathscr{A}^0$-set A satisfying

$$A \subset V, \ \psi(V - A) + \mu(V - A) < \varepsilon' . \tag{2.5.4}$$

Also, our hypotheses ensure that each $\mathfrak{B}_{n_0}$-constituent $V = M_\iota(x)$ includes an $\mathscr{A}$-set A' with $x \in A'$; thus, mindful that $A \cup A'$ is again an $\mathscr{A}$-set as well as an $\mathscr{M}$-set, we may assume that the set A in (2.5.4) belongs to $\mathscr{A} \cap \mathscr{M}$ and includes the point x associated with the set $M_\iota(x) = V$.

To each $\mathfrak{B}_{n_0}$-sequence $(M_\iota(x))$, we determine all possible sequences $(A_\iota(x))$ of $\mathscr{A} \cap \mathscr{M}$-sets such that

$$x \in A_\iota(x) , \ \mu(A_\iota(x)) > \eta\,\mu(M_\iota(x)) , \text{ and } \psi(A_\iota(x)) \geqslant \delta\psi(M_\iota(x)) . \tag{2.5.5}$$

Because of (2.5.4), such associated sequences $(A_\iota(x))$ exist for each $\mathfrak{B}_{n_0}$-sequence $(M_\iota(x))$. Thus we may define a basis $\mathfrak{A}'$ to consist of all such sequences $(A_\iota(x))$, and associate with each such sequence the point x corresponding to $(M_\iota(x))$. It is clear that $\mathfrak{A}'$ has domain X_{n_0}, its constituents are $\mathscr{A} \cap \mathscr{M}$-sets, and Axiom (E) holds. Moreover, condition (3) of Th. 2.2 clearly holds for $\mathfrak{A}'$.

To each $\mathfrak{A}'$-constituent $V' = A_\iota(x)$ there corresponds at least one $\mathfrak{B}_{n_0}$-constituent $V = M_\iota(x)$ satisfying (2.5.5). We choose exactly one of these and denote it by $V = D(V')$ (the *dilation* of V'); we have then

$$V' \subset V , \quad \mu(V') > \eta\,\mu(V) , \quad \psi(V') \geqslant \delta\psi(V) . \tag{2.5.6}$$

We define the disentanglement function Δ' on the spread of $\mathfrak{A}'$ by the relation

$$\Delta'(V') = \Delta(V) , \quad V = D(V') . \tag{2.5.7}$$

We observe that the halo $H'(\Delta', \alpha, V')$, which is a union of $\mathfrak{A}'$-constituents, is included in $H(\Delta, \alpha, V)$, hence

$$\begin{aligned} \rho'(\Delta', \alpha, V') &= \bar{\mu}(H'(\Delta', \alpha, V'))/\mu(V') \\ &\leqslant \frac{\bar{\mu}(H(\Delta, \alpha, V))}{\mu(V)} \cdot \frac{\mu(V)}{\mu(V')} \leqslant \rho(\Delta, \alpha, V)/\eta . \end{aligned} \tag{2.5.8}$$

Combining (2.5.1), (2.5.7), and (2.5.8), we infer that

$$\Delta'(V') + \rho'(\Delta',\alpha,V') \leqslant (\Delta(V) + \rho(\Delta,\alpha,V))/\eta < n_0/\eta .$$

We now see that $\mathfrak{A}'$ possesses the special Morse halo property (Def. 2.2.1) and satisfies the hypotheses of Th. 2.2. Hence, there exists a disjoint M-family $\mathscr{E}'$ of $\mathfrak{A}'$-constituents with $X_{n_0} \subset S' \pmod{\mathscr{N}^\star}$, where $S' = \sigma\mathscr{E}'$. We let $\mathscr{E}$ be the M-family of $\mathfrak{B}_{n_0}$-constituents obtained from $\mathscr{E}'$ by the correspondence $V = D(V')$, $V' \in \mathscr{E}'$, and set $S = \sigma\mathscr{E}$. Since $X_{n_0} \subset S' \subset S \pmod{\mathscr{N}^\star}$, we see from (2.5.2) that $\mu(\bar{X} - \bar{X}\cdot S) \leqslant \mu(\bar{X}) - \mu(\bar{X}_{n_0}) < \varepsilon$.

Since $\mathscr{E}$ is disjoint and $S' \subset S \subset G_N^0$, then (2.5.3) yields

$$\begin{aligned}\omega(\mathscr{E},\psi) &= \sum_{V \in \mathscr{E}} \psi(V) - \psi(S) \leqslant \delta^{-1} \sum_{V' \in \mathscr{E}'} \psi(V') - \psi(S') \\ &= (\delta^{-1} - 1)\psi(S') \leqslant (\delta^{-1} - 1)\psi^0 < \varepsilon .\end{aligned}$$

Hence $\mathfrak{B}$ has the Vitali ψ-property; since ψ is arbitrary, $\mathfrak{B}$ has the Vitali property for Radon measures.

2.6. Corollary. If $\mathfrak{B}$ satisfies the hypotheses of Th. 2.5, then $\mathfrak{B}$ derives every Radon measure.

Remarks. The essential steps in the foregoing proof are (i) the contraction of the $\mathfrak{B}_n$-sets into $\mathscr{A}$-sets of nearly equal ψ-measure, with μ-exhaustion power exceeding η, (ii) the transfer, expressed by (2.5.7), of the function Δ from the original sets to the new ones. The second step shows the power of Morse's methods residing here in the choice of the new disentanglement function. The "concrete" bases in Ch. VI will give point to these assertions.

In Morse's paper [*29*, 206], it is mentioned that the metric axiom

$$\delta(p',p'') = 0 \quad \text{implies} \quad p' = p''$$

is never used. Discarding this axiom, the (metric) closure P, of the set consisting of the single point $\{p\}$, is the set of points x satisfying the relation $\delta(x,p) = 0$. To the various points x of P there may be associated various sequences $(M_\iota(x))$, but any Borel set containing p must include P. That is why the first assumption in Th. 2.5 is satisfied under Morse's relaxed hypotheses.

Morse's halo property in the general case involves the contracting process; this is not so, however, in the special case of uniformity. If, for a blanket (cf. I. 3.3) $\mathfrak{F}$, there exist Δ and $\alpha > 1$ such that $\Delta(V) + \rho(\Delta,\alpha,V)$ is bounded on the spread of $\mathfrak{F}$, then the same is true of any blanket with the same spread; in particular, it holds for the blanket $\mathfrak{F}^\star$ defined so that $\mathfrak{F}^\star(x)$ is the family of $\mathfrak{F}$-constituents containing x, which is a D-basis. Not only does $\mathfrak{F}$ derive any Radon measure, but $\mathfrak{F}^\star$ does, also.

In 1947, O. Nikodym raised the question as to whether or not the property (S.V. mod $\mathscr{N}^\star$) is equivalent to the validity of the derivation theorem for μ-finite μ-integrals. Referring to Ths. III. 2.9, and 2.5 above, Examples 2.4.3, and 2.4.4 show that the answer to this question is negative.

3. Abstract version of the strong Vitali theorem modelled after Carathéodory.

3.1. Theorem. *(The special* C*-Vitali theorem.) We assume that*

(1) *the* $\mathfrak{B}$*-constituents are* $\mathscr{A}$*-sets;*

(2) $\mathfrak{B}$ *satisfies conditions* (S.V. 1) *and* (S.V. 2) *of Def.* II. 6.1;

(3) *there is a bounded function* Δ *and a (finite) number* $\lambda > 1$ *such that for each* $\mathfrak{B}$*-constituent* V, $\rho(H(\Delta,1,V)) < \lambda$ *(this is the halo dilation of* V *corresponding to* $\alpha = 1$*).*

Then for each $\mathfrak{B}$*-fine covering* $\mathscr{V}$ (mod $\mathscr{N}^*$) *of an arbitrary subset* X *of* E *of finite outer measure, each* $\varepsilon > 0$, *and each finite set of pairwise disjoint* $\mathscr{V}$*-sets* $W_1, W_2, \ldots, W_p$ *for which*

$$\mu(F - F \cdot \bar{X}) < \varepsilon, \quad F = \bigcup_{i=1}^{p} W_i,$$

there exists a countable disjoint family of $\mathscr{V}$*-sets that covers* X (mod $\mathscr{N}^*$), *whose* (X, μ)*-overflow is less than* ε*, and which includes the sets* $W_1, W_2, \ldots W_p$.

Proof. We take X, $\mathscr{V}$, ε, and $W_1, W_2, \ldots, W_p$ in accordance with the hypotheses above. We may suppose $\bar{\mu}(X) > 0$, since, if $\bar{\mu}(X) = 0$, there is nothing to prove.

We establish the theorem first on the assumption that there are no given sets $W_1, W_2, \ldots, W_p$. Then condition (2) of our hypotheses guarantees the existence of a sequence $V_1^1, V_1^2, \ldots, V_1^j, \ldots$ of $\mathscr{V}$-sets satisfying

$$X_1 \subset \bigcup_j V_1^j \ (\mathrm{mod}\ \mathscr{N}^*), \qquad \mu\left(\bigcup_j V_1^j - \bar{X}_1 \cdot \left(\bigcup_j V_1^j\right)\right) < \varepsilon' < \varepsilon/2, \tag{3.1.1}$$

where $X_1 = X$ and ε' is any number such that $0 < \varepsilon' < \min(\mu(\bar{X}_1)/4\lambda, \varepsilon/2)$.

From the sequence (V_1^j) we may and do extract a finite subsequence $V_1^1, V_1^2, \ldots, V_1^{p_1}$ for which

$$\mu(\bar{X}_1 - \bar{X}_1 \cdot S_1) < \varepsilon'', \tag{3.1.2}$$

where $S_1 = \bigcup_{j=1}^{p_1} V_1^j$ and $\varepsilon'' = \mu(\bar{X}_1)/4$.

The next step consists of disentanglement of certain nuclei, a process that depends on the halo condition and shows the role played by Δ. We may assume that the sets $V_1^1, V_1^2, \ldots, V_1^{p_1}$ are numbered so that $\Delta(V_1^1) \geqslant \Delta(V_1^2) \geqslant \cdots \geqslant \Delta(V_1^{p_1})$. Since $\alpha = 1$, then $H(\Delta, 1, V_1^1)$ includes all those members of the sequence $V_1^1, V_1^2, \ldots, V_1^{p_1}$ that intersect V_1^1. We let $V_1 = V_1^1$. If not all these sets intersect V_1, then we let V_2 denote the first of them in the sequential order that does not intersect V_1. If there are sets of the sequence following V_2 and intersecting either V_1

or V_2, those sets will be included in either $H(\Delta,1,V_1)$ or $H(\Delta,1,V_2)$, possibly both. In case there are sets following V_2 and not intersecting V_1 or V_2, we denote the first of them by V_3 and continue thus inductively, until the process eventually terminates. We evidently arrive at a finite disjoint subsequence $V_1, V_2, \ldots, V_k$ of the original sequence $V_1^1, V_1^2, \ldots, V_1^{p_1}$ such that each member of the original sequence is contained in the halo of some set of the subsequence; that is, $S_1 \subset \bigcup_{i=V}^{k} H(\Delta,1,V_i)$. We denote by $\mathcal{T}_1$ the family of sets occurring in this subsequence, and by T_1 their union. Using (3.1.1) and (3.1.2), we now see that

$$\begin{aligned}\mu(\bar{X}_1) &= \mu(S_1\cdot\bar{X}_1) + \mu(S_1\cdot\bar{X}_1 - \bar{X}_1\cdot S_1) \leqslant \mu(S_1\cdot\bar{X}_1) + \varepsilon'' \leqslant \mu(S_1) + \varepsilon''\\ &\leqslant \sum_{V\in\mathcal{T}_1}\bar{\mu}(H(\Delta,1,V)) + \varepsilon'' < \lambda\sum_{V\in\mathcal{T}_1}\mu(V) + \varepsilon'' = \lambda\mu(T_1) + \varepsilon''\\ &= \lambda\mu(T_1\cdot\bar{X}_1) + \lambda\mu(T_1 - T_1\cdot\bar{X}_1) + \varepsilon'' < \lambda\mu(T_1\cdot\bar{X}_1) + \lambda\varepsilon' + \varepsilon''.\end{aligned}$$

Taking into account the definitions of ε' and ε'', we infer from these last relations and (3.1.1) that

$$\mu(T_1\cdot\bar{X}_1) > \mu(\bar{X}_1)/2\lambda; \qquad \mu(T_1 - T_1\cdot\bar{X}_1) < \varepsilon' < \varepsilon/2.$$

We now let $X_2 = X_1 - X_1\cdot T_1$. The family of those $\mathcal{V}$-sets that do not intersect T_1 form a $\mathfrak{B}$-fine covering (mod $\mathcal{N}^\star$) of X_2. In case $\mu(\bar{X}_2) \neq 0$, we may repeat on X_2 the same process just carried out on X_1, to find a finite family of $\mathcal{V}$-sets $\mathcal{T}_2$, pairwise disjoint and not intersecting any member of $\mathcal{T}_1$, such that, if $T_2 = \cup\mathcal{T}_2$, then

$$\mu(T_2\cdot\bar{X}_2) > \mu(\bar{X}_2)/2\lambda, \qquad \mu(T_2 - T_2\cdot\bar{X}_2) < \varepsilon/2^2.$$

Repeating this process inductively, we now suppose that for the positive integer n, finite disjoint families of $\mathcal{V}$-sets $\mathcal{T}_1, \mathcal{T}_2, \ldots, \mathcal{T}_n$ have been defined such that, setting $T_i = \cup\,\mathcal{T}_i$, $X_1 = X$, $X_{i+1} = X_i - X_i\cdot\left(\bigcup_{j=1}^{i} T_j\right)$ for $i = 1, 2, \ldots, n$, we have

$$\mu(T_i\cdot\bar{X}_i) > \mu(\bar{X}_i)/2\lambda, \quad \mu(T_i - T_i\cdot\bar{X}_i) < \varepsilon/2^i \quad\text{and}\quad T_i\cdot T_j = \emptyset \qquad (3.1.3)$$

for all $i, j, = 1, 2, \ldots, n$, $i \neq j$. Then, in case $\bar{\mu}(X_{n+1}) \neq 0$, we may repeat the process to define still another finite family satisfying the relation (3.1.3) for $i = n+1$. If $\bar{\mu}(X_{n+1}) = 0$, the process stops. We let $\mathcal{T} = \bigcup_i \mathcal{T}_i$, $T = \bigcup_i T_i$. Clearly, $\mathcal{T}$ is a countable disjoint subfamily of $\mathcal{V}$.

It is easily seen that for each positive integer n prior to termination of the inductive process, should that occur, or for all n otherwise, we have

$$X_{n+1} = X_n - X_n\cdot\left(\bigcup_{i=1}^{n} T_i\right) = X_n - X_n\cdot T_n = X - X\cdot\left(\bigcup_{i=1}^{n} T_i\right). \qquad (3.1.4)$$

Thus, if the inductive process stops after N steps, say, then $\bar{\mu}(X - X \cdot T) = \bar{\mu}(X_{N+1}) = 0$ so that $\mathscr{T}$ covers X (mod $\mathscr{N}^\star$). Otherwise, we may set $\zeta = (2\lambda)^{-1}$, note that $0 < \zeta < 1$, and use (3.1.4) and (3.1.3) to infer that, for each positive integer n,

$$\bar{\mu}(X_{n+1}) = \bar{\mu}(X_n) - \bar{\mu}(X_n \cdot T_n) \leqslant (1-\zeta)\bar{\mu}(X_n),$$

whence $\bar{\mu}(X_{n+1}) \leqslant (1-\zeta)^n \bar{\mu}(X_1)$. Since $(X - T) \subset \left(X - X \cdot \left(\bigcup_{i=1}^{n} T_i\right)\right)$ for each such n, this leads to $\bar{\mu}(X - T) \leqslant \bar{\mu}(X_{n+1}) \leqslant (1-\zeta)^n \bar{\mu}(X_1)$ for all n, and thus $\bar{\mu}(X - T) = 0$. In any case then, $\mathscr{T}$ covers X (mod $\mathscr{N}^\star$).

Regardless of whether or not the inductive process terminates, from the second relation in (3.1.3) and the fact that $T = \bigcup_i T_i$, we obtain $\mu(T - T \cdot \bar{X}) \leqslant \sum_i \mu(T_i - T_i \cdot \bar{X}) < \varepsilon$. Thus the theorem is true if no sets $W_1, W_2, \ldots, W_p$ are prescribed.

We now assume that such a finite disjoint sequence is prescribed; $F = \bigcup_{i=1}^{p} W_i$. We let $X_1 = X - X \cdot F$, set $\varepsilon' = \varepsilon - \mu(F - F \cdot \bar{X}) > 0$, and follow the procedure just described to find a countable disjoint subfamily $\mathscr{T}'$ of $\mathscr{V}$ such that, putting $T' = \cup \mathscr{T}'$, we have

$$\mu(T' - T' \cdot \bar{X}_1) < \varepsilon', \quad \mu(\bar{X}_1 - \bar{X}_1 \cdot T') = 0. \tag{3.1.5}$$

Because F is an $\mathscr{A}$-set, and because those members of $\mathscr{V}$ that do not intersect F comprise a $\mathfrak{B}$-fine covering of X_1, there is no loss of generality in assuming that the sets belonging to $\mathscr{T}'$ were selected from them, so that $T' \cdot F = \emptyset$.

We let $\mathscr{T}$ denote the obviously countable disjoint subfamily of $\mathscr{V}$ obtained by adjoining $W_1, W_2, \ldots, W_p$ to $\mathscr{T}'$, and we set $T = \cup \mathscr{T}$. Since $T = T' + F$ and $X = X_1 + F \cdot X$, it follows from (3.1.5) that

$$\mu(\bar{X} - \bar{X} \cdot T) \leqslant \mu(\bar{X}_1 - \bar{X}_1 \cdot T') = 0;$$
$$\mu(T - T \cdot \bar{X}) \leqslant \mu(T' - T' \cdot \bar{X}_1) + \mu(F - F \cdot \bar{X}) < \varepsilon' + \mu(F - F \cdot \bar{X}) = \varepsilon,$$

and our proof is complete.

Just as we defined a sharp (R. S. V.) property for a family of sets $\mathscr{V} \subset \mathscr{M}$, we may define a sharp (S. V.) property for such families. The following lemma corresponds to Lemma 2.3.

3.2. Lemma. *We assume that*

(1) $X = \lim_i X_i$ (mod $\mathscr{N}^\star$), *where* (X_i) *is an expanding sequence of sets each of finite outer measure;*

(2) $\mathscr{V}$ *is a family of* $\mathscr{M}$*-sets possessing the sharp (S. V.) property with respect to each set* X_i, $i = 1, 2, \ldots$; *i. e., for each such* i, *each* $\varepsilon > 0$, *and each*

finite disjoint subfamily $W_1, W_2, \dots W_p$ *of* $\mathscr{V}$, *with* $F = \bigcup_{i=1}^{p} W_i$, *and satisfying* $\mu(F - F \cdot \bar{X}_i) < \varepsilon$, *there exists a countable disjoint subfamily* $\mathscr{T}$ *of* $\mathscr{V}$ *which contains the sets* $W_1, W_2, \dots, W_p$, *covers* X_i (mod $\mathscr{N}^\star$), *and whose* (X_i, μ)-*overflow is less than* ε.

Then $\mathscr{V}$ *has the sharp (S.V.) property with respect to* X.

Proof. We suppose that $\varepsilon > 0$, $W_1, W_2, \dots, W_p$ are pairwise disjoint and belong to $\mathscr{V}$, $F = \bigcup_{i=1}^{p} W_i$, and $\mu(F - F \cdot \bar{X}) < \varepsilon$. We select any strictly decreasing nullsequence of positive numbers $\varepsilon_1, \varepsilon_2, \dots, \varepsilon_n, \dots$ and choose any positive number $\varepsilon^\star$ such that

$$\mu(F - F \cdot \bar{X}) < \varepsilon^\star < \varepsilon. \tag{3.2.1}$$

Because $(F \cdot X_i)$ is a non-decreasing sequence of sets whose limit is $F \cdot X$ (mod $\mathscr{N}^\star$) and $\mu(F) < \infty$, we have $\mu(F - F \cdot \bar{X}) = \lim_i \mu(F - F \cdot \bar{X}_i)$; thus, from (3.2.1), we may determine a positive integer n_0 such that

$$\mu(F - F \cdot \bar{X}_n) < \varepsilon^\star \tag{3.2.2}$$

whenever $n \geqslant n_0$. Without loss of generality, we may discard the first $n_0 - 1$ terms of the sequence (X_i) and assume that $n_0 = 1$.

We now use condition (2) with $i = 1$ and the prescribed finite disjoint sequence $W_1, W_2, \dots, W_p$, satisfying (3.2.2), to obtain a disjoint sequence $V_1^1, V_1^2, \dots, V_1^j, \dots$ of $\mathscr{V}$-sets with $S_1 = \bigcup_j V_1^j$, such that $V_1^j = W_j$, $j = 1, 2, \dots, p$, and

$$X_1 \subset S_1 \,(\mathrm{mod}\, \mathscr{N}^\star), \qquad \mu(S_1 - S_1 \cdot \bar{X}_1) < \varepsilon^\star. \tag{3.2.3}$$

Since $\bar{\mu}(X_1) < \infty$, we may use (3.2.3) to choose $p_1 \geqslant p$ so that, setting $S_1^{p_1} = \bigcup_{j=1}^{p_1} V_1^j \subset S_1$, we have

$$\mu(\bar{X}_1 - \bar{X}_1 \cdot S_1^{p_1}) < \varepsilon, \qquad \mu(S_1^{p_1} - S_1^{p_1} \cdot \bar{X}) < \varepsilon^\star. \tag{3.2.4}$$

Next, we invoke condition (2) with $i = 2$ and $V_1^1, V_1^2, \dots, V_1^{p_1}$ as the prescribed finite disjoint sequence of $\mathscr{V}$-sets to find a disjoint sequence $V_2^1, V_2^2, \dots, V_2^j, \dots$ of $\mathscr{V}$-sets with $S_2 = \bigcup_j V_2^j$, such that $V_2^j = V_1^j$, $j = 1, 2, \dots, p_1$ and

$$X_2 \subset S_2 \,(\mathrm{mod}\, \mathscr{N}^\star), \qquad \mu(S_2 - S_2 \cdot \bar{X}_2) < \varepsilon^\star.$$

From (3.2.4) and the fact that $\bar{\mu}(X_2) < \infty$, we may and do select $p_2 \geqslant p_1$ so that, setting $S_2^{p_2} = \bigcup_{j=1}^{p_2} V_2^j$,

$$\mu(\bar{X}_2 - \bar{X}_2 \cdot S_2^{p_2}) < \varepsilon_2, \qquad \mu(S_2^{p_2} - S_2^{p_2} \cdot \bar{X}_2) < \varepsilon^\star.$$

We continue this procedure inductively and obtain, exactly as in the proof of Lemma 2.3, nested finite sequences of pairwise disjoint $\mathscr{V}$-sets, each containing $W_1, W_2, \ldots, W_p$, whose union is a countable disjoint subfamily that covers X (mod $\mathcal{N}^\star$). In the present theorem, the unions $S_\nu^{p_\nu}$ of each of these nested finite sequences satisfy the overflow condition

$$\mu(S_\nu^{p_\nu} - S_\nu^{p_\nu} \cdot \bar{X}_\nu) < \varepsilon^\star, \qquad \nu = 1, 2, \ldots. \tag{3.2.5}$$

From (3.2.5) and the expanding nature of the sequence $(S_\nu^{p_\nu} - S_\nu^{p_\nu} \cdot X)$, we obtain, putting $S = \bigcup_\nu S_\nu^{p_\nu}$,

$$\mu(S - S \cdot \bar{X}) = \lim_\nu \mu(S_\nu^{p_\nu} - S_\nu^{p_\nu} \cdot \bar{X}) \leqslant \varepsilon^\star < \varepsilon .$$

Thus our sequence has all the properties required of it.

3.3. Theorem. *(The general* C*-Vitali theorem.) We assume the hypotheses* (1) *and* (2) *of Th.* 3.1, *also the existence of a non-negative function* Δ *such that*

$$\sup \{\lim_\iota \sup [\Delta(M_\iota(x)) + \rho(\Delta, 1, M_\iota(x))]\} < \infty$$

holds for $\mu^\star$*-almost all* $x \in X$.

Then $\mathfrak{B}$ *has the sharp (S.V.) property.*

Proof. We take any set $X \subset E$ with $\bar{\mu}(X) < \infty$, any $\mathfrak{B}$-fine covering $\mathcal{V}$ of X (mod $\mathcal{N}^\star$), and choose Δ in accordance with our hypotheses.

For $n = 1, 2, \ldots$, we define X_n as the set of those points $x \in X$ for which

$$\sup \{\lim_\iota \sup [\Delta(M_\iota(x)) + \rho(\Delta, 1, M_\iota(x)]\} < n . \tag{3.3.1}$$

Now (X_n) is an expanding sequence and $X = \lim_n X_n$ (mod $\mathcal{N}^\star$) because of our hypotheses. We take an arbitrary but fixed positive integer n. We further suppose that $\varepsilon > 0$ and $W_1, W_2, \ldots, W_p$ is any finite disjoint collection of $\mathcal{V}$-sets with $F = \bigcup_{i=1}^{p} W_i$ satisfying $\mu(F - F \cdot \bar{X}_n) < \varepsilon$. We let $X_n' = X_n - X_n \cdot F$.

$\mathcal{V}$ is *a fortiori* a $\mathfrak{B}$-fine covering of X_n (mod $\mathcal{N}^\star$) and so of X_n' (mod $\mathcal{N}^\star$). Hence, because of (3.3.1), for each $x \in X_n'$ there exists at least one x-converging sequence $(M_\iota(x))$, each set V of which satisfies

$$\Delta(V) + \rho(\Delta, 1, V) < n, \qquad V \in \mathcal{V}. \tag{3.3.2}$$

Thus we may define a subbasis $\mathfrak{B}_n$ of $\mathfrak{B}$ with spread $\mathcal{V}_n$ and domain X_n' such that $\mathcal{V}_n \subset \mathcal{V}$ and each set $V \in \mathcal{V}_n$ satisfies (3.3.2). Since $G \in \mathcal{G}$, there is no loss of generality in assuming that all members of $\mathcal{V}_n$ are included in G. Also, because F is an $\mathcal{A}$-set, we may assume that the sets of $\mathcal{V}_n$ do not intersect F. It follows that $\mathfrak{B}_n$ is a basis satisfying the hypotheses of Th. 3.1. Thus there exists a countable disjoint subfamily $\mathcal{F}'$ of $\mathcal{V}_n$, with $S' = \cup \mathcal{F}'$, such that

$$\mu(\bar{X}_n' - \bar{X}_n' \cdot S') = 0, \quad \mu(S' - S' \cdot \bar{X}_n') < \varepsilon', \quad \varepsilon' = \varepsilon - \mu(F - F \cdot \bar{X}_n) > 0 . \tag{3.3.3}$$

We let $\mathscr{F}$ denote the clearly countable disjoint subfamily of $\mathscr{V}$ obtained by adjoining $W_1, W_2, \ldots, W_p$ to $\mathscr{F}'$, and set $S = \cup \mathscr{F}$. Since $S = S' + F$ and $X'_n = X_n - X_n \cdot F$, it follows from the first relation in (3.3.3) that $\mathscr{F}$ covers $X_n (\bmod \mathscr{N}^\star)$. Since $(S - S \cdot \bar{X}_n) \subset (S' - S' \cdot X'_n) \cup (F - F \cdot \bar{X}_n)$, the second and third relations of (3.3.3) yield $\mu(S - S \cdot \bar{X}_n) < \varepsilon$. Thus $\mathscr{V}$ has the sharp (S.V.) property of Lemma 3.2 with respect to X_n, and consequently with respect to X. From this it follows that $\mathfrak{B}$ has the sharp (S.V.) property.

Remarks on the general C-*Vitali theorem.* The hypothesis (2) marks the principal difference between the Banach and Carathéodory settings of the Vitali theorem. This condition is satisfied in a metric space with a countable basis (a *separable* metric space) when μ is a Radon measure, the $\mathfrak{B}$-constituents are Jordan measurable, and the x-converging sequences are ordinary sequences containing x in their interiors, with diameters tending to zero. This assertion follows from LINDELÖF'S topological theorem and the condition $(U\,G)$ (recall Def. II. 1.6). For this reason, the hypothesis (2) is known as Property (L).

3.4. Theorem. *We suppose that*

(1) *Haupt's adaptation property holds (Def. II. 1.6);*

(2) $\mathfrak{B}$ *satisfies the conditions (S.V. 1) and (S.V. 2) of Def. II. 6.1 (Property* (L)*);*

(3) *there exists a disentanglement function* Δ *such that*

$$\sup \{\limsup_{\iota} [\Delta(M_\iota(x)) + \rho(\Delta, 1, M_\iota(x))]\} < \infty$$

for $\mu^\star$*-almost all* $x \in E$.

Then $\mathfrak{B}$ *possesses the Vitali property for Radon measures.*

Proof. We take an arbitrary Radon measure ψ, an arbitrary bounded subset X of E, any $\mathfrak{B}$-fine covering $\mathscr{V}$ of X, and any $\varepsilon > 0$. We may assume $X \subset G_N^0$, where (G_n^0) is the sequence associated with (G_σ). Because of Prop. II. 1.8, we need to prove only that there exists an M-family $\mathscr{E}$ of $\mathscr{V}$-sets such that $\mu(\bar{X} - \bar{X} \cdot \sigma\mathscr{E}) < \varepsilon$ and $\omega(\mathscr{E}, \psi) < \varepsilon$.

For $n = 1, 2, \ldots$, we denote by $\mathfrak{B}_n$ the set of those $\mathfrak{B}$-sequences $(M_\iota(x))$ with $x \in X$, whose constituents belong to $\mathscr{V}$, are included in G_N^0 and satisfy, for $V = M_\iota(x)$,

$$\Delta(V) + \rho(\Delta, 1, V) < n. \tag{3.4.1}$$

We let X_n denote the domain of $\mathfrak{B}_n$. Since the sequence (X_n) is increasing and $\lim_n X_n = X (\bmod \mathscr{N}^\star)$ due to the hypothesis (3), we may and do choose ν so that

$$0 \leqslant \mu(\bar{X}) - \mu(\bar{X}_\nu) < \varepsilon, \tag{3.4.2}$$

and we let $Z = X_\nu$.

We denote by $\mathscr{V}^1$ the spread of $\mathfrak{B}_\nu$, by η a fixed number, $0 < \eta < 1$, and by δ a positive number such that

$$0 < (\delta^{-1} - 1)\psi^0 < \varepsilon, \tag{3.4.3}$$

where $\psi^0 = \psi(G_N^0)$.

Condition (2) of our hypotheses permits us to select an M-family $V_1, V_2, \ldots, V_j, \ldots$ of $\mathscr{V}^1$-sets such that

$$Z \subset T(\operatorname{mod} \mathscr{N}^\star), \qquad \mu(T - T \cdot \bar{Z}) < \varepsilon_1, \tag{3.4.4}$$

where $T = \bigcup_j V_j$ and $\varepsilon_1 = \eta \bar{\mu}(Z)/2(\eta + \nu)$.

Setting $T_q = \bigcup_{j=1}^{q} V_j$, $q = 1, 2, \ldots$, we may choose Q so that

$$\mu(T - T_Q) < \varepsilon_1 . \tag{3.4.5}$$

Now, exactly as in the proof of Th. 3.1, we use the condition (3.4.1) with $n = \nu$ to select a disjoint subfamily $V_1^1, V_1^2, \ldots, V_1^{q_1}$ of $V_1, V_2, \ldots V_Q$ such that

$$\mu(T^1) > \mu(T_Q)/\nu, \tag{3.4.6}$$

where $T^1 = \bigcup_{k=1}^{q_1} V_1^k$. This is the disentanglement step.

With each set V_1^k we associate an $\mathscr{A}^0$-set $A_1^k \subset V_1^k$ satisfying

$$\mu(A_1^k) > \eta\,\mu(V_1^k), \qquad \psi(A_1^k) \geqslant \delta\psi(V_1^k), \qquad k = 1, 2, \ldots, q_1, \tag{3.4.7}$$

which is possible because of the consequences of Haupt's adaptation property (cf. Remarks after Prop. II. 1.7). We now evaluate the *Z-exhaustion* of $S^1 = \bigcup_{i=1}^{q_1} A_1^k$, by which we mean the value of $\mu(S^1 \cdot \bar{Z}) = \bar{\mu}(S^1 \cdot Z)$. Since $S^1 \subset T$ and $Z \subset T(\operatorname{mod} \mathscr{N}^\star)$, we obtain, with the help of (3.4.4), (3.4.5), (3.4.6), and (3.4.7),

$$\begin{aligned}
\mu(S^1 \cdot \bar{Z}) &\geqslant \mu(\bar{Z}) + \mu(S^1) - \mu(T) \\
&> \bar{\mu}(Z) + \eta\,\mu(T^1) - \mu(T) \\
&> \bar{\mu}(Z) + \tfrac{\eta}{\nu}\mu(T_Q) - (\mu(T \cdot \bar{Z}) + \varepsilon_1) \\
&\geqslant \bar{\mu}(Z) + \tfrac{\eta}{\nu}(\mu(T) - \varepsilon_1) - (\bar{\mu}(Z) + \varepsilon_1) \\
&\geqslant \tfrac{\eta}{\nu}\bar{\mu}(Z) - \varepsilon_1(\tfrac{\eta}{\nu} + 1) \geqslant \tfrac{\eta}{2\nu}\bar{\mu}(Z) .
\end{aligned}$$

Thus we have obtained a finite disjoint family $V_1^1, V_1^2, \ldots, V_1^{q_1}$ of the $\mathscr{V}^1$-sets, and hence of the $\mathscr{V}$-sets, and a corresponding finite disjoint sequence of $\mathscr{A}$-sets $A_1^1, A_1^2, \ldots, A_1^{q_1}$, whose union S^1 has a Z-power of exhaustion exceeding $\eta/2\nu$. Moreover, $A_1^k \subset V_1^k$, $k = 1, 2, \ldots, q_1$.

We repeat this process with the set $Z_2 = Z - Z \cdot S^1$ and the family $\mathscr{V}^2$ consisting of those $\mathscr{V}^1$-sets that do not intersect the $\mathscr{A}$-set S^1, thus producing two finite disjoint families $V_2^1, V_2^2, \ldots, V_2^{q_2}$ and $A_2^1, A_2^2, \ldots, A_2^{q_2}$ that satisfy, for $i = 1, 2, \ldots, q_2$, $j = 1, 2, \ldots, q_1$ the relations $V_2^i \in \mathscr{V}^2$,

$A_2^i \subset V_2^i$, $A_2^i \in \mathscr{A}^0$, $A_1^k \cdot V_2^i = \emptyset$, $\psi(A_2^k) \geqslant \delta\psi(V_2^k)$ and $\mu(S^2 \cdot \bar{Z}_2) > \frac{\eta}{2\nu}\bar{\mu}(Z_2)$, where $S^2 = \bigcup_{k=1}^{q_2} A_2^k$.

The iteration of this process leads to the construction of two M-families, namely $\mathscr{E}$ consisting of the $\mathscr{V}$-sets $V_1^1, V_1^2, \ldots, V_1^{q_1}, V_2^1, V_2^2, \ldots, V_2^{q_2}, \ldots$ and $\mathscr{C}$, consisting of the $\mathscr{A}$-sets $A_1^1, A_1^2, \ldots, A_1^{q_1}, A_2^1, A_2^2, \ldots, A_2^{q_2}, \ldots$. The sets of $\mathscr{C}$ are pairwise disjoint and satisfy, for $i = 1,2,\ldots q_j$, $j = 1,2,\ldots$, the relations

$$A_j^i \subset V_j^i, \quad \psi(A_j^i) \geqslant \delta\psi(V_j^i), \quad \mu(S^j \cdot \bar{Z}_j) > \frac{\eta}{2\nu}\bar{\mu}(Z_j), \tag{3.4.8}$$

$$\text{where} \quad S^j = \bigcup_{k=1}^{q_j} A_k, \quad Z_{j+1} = Z - Z \cdot \left(\bigcup_{k=1}^{j} S^k\right).$$

Using the exhaustion argument of Th. 3.1, it follows that $Z \subset \sigma\mathscr{E} \pmod{\mathscr{N}^\star}$; thus, because of (3.4.2), $\mu(\bar{X} - \bar{X} \cdot \sigma\mathscr{E}) < \varepsilon$. Finally, since $\mathscr{C}$ is disjoint and $\sigma\mathscr{C} \subset \sigma\mathscr{E} \subset G_N^0$, then (3.4.8) and (3.4.3) yield

$$\begin{aligned}\omega(\mathscr{E},\psi) &= \sum_{i,j} \psi(V_j^i) - \psi(\sigma\mathscr{E}) \leqslant \delta^{-1} \sum_{i,j} \psi(A_j^i) - \psi(\sigma\mathscr{C}) \\ &= (\delta^{-1} - 1)\psi(\sigma\mathscr{C}) \leqslant (\delta^{-1} - 1)\psi^0 < \varepsilon .\end{aligned}$$

Accordingly, $\mathfrak{B}$ has the Vitali ψ-property in the ε-version, and so possesses the Vitali property for Radon measures.

Remarks. Haupt's adaptation principle is used only in the contracting process. If the weaker condition (UG) is substituted, then the assertion of the theorem remains true for Radon integrals.

In the proof of Th. 2.5, Morse's function Δ is used to disentangle the infinite family of constituents to produce the desired countable family, hence the necessity of a choice condition based on the boundedness of Δ. In the proof just completed, we disentangle a finite family and repeat the process, obtaining the desired family by juxtaposition of sections.

In the proof of Th. 3.4, we do not deal with Morse's halo $H(\Delta,\alpha,V_0)$ itself, but only with a finite family of halo constituents, which may be aptly called a *partial halo*. For this reason we define $H'(\Delta,\alpha,V_0)$ as the *essential union* (cf. Def. IV.4.1.1) of the $\mathscr{V}$-constituents V intersecting V_0 with $\Delta(V) \leqslant \alpha\Delta(V_0)$. In the formulation of Ths. 3.3 and 3.4 we can replace the halo dilation $\rho(\Delta,\alpha,V_0)$ by

$$\rho'(\Delta,\alpha,V_0) = \mu(H'(\Delta,\alpha,V_0))/\mu(V_0) ;$$

clearly, $\rho' \leqslant \rho$.

In defining H' or H we accept all constituents V intersecting V_0 and satisfying $\Delta(V) \leqslant \alpha\Delta(V_0)$. The *incidence requirement* is $V \cdot V_0 \neq \emptyset$. Correspondingly, disentanglement requires the determination of a strictly disjoint family of constituents. The incidence requirement may be altered to *essential intersection*, that is, $\mu(V \cdot V_0) \neq 0$, and, in turn, the disentanglement changed to require the production of a family of pairwise (mod $\mathcal{N}$) disjoint constituents. This point of view may, if we wish, be adopted in Th. 3.3 if we want to achieve the Vitali property (mod $\mathcal{N}$), and in Th. 3.4, if we restrict the assertion to Radon integrals. The stronger we make the incidence requirements, the weaker the halo condition becomes.

4. Weak halo evanescence condition.

4.1. Preliminaries. The following definition involves only the σ-finite measure space $(R, \mathcal{M}, \mu)$.

4.1.1. Definition. Let $\mathcal{M}/\mathcal{N}$ denote the quotient of the Boolean σ-algebra $\mathcal{M}$ by $\mathcal{N}$, H the corresponding homomorphism. The order relation in $\mathcal{M}/\mathcal{N}$ is defined so that $H(Z') \geqslant H(Z'')$ iff $Z' \supset Z''$ (mod $\mathcal{N}$) holds.

Thus established, $\mathcal{M}/\mathcal{N}$ is a complete lattice [46, 378–380). For the *essential union* $\bigcup^\star \mathcal{Z}$ is defined as $H^{-1}(\vee \mathcal{Z})$; the *essential intersection* $\bigcap^\star \mathcal{Z}$ is defined as $H^{-1}(\wedge \mathcal{Z})$. The *essential union* $\bigcup^\star \mathcal{Z}$ is thus characterized by the property that every $\mathcal{M}$-set which includes each $\mathcal{Z}$-set (mod $\mathcal{N}$) includes $\bigcup^\star \mathcal{Z}$ (mod $\mathcal{N}$). There exists a countable subfamily of $\mathcal{Z}$-sets $Z_1, Z_2, \ldots, Z_n, \ldots$ such that $\bigcup_n Z_n = \bigcup^\star \mathcal{Z}$ (mod $\mathcal{N}$). If all $\mathcal{Z}$-sets are included (mod $\mathcal{N}$) in an $\mathcal{M}$-set of finite μ-measure, then $\bigcup^\star \mathcal{Z}$ is any $\mathcal{M}$-set of minimal μ-measure including each $\mathcal{Z}$-set (mod $\mathcal{N}$).

Remark. Essential unions are always $\mathcal{M}$-measurable; the ordinary union of a non-countable family $\mathcal{Z}$ need not be.

Throughout this section we assume that $\mathfrak{B}$ is a D-basis $[\mathcal{U}, \delta]$.

4.1.2. Definition. If M denotes a δ-bounded $\mathcal{M}$-set of finite μ-measure, η is a positive number, and $0 < \alpha < 1$, then we define the *η-weak halo* $S(\alpha, \eta, M)$ as an essential union of those $\mathfrak{B}$-constituents V for which

$$\mu(M \cdot V) > \alpha \mu(V), \qquad \delta(V) < \eta,$$

and the *weak halo* $S(\alpha, M)$ as $\bigcup^\star_{\eta>0} S(\alpha, \eta, M)$, that is, an essential union of all $\mathfrak{B}$-constituents V for which $\mu(M \cdot V) > \alpha \mu(V)$. The set M is called the *nucleus*. The ratio $\mu(M \cdot V)/\mu(V)$ is called the *mean density of M on V*. If we restrict the union to a finite number of $\mathfrak{B}$-constituents we obtain, respectively, the *η-weak partial halo* and the *weak partial halo*. The η-weak halo and weak halo are defined to within an indeterminate set $N \in \mathcal{N}$.

If the $\mathcal{U}$-sets ($\mathfrak{B}$-constituents) are open sets in a Hausdorff locally compact space provided with a Radon measure μ, then the ordinary union of the $\mathcal{U}$-sets is an essential union.

Morse's halo $H(V_0)$ (§ 1) has been defined as an ordinary union because Ths. 2.4 and 3.3 aim at securing the validity of the strong

Vitali property; that is, the strict disjunction of the extracted family $\mathscr{E}$. If we define Morse's halo as an essential union, then the corresponding Theorems 2.4.2 and 3.3 assert only the properties (R.S.V. mod $\mathscr{N}$) and (S.V. mod $\mathscr{N}$).

4.2. A first criterion for the validity of the Density Theorem.

4.2.1. Definition. A *nullsequence (μ-nullsequence)* of sets in a measure space $(R,\mathscr{M},\mu)$ is a sequence of $\mathscr{M}$-sets $M_1, M_2, \ldots, M_n, \ldots$ such that $\lim_n M_n = \emptyset$ ($\lim_n \mu(M_n) = 0$).

4.2.2. Definition. We say that the *weak halo evanescence condition* holds for $[\mathscr{U},\delta]$ iff for each α, $0 < \alpha < 1$, each non-increasing nullsequence $M_1, M_2, \ldots, M_n, \ldots$ of bounded $\mathscr{M}$-sets, and each non-increasing nullsequence of positive numbers $\delta_1, \delta_2, \ldots, \delta_n, \ldots$, the relation $\lim_n \mu(S(\alpha, \delta_n, M_n)) = 0$ holds.

Remark. This is equivalent to demanding that the μ-measure of any partial δ_n-halo tends to zero as n tends to infinity.

4.2.3. In this subsection, we assume that R is a locally compact metric space with metric δ, μ is a classical Radon measure, and $\mathscr{U}$ is a family of open sets of positive finite μ-measure. $\mathscr{G}$ is the family of all open sets.

4.2.3.1. Lemma. *Under the hypotheses just stated, the domain of the basis $[\mathscr{U},\delta]$ is a $\mathscr{G}_\delta$-set, hence an $\mathscr{M}$-set. Moreover, if ψ is an extended real-valued function defined on $\mathscr{U}$, then the extreme $[\mathscr{U},\delta]$-derivates of ψ are $\mathscr{M}$-measurable; in particular, Baire functions of the second class.*

Proof. If $\mathscr{U}_\eta$ denotes the family of those $\mathscr{U}$-sets of diameter less than η, where $\eta > 0$, then for any decreasing nullsequence (δ_n) of real numbers, the domain of $[\mathscr{U},\delta]$ coincides with $\bigcap_{n=1}^{\infty} (\bigcup \mathscr{U}_{\delta_n})$, and so is a $\mathscr{G}_\delta$-set, hence an $\mathscr{M}$-set.

Next for any (finite) real number β, we let $P = [D^\star \psi > \beta]$. For any positive integers h and k, we denote by $R_{h,k}$ the union of those $\mathscr{U}$-sets U for which $\delta(U) < k^{-1}$ and $\psi(U)/\mu(U) > \beta + h^{-1}$. From the nature of the basis $[\mathscr{U},\delta]$ it follows that $P = \bigcup_{h=1}^{\infty} \bigcap_{k=1}^{\infty} P_{h,k}$. Hence P is a $\mathscr{G}_{\delta\sigma}$-set, so belongs to $\mathscr{M}$. The $\mathscr{M}$-measurability of $D_\star \psi$ can be proved by an analogous argument.

4.2.3.2. Theorem. *Under the hypotheses above, $[\mathscr{U}, \delta]$ has the density property iff the weak halo evanescence condition holds.*

Proof. We begin by assuming the validity of the density property. Using the notation of Def. 4.2.2, it follows from the density property that for any bounded set $M \in \mathscr{M}$,

$$\bigcap_{\eta>0} S(\alpha,\eta,M) = M \cdot E \,(\mathrm{mod}\,\mathscr{N}) \,. \qquad (4.2.3.2.1)$$

For $v > n$, the relation

$$\mu(S(\alpha,\delta_v,M_v)) \leqslant \mu(S(\alpha,\delta_v,M_n))$$

holds. From (4.2.3.2.1) we see that for each n,

$$\lim_v \mu(S(\alpha,\delta_v,M_n)) = \mu(M_n \cdot E)$$

hence, since (M_v) is a nullsequence, we have

$$0 \leqslant \lim_v \mu(S(\alpha,\delta_v,M_v)) \leqslant \lim_n \mu(M_n \cdot E) = 0 \,,$$

which is the desired weak halo evanescence condition.

We now assume the weak halo evanescence condition. For convenience, we shall fix, as an essential union of a family of open sets, the ordinary union of these sets. We want to show that, for a bounded measurable set M, the set of those points of M at which M has not the density 1 with respect to $[\mathscr{U},\delta]$ is an $\mathscr{N}$-set. We let L denote a subset of M consisting of those points of M where the lower density of M is less than $1-\alpha$, $0<\alpha<1$. We wish to show that every closed subset $F \subset L$ is an $\mathscr{N}$-set.

Evidently, the family $\mathscr{V}$ of the sets V for which

$$\mu(F \cdot V) \leqslant \mu(L \cdot V) < (1-\alpha)\mu(V)$$

is a $[\mathscr{U},\delta]$-fine covering of F. Since F is compact and the sets $V \in \mathscr{V}$ are open, we can select from $\mathscr{V}$ a finite subfamily $V_1^1, V_2^1, \ldots, V_{r_1}^1$, each set of which is of diameter less than 1, and the union of which covers F. We set $G^1 = \bigcup_{j=1}^{r_1} V_j^1$.

Proceeding inductively, we suppose families $G^1, G^2, \ldots, G^{n-1}$ have been defined. Among the $\mathscr{V}$-sets, we determine a finite subfamily $V_1^n, V_2^n, \ldots V_{r_n}^n$ that cover F, are all of diameter less than n^{-1}, and are included in G^{n-1}. We let $G^n = \bigcup_{p=1}^{r_n} V_p^n$. Then $G^n \subset G^{n-1}$.

Since F is compact and the diameters of the sets V_p^n, $p = 1,2,\ldots,r_n$ are all less than n^{-1}, then $F = \bigcap_{n=1}^{\infty} G^n$; hence the sequence of sets $(G^n - F)$ is non-increasing and has an empty intersection. For $n = 1,2,\ldots$ we let $M_n = G^n - F$. Since we have $V_p^n \subset G^n$ for each positive integer n and $p = 1,2,\ldots,r_n$, then

$$\mu(V_p^n) = \mu(M_n \cdot V_p^n) + \mu(F \cdot V_p^n). \tag{4.2.3.2.2}$$

However, for each set V_p^n, we have

$$\mu(F \cdot V_p^n) < (1-\alpha)\mu(V_p^n),$$

hence with the help of (4.2.3.2) we derive

$$\mu(M_n \cdot V_p^n) > \alpha\mu(V_p^n);$$

consequently,

$$F \subset \bigcup_{p=1}^{r_n} V_p^n \subset S(\alpha, n^{-1}, M_n). \tag{4.2.3.2.3}$$

From the weak halo evanescence condition it follows that

$$\lim_n \mu(S(\alpha, n^{-1}, M_n)) = 0;$$

thus from (4.2.3.2.3), $\mu(F) = 0$, and the density property is valid.

Remark. If $S(\alpha, n^{-1}, M_n)$ is taken to be an arbitrary essential union, then to the right hand side of (4.2.3.2.3) we must add '(mod $\mathcal{N}$)'.

5. Further criteria for the validity of the Density Theorem involving the weak halo.

The weak halo evanescence condition, and consequently the density property, holds whenever, for any bounded $\mathcal{M}$-set M,

$$(WH) \quad \mu(S(\alpha, M)) \leqslant \gamma(\alpha)\mu(M),$$

where γ (the *halo function*) is defined and *finite* for each α, $0 < \alpha < 1$, and γ does not depend on M. It is defined with reference to $\mathcal{U}$. We shall show that (WH) and the density property are equivalent for special bases. However, (WH) itself is defined in any measure space provided with a family $\mathcal{U}$ of sets of finite positive measure.

For a set X in Euclidean space of any dimension, we shall let $\mathcal{H}(X)$ denote the family of sets homothetical to X or obtainable from a homothety of X by translation.

5.1. Lemma. *We assume that R is the Euclidean space $\mathbf{R}^m$, μ is Borel measure in R, M is a bounded measurable set of positive measure, G is an open set, ε and η are arbitrary positive numbers, and δ is the metric of $\mathbf{R}^m$.*

Then there exist countably many sets $M_1, M_2, \ldots, M_n \ldots$ belonging to $\mathcal{H}(M)$ and included in G, for which

$$\mu\Big(\bigcup_n M_n\Big) = \mu(G), \quad \sum_n \mu(M_n) < \mu(G) + \varepsilon, \quad \delta(M_n) < \eta.$$

Proof. We consider the basis $\mathfrak{B}_1 = [\mathcal{H}(M), \delta]$ and the basis $\mathfrak{B}_2 = [\mathcal{H}(I), \delta]$, where I denotes the smallest closed interval including

M. We put $\tau = \mu(M)/\mu(I)$. Morse's halo $H_1(\delta,1,M)$ of M with respect to the basis $\mathfrak{B}_1$ is included in the halo $H_2(\delta,1,I)$ of I with respect to the basis $\mathfrak{B}_2$. $H_2(\delta,1,I)$ is a closed interval homothetic to and concentric with I in the linear ratio 3. The halo dilation is, therefore, 3^m and

$$\frac{\mu(H_1(\delta,1,M))}{\mu(M)} \leqslant \frac{\mu(H_2(\delta,1,I))}{\mu(I)} \cdot \frac{\mu(I)}{\mu(M)} = \frac{3^m}{\tau} .$$

This result holds for every set of $\mathscr{H}(M)$ since they are all homothetic or equipollent. The hypotheses of Th. 2.5 are satisfied by the basis $\mathfrak{B}$. Consequently, $\mathfrak{B}_1$ possesses the Vitali property for Radon measures ψ; in particular, for $\psi = \mu$. The family of the $\mathscr{H}(M)$-sets of diameter less than η and included in G is a $\mathfrak{B}_1$-fine covering of G. The application of the Vitali μ-property (weak Vitali property) to this family yields the desired M-family.

Remark. The μ-overflow is zero in this example.

5.2. Theorem. *We take for R the Euclidean space $\mathbf{R}^m$, let μ denote Borel measure in R, and let $\mathscr{U}$ be a family of open sets such that if $G \in \mathscr{U}$, then $\mathscr{H}(G) \subset \mathscr{U}$.*

Then (WH) is a necessary and sufficient condition for the validity of the Density Theorem in $[\mathscr{U},\delta]$.

Proof. (1) We establish the sufficiency first.

It is clear that (WH) implies the weak halo evanescence condition which, by Th. 4.2.3, in turn implies the density property.

(2) We now prove the necessity. To this end, we shall assume the negation of (WH) and show that this implies that every open interval J includes a measurable set M such that $\mu(M) < (2/3)\mu(J)$ and M has positive upper density at all points of J with respect to $[\mathscr{U},\delta]$, thereby invalidating the density property.

The negation of (WH) implies the existence of $\alpha_0 > 0$ such that for each positive number γ_0, there is a bounded set $B_0 \in \mathscr{M}$ depending on γ_0, for which

$$\mu(S(\alpha_0,B_0)) > \gamma_0 \mu(B_0) . \tag{5.2.1}$$

Thus, for each positive integer n there exists a set B_n such that

$$\mu(S(\alpha_0,B_n)) > 4^n \mu(B_n) . \tag{5.2.2}$$

By definition, $S(\alpha_0,B_n)$ is an essential union of those sets $U \in \mathscr{U}$ for which

$$\mu(B_n \cdot U) > \alpha_0 \mu(U) . \tag{5.2.3}$$

We can extract a finite family of these $\mathscr{U}$-sets having a bounded union (partial halo) S_n and satisfying the inequality

$$\mu(S_n) > 4^n \mu(B_n) . \tag{5.2.4}$$

Given any open interval J, according to Lemma 5.1 there exists, for each positive integer n, a sequence (S_n^ν) of $\mathscr{H}(S_n)$-sets with diameters $\delta(S_n^\nu)$ less than n^{-1} for $\nu = 1, 2, \ldots$ and satisfying for $\varepsilon = \mu(J)$ the relation

$$\mu(J) = \mu\left(\bigcup_\nu S_n^\nu\right) \leqslant \sum_\nu \mu(S_n^\nu) < 2\mu(J). \tag{5.2.5}$$

We denote by T_n^ν the homothety or translation carrying S_n into S_n^ν, by B_n^ν the set $T_n^\nu(B_n)$, by S^n the union $\bigcup_\nu S_n^\nu$, and by B^n the union $\bigcup_\nu B_n^\nu$. Then we have

$$\mu(S^n) = \mu(J) \;;\; \mu(B^n) \leqslant \sum_\nu \mu(B_n^\nu) < 4^{-n} \sum_\nu \mu(S_n^\nu) < \frac{2}{4^n}\mu(J). \tag{5.2.6}$$

For the set $B = \bigcup_n B^n$, we see from (5.2.6) that

$$\mu(B) < (2/3)\mu(J). \tag{5.2.7}$$

Also, for each positive integer n and ν, S_n^ν is the union of certain $\mathscr{U}$-sets in each of which B_n^ν has mean density greater than α_0. Since

$$B_n^\nu \subset B^n \subset B,$$

then B has mean density greater than α_0 in each of the named $\mathscr{U}$-sets.

For each positive integer n, almost all points x of J belong to some set S_n^ν, ν depending on n and x. Since $\delta(S_n^\nu) < n^{-1}$, then we see that for almost all $x \in J$, there is a sequence of sets converging to x in each of which the mean density of B exceeds α_0; hence B has an upper density with respect to $[\mathscr{U}, \delta]$ that exceeds α_0 at almost all points of J. Putting $M = B \cdot J$, $\mu(M) < (2/3)\,\mu(J)$ by (5.2.7) and, at each point of J, the upper densities of B and M coincide. For M, the Density Theorem clearly fails to hold.

5.3. Theorem. *Under the hypotheses of Th. 5.2, a necessary and sufficient condition for the validity of the Density Theorem is*

$(WH)'$ for each $\alpha, 0 < \alpha < 1$, *and each set $M \in \mathscr{M}$ of finite μ-measure, the weak halo $S(\alpha, M)$ has finite measure.*

Proof. (1) We show first the necessity of the condition $(WH)'$. To accomplish this, we assume that $(WH)'$ is false. Then there exists α, $0 < \alpha < 1$, and a μ-measurable set B with $\mu(B) < \infty$ and $\mu(S(\alpha, B)) = \infty$. For any prescribed positive number γ, we can find a finite family of $\mathscr{U}$-sets $U_1, U_2, \ldots, U_n$ such that

$$\mu(U_i \cdot B) > \alpha\mu(U_i), \quad \mu\left(\bigcup_{i=1}^n U_i\right) > \gamma\mu(B). \tag{5.3.1}$$

We denote by B' the set $B \cdot \left(\bigcup_{i=1}^{n} U_i\right)$. B' is bounded and $\bigcup_{i=1}^{n} U_i \subset S(\alpha, B')$, hence

$$\mu(S(\alpha,B')) \geqslant \mu\left(\bigcup_{i=1}^{n} U_i\right) \geqslant \gamma\mu(B) \geqslant \gamma\mu(B') . \tag{5.3.2}$$

Consequently (WH) is not valid and, by Th. 5.2, the Density Theorem does not hold.

(2) We now prove the sufficiency of $(WH)'$. To do so, we shall show that if (WH) is false, so is $(WH)'$.

Assuming (WH) to be false, for a suitable α, $0 < \alpha < 1$, there exists for each positive integer n a bounded measurable set B_n such that

$$\mu(S(\alpha,B_n)) > n^2\mu(B_n) . \tag{5.3.3}$$

We apply to each set B_n, $n = 1,2,\ldots$ such a homothety or translation T_n' that if $B_n' = T_n'(B_n)$, then $\mu(B_n') = 1/n^2$, and we denote by S_n' a bounded partial weak halo of B_n' satisfying the inequality

$$\mu(S_n') > n^2\mu(B_n') = 1 , \tag{5.3.4}$$

which may be accomplished because of (5.3.3).

By applying suitable translations T_n'', $n = 1,2,\ldots$, we can achieve the disjunction of the sets $S_n'' = T_n''(S_n')$, $n = 1,2,\ldots$. For any positive integer n, we let $B_n'' = T_n''(B_n')$. Then we use (5.3.4) to see that

$$\mu\left(\bigcup_n (B_n'')\right) \leqslant \sum_n \mu(B_n'') = \sum_n n^{-2} < \infty ; \quad \mu\left(\bigcup_n S_n''\right) = \infty$$

and

$$S\left(\alpha, \bigcup_n B_n''\right) \supset \bigcup_n S(\alpha, B_n'') \supset \bigcup_n S_n'' ;$$

thus $(WH)'$ does not hold.

6. An individual derivability condition of Busemann-Feller type.

6.1. Definition. If ψ is an $\mathscr{M}$-measure, $M \in \mathscr{M}$, $\alpha > 0$, and $\eta > 0$, then we denote by $S(\alpha,\psi,\eta,M)$ an essential union of those $\mathfrak{B}$-constituents V such that

$$\psi(M \cdot V) > \alpha\mu(V) \quad \text{and} \quad \delta(V) \leqslant \eta .$$

The set $S(\alpha,\psi,\eta,M)$ is called a *ψ-halo*.

Remarks. The modification of the BUSEMANN-FELLER definition, which involves the strict union, has been made here because of the possible nonmeasurability in our setting.

The relation between $S(\alpha,r,\mathscr{E})$, defined in (III. 4), and the concept just defined, may be written as $S(\alpha,r,\mathscr{E}) = S(\alpha,\psi,\eta,M)$ provided

$M = \theta\mathscr{E}$, ψ is the indefinite μ-integral of that function coinciding with $(\in_{\mathscr{E}})^r$ on $\theta\,\mathscr{E}$, zero elsewhere, and $\eta = \nu(\mathscr{E})$.

6.2. Theorem. *If $\mathfrak{B}$ is a D-basis and an S^1-basis, ψ is a nonnegative μ-integral, and (G_σ) holds, then a necessary and sufficient condition that $\mathfrak{B}$ derive ψ is the following ψ-halo evanescence condition: if (M_n) is any bounded non-increasing sequence of sets in $\mathscr{M}$ with $\lim_n \mu(M_n) = 0$, (η_n) is any non-increasing sequence of positive numbers with $\lim_n \eta_n = 0$, and α is any positive number, then $\lim_n \mu(S(\alpha,\psi,\eta_n,M_n)) = 0$.*

Proof. We first establish the sufficiency of the condition. We shall show that the ψ-halo evanescence condition implies the Vitali ψ-property in the ε-version. We take an arbitrary bounded subset X of E, a $\mathfrak{B}$-fine covering $\mathscr{V}$ of X, and a positive number ε.

For a suitable positive integer N, we have $X \subset G_N^0$ (cf. Def. II. 1). Pruning $\mathscr{V}$ if necessary, we may assume that all the $\mathscr{V}$-sets are included in G_N^0. Since $\mathfrak{B}$ is a D-basis and an S^1-basis, then corresponding to each positive integer n, there exists an M-family $\mathscr{E}_n$ of $\mathscr{V}$-sets such that, putting $S_n = \sigma\mathscr{E}_n$, we have

$$X \subset S_n \,(\mathrm{mod}\, \mathscr{N}^\star), \qquad \mu(S_n - S_n \cdot \bar{X}) < 2^{-n-1},$$
$$\omega(\mathscr{E}_n,\mu) < 2^{-n-1}, \qquad \nu(\mathscr{E}_n) \leqslant 1/n. \tag{6.2.1}$$

Setting

$$O_n = \theta\mathscr{E}_n, \qquad D_n = \bigcup_{k=n}^{\infty} O_k,$$

we have

$$\mu(O_n) < 2^{-n-1}, \qquad \mu(D_n) < 2^{-n}, \qquad n = 1,2,\ldots$$

Next, we define $\alpha = \varepsilon/(\bar{\mu}(X)+1)$. For $n = 1,2,\ldots$ we denote by $\mathscr{H}_n$ the family of those sets $V \in \mathscr{E}_n$ for which

$$\psi(V \cdot O_n) \leqslant \alpha\mu(V). \tag{6.2.2}$$

For each positive integer n we have

$$\begin{aligned}\omega(\mathscr{H}_n,\psi) &= \int_{\sigma\mathscr{H}_n} \in_{\mathscr{H}_n} d\psi \leqslant \int_{O_n} \varphi_{\mathscr{H}_n} d\psi \\ &= \sum_{V\in\mathscr{H}_n} \psi(V\cdot O_n) \leqslant \alpha \sum_{V\in\mathscr{H}_n} \mu(V) \leqslant \alpha(\mu(S_n) + \omega(\mathscr{E}_n,\mu)) \\ &< \alpha(\bar{\mu}(X) + 2^{-n-1} + 2^{-n-1}) \leqslant \alpha(\bar{\mu}(X)+1) = \varepsilon.\end{aligned} \tag{6.2.3}$$

Now $\mathscr{E}_n$ is a covering of X (mod $\mathscr{N}^\star$). The sets belonging to $\mathscr{E}_n - \mathscr{H}_n$ are included in $S(\alpha,\psi,1/n,O_n)$(mod $\mathscr{N}^\star$); therefore $\mathscr{H}_n$ covers

$$X - X \cdot S(\alpha,\psi,1/n,O_n)\,(\mathrm{mod}\,\mathscr{N}^\star),$$

and consequently

$$X - X \cdot S(\alpha, \psi, 1/n, O_n) \subset \sigma \mathscr{H}_n \,(\mathrm{mod}\, \mathscr{N}^\star), \qquad n = 1, 2, \ldots \tag{6.2.4}$$

Since $D_{n+1} \subset D_n \subset G_N^0$ for each positive integer n, $\mu(G_N^0) < \infty$, and $\lim_n \mu(D_n) = 0$, then from the ψ-halo evanescence condition we conclude that

$$\lim_n \mu(S(\alpha, \psi, n^{-1}, D_n)) = 0.$$

There exists a positive number η such that for any $\mathscr{M}$-set $M \subset G_N^0$ with $\mu(M) < \eta$, we have $\psi(M) < \varepsilon$. We may clearly assume that $\eta < \varepsilon$. We fix n_0 so that

$$\mu(S(\alpha, \psi, 1/n_0, D_{n_0})) < \eta, \qquad 2^{-n_0 - 1} < \eta. \tag{6.2.5}$$

Thus the family $\mathscr{E} = \mathscr{H}_{n_0}$ satisfies the relation

$$\psi(\bar{X} - \bar{X} \cdot \sigma \mathscr{E}_{n_0}) < \varepsilon;$$

that is, $\mathscr{E}$ satisfies the ε-covering condition of Def. II. 1.3. From the second relation in (6.2.1) and the fact that $\sigma\mathscr{E} \subset S_{n_0}$ we infer that $\psi(\sigma\mathscr{E} - \sigma\mathscr{E} \cdot \bar{X}) < \varepsilon$; the ψ-overflow of $\mathscr{E}$ with respect to X is less than ε. Finally, the ψ-overlap of $\mathscr{E}$ is less than ε by virtue of (6.2.3).

We turn now to the proof of the necessity of the condition. We consider an arbitrary non-increasing sequence of $\mathscr{M}$-sets $M_1, M_2, \ldots, M_n \ldots$ with $\lim_n \mu(M_n) = 0$, an arbitrary non-increasing sequence of positive numbers $\delta_1, \delta_2, \ldots, \delta_n, \ldots$ with $\lim_n \delta_n = 0$, and an arbitrary $\alpha > 0$. We let

$$H_n = S(\alpha, \psi, \delta_n, M_n), \qquad n = 1, 2, \ldots; \qquad H = \bigcap_{n=1}^{\infty} H_n.$$

Since the halo $S(\alpha, \psi, \delta, M)$ is a non-decreasing function of δ, it follows readily that for any pair of positive integers n, ν we have

$$H \subset S(\alpha, \psi, \delta_n, M_\nu). \tag{6.2.6}$$

For each such pair of positive integers, there exist enumerably many $\mathfrak{B}$-constituents $V_{n,\nu}^1, V_{n,\nu}^2, \ldots, V_{n,\nu}^j, \ldots$ such that

$$S_{n,\nu} = \bigcup_{j=1}^{\infty} V_{n,\nu}^j = S(\alpha, \psi, \delta_n, M_\nu) \,(\mathrm{mod}\, \mathscr{N}) \tag{6.2.7}$$

$$\delta(V_{n,\nu}^j) < \delta_n, \psi(V_{n,\nu}^j \cdot M_\nu) > \alpha \mu(V_{n,\nu}^j),$$

for $j = 1, 2, \ldots$. Corresponding to each point x of the set

$$H_\nu^\star = \bigcap_{n=1}^{\infty} S_{n,\nu},$$

$\nu = 1,2,\dots$, there exists a sequence of $\mathfrak{B}$-constituents $W_1, W_2, \dots, W_n, \dots$ satisfying the relations

$$x \in W_n, \quad \delta(W_n) \leqslant \delta_n, \quad \psi(W_n \cdot M_\nu) > \alpha\mu(W_n). \tag{6.2.8}$$

We denote by f an integrand of ψ, define r_ν so that

$$r_\nu(x) = f(x) \quad \text{if} \quad x \in M_\nu, \quad r_\nu(x) = 0 \quad \text{if} \quad x \notin M_\nu,$$

and let

$$\rho_\nu(M) = \int_M r_\nu \, d\mu$$

for $M \in \mathscr{M}$. We deduce from (6.2.8) that

$$D^\star \rho_\nu(x) \geqslant \alpha, \tag{6.2.9}$$

for each $x \in H_\nu^\star$.

Since $\mathfrak{B}$ is an S^1-basis and derives ψ, then by Th. III. 2.3, $\mathfrak{B}$ derives the non-negative Radon integrals dominated by ψ for μ-almost all $x \in E$. Hence, from (6.2.9), $r_\nu(x) \geqslant \alpha$ for μ-almost all $x \in H_\nu^\star$. However, $r_\nu(x) = 0$ for each $x \notin M_\nu$, thus $H_\nu^\star \subset M_\nu \pmod{\mathscr{N}^\star}$, which, by (6.2.7), means that

$$\bigcap_{n=1}^{\infty} (S(\alpha, \psi, \delta_n, M_\nu)) \subset M_\nu \pmod{\mathscr{N}^\star};$$

using (6.2.6), we conclude that $H \subset M_\nu \pmod{\mathscr{N}^\star}$, $\nu = 1,2,\dots$. Since $\lim_\nu \mu(M_\nu) = 0$, then $\mu(H) = 0$, which completes the proof.

7. The weak halo property in general bases. Here we consider a general derivation basis $\mathfrak{B}$ with the following *weak halo property*, namely,

$$(WH)'' \qquad \mu(S(\alpha, \mu, K)) \leqslant \gamma(\alpha)\mu(K)$$

whenever K is the union of an arbitrary finite subfamily of the spread of $\mathfrak{B}$, α is any number satisfying $0 < \alpha < 1$, and γ is a finite-valued function of α that does not depend on K. Obviously $\gamma(\alpha) \geqslant 1$ for each admissible value of α.

7.1. Theorem. *We assume that*

(1) *If $X \subset E$, $0 < \bar{\mu}(X) < \infty$, $\varepsilon > 0$, and $\mathscr{V}$ is any $\mathfrak{B}$-fine covering of X, then there exists a countable subfamily $\mathscr{E}$ of $\mathscr{V}$ such that*

$$X \subset S \pmod{\mathscr{N}^\star} \quad \textit{and} \quad \mu(S - S \cdot \bar{X}) < \varepsilon,$$

where $S = \bigcup \mathscr{E}$ (this is Property (L); cf. Remarks following Th. 3.3).

(2) *$(WH)''$ holds.*

Then $\mathfrak{B}$ has the density property.

Proof. It is sufficient to show that $\mathfrak{B}$ has the Vitali μ-property in the ε-version. To this end, we take an arbitrary set $X \subset E$ such that

$\bar{\mu}(X) < \infty$, an arbitrary $\mathfrak{B}$-fine covering $\mathscr{V}$ of X, and an arbitrary positive number ε. We may clearly assume $0 < \bar{\mu}(X)$ and $\varepsilon < 1$ without loss of generality.

We choose a sequence of positive numbers $\varepsilon_1, \varepsilon_2, \ldots, \varepsilon_n, \ldots$ such that $\sum_{i=1}^{\infty} \varepsilon_i < \varepsilon/2$ and set $X_1 = X$. We further take any positive number α such that

$$0 < \alpha < \frac{\varepsilon_1}{2(\bar{\mu}(X)+1)} < \frac{1}{2}. \tag{7.1.1}$$

We let γ denote the number $\gamma(\alpha)$ associated with α by $(WH)''$, define

$$\varepsilon_1' = \min\left(\varepsilon_1, \frac{\bar{\mu}(X_1)}{4\gamma}\right)$$

and select any number ε_2' such that

$$0 < \varepsilon_2' < \frac{\bar{\mu}(X_1) - 2\varepsilon_1'\gamma}{2} \tag{7.1.2}$$

which is possible due to the definition of ε_1'.

According to the condition (1) of our hypotheses, there exists a countable subfamily $\mathscr{E}$ of $\mathscr{V}$ whose members are $V_1, V_2, \ldots, V_n, \ldots$ such that

$$\mu(\bar{X}_1 - T \cdot \bar{X}_1) = 0, \quad \mu(T - T \cdot \bar{X}_1) < \varepsilon_1', \tag{7.1.3}$$

where $T = \bigcup \mathscr{E}$. For any positive integer n we let $T_n = \bigcup_{i=1}^{n} V_i$, and since $\mu(T) < \infty$ because of (7.1.3), we may select N so that

$$\mu(T - T_N) < \varepsilon_2'. \tag{7.1.4}$$

Next, we let $U_1 = V_1$ and consider the ratios

$$\mu(U_1 \cdot V_i)/\mu(V_i)$$

for $i = 1, 2, \ldots, N$. In case all these ratios exceed α, we stop at this point. If not, we select one of the sets V_i for which the corresponding ratio does not exceed α and denote it by U_2. Since $0 < \alpha < 1$, it is clear that $U_1 \neq U_2$ and

$$\mu(U_1 \cdot U_2) \leqslant \alpha \mu(U_2).$$

Next, we consider the ratios

$$\mu((U_1 \cup U_2) \cdot V_i)/\mu(V_i)$$

for $i = 1, 2, \ldots, N$, and if they all exceed α, we stop at this point. Otherwise, we pick one of the sets V_i whose corresponding ratio does not exceed α and name it U_3.

Evidently

$$U_3 \neq U_1, \quad U_3 \neq U_2, \quad \text{and} \quad \mu((U_1 \cup U_2) \cdot U_3) \leqslant \alpha \mu(U_3).$$

Proceeding inductively in this manner, we are led to a finite non-repetitive sequence of sets (U_j), $j = 1,2,\ldots,k_1 \leqslant N$, drawn from the original sequence (V_i), $i = 1,2,\ldots,N$, and a corresponding sequence of sets (S_j), $j = 1,2,\ldots,k_1$, satisfying the relations

$$\begin{aligned} S_j &= \bigcup_{i=1}^{j} U_i \subset T_N, \quad j = 1,2,\ldots,k_1; \\ \mu(U_j \cdot S_{j-1}) &\leqslant \alpha \mu(U_j), \quad j = 2,3,\ldots,k_1. \end{aligned} \tag{7.1.5}$$

We let $\mathscr{E}_1$ denote the subfamily of $\mathscr{E}$ consisting of the sets $U_1, U_2, \ldots, U_{k_1}$. Since the inductive process stops after k_1 steps, it follows that all the sets V_i, $i = 1,2,\ldots,N$ satisfy the relation

$$\mu(S_{k_1} \cdot V_i) > \alpha \mu(V_i), \tag{7.1.6}$$

and, since $\bigcup_{i=1}^{N} V_i = T_N$, it follows that

$$T_N \subset S(\alpha, \mu, S_{k_1}).$$

Consequently, from 7.1.4 and $(WH)''$, we obtain

$$\mu(T) < \varepsilon_2' + \gamma \mu(S_{k_1}). \tag{7.1.7}$$

From (7.1.3) and (7.1.5), we see that $(\bar{X}_1 \cup S_{k_1}) \subset T (\mathrm{mod}\, \mathscr{N})$; using (7.1.3) and (7.1.7), we conclude that

$$\begin{aligned} \mu(\bar{X}_1 \cdot S_{k_1}) &= \mu(\bar{X}_1) + \mu(S_{k_1}) - \mu(\bar{X}_1 \cup S_{k_1}) \\ &> \mu(\bar{X}_1) + \mu(S_{k_1}) - (\mu(\bar{X}_1) + \varepsilon_1') \\ &> (1/\gamma)(\mu(T) - \varepsilon_2') - \varepsilon_1' \\ &= \frac{1}{\gamma} \mu(T) - \frac{\varepsilon_2'}{\gamma} - \varepsilon_1'. \end{aligned} \tag{7.1.8}$$

Using (7.1.2) in conjunction with (7.1.8) yields

$$\mu(\bar{X}_1 \cdot (\bigcup \mathscr{E}_1)) = \mu(\bar{X}_1 \cdot S_{k_1}) > \tfrac{1}{2\gamma} \mu(\bar{X}_1). \tag{7.1.9}$$

Now, since the sets U_1 and $U_j - U_j \cdot S_{j-1}$, $j = 2,3,\ldots,k_1$ are pairwise disjoint and

$$S_{k_1} = U_1 \cup \bigcup_{j=2}^{k_1} (U_j - U_j \cdot S_{j-1}),$$

it follows with the help of the second relation in (7.1.5) that

$$\begin{aligned}\mu(\bigcup \mathscr{E}_1) = \mu(S_{k_1}) &= \mu(U_1) + \sum_{j=2}^{k_1} \mu(U_j - U_j \cdot S_{j-1}) \\ &\geqslant \mu(U_1) + (1-\alpha)\sum_{j=2}^{k_1} \mu(U_j) \\ &> (1-\alpha)\sum_{j=1}^{k_1} \mu(U_j).\end{aligned} \tag{7.1.10}$$

From (7.1.10), (7.1.1), and (7.1.3), we now see that

$$\begin{aligned}\omega(\mathscr{E},\mu) &= \sum_{j=1}^{k_1} \mu(U_j) - \mu(S_{k_1}) \\ &< \mu(S_{k_1})\left(\frac{1}{1-\alpha} - 1\right) = \frac{\alpha}{1-\alpha}\mu(S_{k_1}) \\ &< \frac{\varepsilon_1}{2(\bar{\mu}(X)+1)}\mu(\bigcup \mathscr{E}) \leqslant \varepsilon_1 ;\end{aligned} \tag{7.1.11}$$

finally, since $\bigcup \mathscr{E}_1 \subset \bigcup \mathscr{E} = T$ we infer from (7.1.3) that

$$\mu(\bigcup \mathscr{E}_1 - \bigcup \mathscr{E}_1 \cdot \bar{X}_1) < \varepsilon_1' \leqslant \varepsilon_1 . \tag{7.1.12}$$

Thus we have succeeded in finding a finite subfamily $\mathscr{E}_1$ of $\mathscr{V}$ (i) whose μ-exhaustion power of X exceeds $1/2\,v$ (7.1.9); (ii) whose μ-overlap is less than ε_1 (7.1.11); and (iii) whose (X,μ)-overflow is less than ε_1 (7.1.12).

It is apparent that by repeating this process inductively, we can find a sequence of finite families $\mathscr{E}_n \subset \mathscr{V}$, $n = 1,2,\ldots$, such that

$$\begin{gathered}\mu(\bar{X}_n \cdot S_n') > (1/2\,v)\mu(\bar{X}_n), \quad \omega(\mathscr{E}_n,\mu) < \varepsilon_n , \\ \mu(S_n' - S_n' \cdot \bar{X}_n) < \varepsilon_n ,\end{gathered} \tag{7.1.13}$$

where $S_n' = \bigcup \mathscr{E}_n$, $X_1 = X$, and $X_{n+1} = X_n - X_n \cdot S_n'$. The inductive process terminates if, for some positive integer K, $\mu(\bar{X}_{K+1}) = 0$; otherwise it continues for each positive integer n. It is easily seen that for each value of n prior to termination of the induction, should that occur, or for all n otherwise,

$$X_{n+1} = X - X \cdot \left(\bigcup_{i=1}^{n} S_i'\right). \tag{7.1.14}$$

We let $\mathscr{F}_n = \bigcup_{i=1}^{n} \mathscr{E}_i$ for each n for which $\mathscr{E}_n$ is defined, and note that $\bigcup \mathscr{F}_n = \bigcup_{i=1}^{n} S_i'$. In case $\mu(\bar{X}_{N+1}) = 0$ for some positive integer N, we see from (7.1.14) that $X \subset \bigcup \mathscr{F}_N \pmod{\mathscr{N}^\star}$. If, on the other hand, $\mu(\bar{X}_n) > 0$ for each positive integer n, then we see from (7.1.13) that for each such n,

$$\mu(\bar{X}_{n+1}) = \mu(\bar{X}_n) - \mu(\bar{X}_n \cdot S_n') < (1 - \tfrac{1}{2\gamma})\mu(\bar{X}_n)$$

whence, by induction,

$$\mu(\bar{X}_{n+1}) < (1-\tfrac{1}{2\gamma})^n \mu(\bar{X}).$$

Using (7.1.14) we can, therefore, select N so that

$$\mu\left(\bar{X} - \bar{X} \cdot \bigcup_{i=1}^{N} S_i\right) < \varepsilon. \tag{7.1.15}$$

Thus $\mathscr{F}_N$ is an ε-covering of X. In any case, we can find a finite subfamily $\mathscr{F}_N$ of $\mathscr{V}$ such that $\mathscr{F}_N = \bigcup_{i=1}^{N} \mathscr{E}_i$ and $\mathscr{F}_N$ is an ε-covering of X.

We wish to show that $\mathscr{F}_N$ has (X, μ)-overflow and μ-overlap both less than ε. In case $N = 1$, there is nothing to prove because of (7.1.13). If $N > 1$, we consider any positive integer n such that $n + 1 \leqslant N$ and observe that

$$\omega(\mathscr{F}_{n+1}, \mu) = \sum_{V \in \mathscr{F}_{n+1}} \mu(V) - \mu(\cup \mathscr{F}_{n+1}) = \sum_{V \in \mathscr{F}_{n+1}} \mu(V) - \mu\left(\bigcup_{i=1}^{n+1} S'_i\right)$$

$$\leqslant \sum_{V \in \mathscr{F}_n} \mu(V) + \sum_{V \in \mathscr{E}_{n+1}} \mu(V) - \left(\mu\left(\bigcup_{i=1}^{n} S'_i\right) + \mu(S'_{n+1}) - \mu\left(S'_{n+1} \cdot \left(\bigcup_{i=1}^{n} S'_i\right)\right)\right)$$

$$= \left(\sum_{V \in \mathscr{F}_n} \mu(V) - \mu\left(\bigcup_{i=1}^{n} S'_i\right)\right) + \left(\sum_{V \in \mathscr{E}_{n+1}} \mu(V) - \mu(S'_{n+1}) + \mu\left(S'_{n+1} \cdot \left(\bigcup_{i=1}^{n} S'_i\right)\right)\right)$$

$$= \omega(\mathscr{F}_n, \mu) + \omega(\mathscr{E}_{n+1}, \mu) + \mu\left(S'_{n+1} \cdot \left(\bigcup_{i=1}^{n} S'_i\right)\right). \tag{7.1.16}$$

Since $\bar{X}_{n+1} = \bar{X} - \bar{X} \cdot \left(\bigcup_{j=1}^{n} S'_j\right)$, it follows that for each i, $1 \leqslant i \leqslant n$, $S'_i \cdot \bar{X}_{n+1} = \emptyset$, and therefore $S'_i \cdot S'_{n+1} \subset (S'_{n+1} - S'_{n+1} \cdot \bar{X}_{n+1})$. Thus, using (7.1.13), we obtain

$$\mu\left(S'_{n+1} \cdot \left(\bigcup_{i=1}^{n} S'_i\right)\right) \leqslant \mu(S'_{n+1} - S'_{n+1} \cdot \bar{X}_{n+1}) < \varepsilon_{n+1}.$$

Placing this last result in (7.1.16) yields

$$\omega(\mathscr{F}_{n+1}, \mu) < \omega(\mathscr{F}_n, \mu) + \omega(\mathscr{E}_{n+1}, \mu) + \varepsilon_{n+1}.$$

Using (7.1.13) and simple induction on this last inequality, we derive

$$\omega(\mathscr{F}_N, \mu) < \sum_{i=1}^{N} \omega(\mathscr{E}_i, \mu) + \sum_{i=2}^{N} \varepsilon_i < 2 \sum_{i=1}^{N} \varepsilon_i < \varepsilon.$$

Hence $\mathscr{F}_N$ has the required overlap property.

Finally, since

$$\cup \mathscr{F}_N - \bar{X} \cdot (\cup \mathscr{F}_N) \subset \bigcup_{i=1}^{N} (S'_i - S'_i \cdot \bar{X}_i)$$

we infer from (7.1.13) that

$$\mu(\cup \mathscr{F}_N - \bar{X} \cdot (\cup \mathscr{F}_N)) \leqslant \sum_{i=1}^{N} \mu(S_i' - S_i' \cdot \bar{X}_i) < \sum_{i=1}^{N} \varepsilon_i < \varepsilon,$$

and $\mathscr{F}_N$ has the required overflow property. The proof is now complete.

Remarks. (1) Th. 7.1 is a weak counterpart to Th. 3.3.

(2) The halo function γ plays the role of a disentanglement function.

(3) We note that the condition $(W\,H)''$ involves the spread and not the pretopology. The latter appears in condition (L) which involves the covering.

(4) In $(W\,H)$ the nucleus M can be any bounded measurable set, whereas in $(W\,H)''$ it is a finite union of $\mathscr{U}$-sets. Because of condition (L), every bounded measurable set can be approximated arbitrarily closely by a finite union of $\mathscr{U}$-sets.

8. Product invariance of a weak halo property [*6*, 239–242; *17*, 202–207). The weak halo property of § 7 can be formulated in an obvious way from families of sets, without reference to a basis or even to points. We now define this concept and a strengthening of it that we need shortly.

8.1. Definition. A family $\mathscr{U}$ of sets U with $0 < \mu(U) < \infty$ possesses the *weak halo property* $(W\,H)''$ iff there exists a finite function γ such that

$$\mu(S(\alpha, K)) < \gamma(\alpha)\,\mu(K)$$

holds whenever K is the union of a finite subfamily of $\mathscr{U}$ and $0 < \alpha < 1$. This weak halo property is called *sharp* if the stated inequality holds whenever K is a set in the Boolean extension of $\mathscr{U}$.

Preliminaries. We shall assume in this section that $(R_i, \mathscr{M}_i, \mu_i)$, $i = 1, 2, \ldots, n$ are each σ-finite measure spaces provided with a family $\mathscr{U}_i \subset \mathscr{M}_i$ of sets of positive finite μ_i-measure. The finite unions of sets of the form $I = M_1 \times M_2 \times \ldots \times M_n$ (called *intervals*) where $M_i \in \mathscr{M}_i$, $i = 1, 2, \ldots, n$, constitute a Boolean algebra $\mathscr{B}$, the Boolean extension of the family $\mathscr{I}$ of the intervals. The function μ defined on $\mathscr{I}$ by

$$\mu(I) = \prod_{i=1}^{n} \mu_i(M_i)$$

for $I \in \mathscr{I}$ can be extended to a σ-additive content (Jordan measure) $\mu|\mathscr{B}$ defined on $\mathscr{B}$, and also to a σ-additive measure $\mu|\mathscr{M}$, the Borel extension of $\mu|\mathscr{I}$ and of $\mu|\mathscr{B}$. The set $\mathscr{U} = \prod_{i=1}^{n} \mathscr{U}_i$ is the family of all sets U of the form $U = \prod_{i=1}^{n} U_i$, where $U_i \in \mathscr{U}_i$, $i = 1, 2, \ldots, n$.

8.2. Theorem. *If each of the families $\mathscr{U}_i$, $i = 1, 2, \ldots, n$ has the property $(W\,H)''$, and if all but at most one of them has the sharp weak halo property, then the class $\mathscr{U}$ also possesses the property $(W\,H)''$.*

Proof. It is clear that if we can prove the assertion of the theorem in the case $n = 2$, with one of the families $\mathscr{U}_1$ or $\mathscr{U}_2$ having the sharp weak halo property and the other merely the property $(W\,H)''$, then the theorem will be true for arbitrary n by induction.

Accordingly, we assume that $\mathscr{U}_1$ has the sharp weak halo property and that $\mathscr{U}_2$ has the property $(W\,H)''$. We take any set $K = \bigcup_{i=1}^{n} B_i$ (*nucleus*), where $B_1, B_2, \ldots, B_n$ are members of the family $\mathscr{U} = \mathscr{U}_1 \times \mathscr{U}_2$. There exist sets $U_{1i} \in \mathscr{U}_1$, $U_{2i} \in \mathscr{U}_2$ such that $B_i = U_{1i} \times U_{2i}$ for $i = 1, 2, \ldots, n$. We also introduce any finite collection of sets $V_1, V_2, \ldots, V_m$ in $\mathscr{U}$ such that

$$\alpha\mu(V_j) < \mu(V_j \cdot K), \qquad 0 < \alpha < 1, \qquad \text{for} \quad j = 1, 2, \ldots, m. \quad (8.2.1)$$

We let $\mathscr{B}_1$ and $\mathscr{B}_2$ denote the smallest Boolean algebras generated by the finite collections of sets $U_{11}, U_{12}, \ldots, U_{1n}, W_{11}, W_{12}, \ldots, W_{1m}$ and $U_{21}, U_{22}, \ldots, U_{2n}, W_{21}, W_{22}, \ldots, W_{2m}$, respectively, where $V_j = W_{1j} x\, W_{2j}$, $j = 1, 2, \ldots, m$. The union of the sets V_j is denoted by H (the partial halo). Because $\mathscr{B}_1$ and $\mathscr{B}_2$ are generated by finite collections of elements, there exist, as is well known, finite disjoint classes $\mathscr{P}_1 \subset \mathscr{B}_1$ and $\mathscr{P}_2 \subset \mathscr{B}_2$ of *atomic* elements, that is, any element of $\mathscr{B}_1$ or $\mathscr{B}_2$ may be expressed as a union of members of $\mathscr{P}_1$ or $\mathscr{P}_2$, respectively. Correspondingly, each of the sets $B_1, B_2, \ldots, B_n, V_1, V_2, \ldots, V_m$, and K is the union of finitely many pairwise disjoint sets of the form $P = P_1 \times P_2$, where $P_1 \in \mathscr{P}_1$ and $P_2 \in \mathscr{P}_2$.

For any set B of the form

$$B = \bigcup_{k=1}^{r} (B_{1k} \times B_{2k}),$$

where $B_{1k} \in \mathscr{B}_1$ and $B_{2k} \in \mathscr{B}_2$ for $k = 1, 2, \ldots, r$, and any set $P_1 \in \mathscr{P}_1$ we let $B(P_1)$ denote the union (possibly empty) of those sets B_{2k}, $k = 1, 2, \ldots, r$ whose corresponding set B_{1k} satisfies the relation $P_1 \subset B_{1k}$. In precisely analogous fashion we define $B(P_2)$, for any set $P_2 \in \mathscr{P}_2$. It is evident that $B(P_1) \in \mathscr{B}_2$, $B(P_2) \in \mathscr{B}_1$, $P_1 \times B(P_1) \subset B$, and $B(P_2) \times P_2 \subset B$. It is easy to see that

$$B = \bigcup_{P_1 \in \mathscr{P}_1} (P_1 \times B(P_1)) = \bigcup_{P_2 \in \mathscr{P}_2} (B(P_2) \times P_2);$$

also, the sets occurring in each of these unions are pairwise disjoint and finite in number.

For any value $j = 1, 2, \ldots, m$ we denote by $\mathscr{P}_{1j}$ the set of those members $P_1 \in \mathscr{P}_1$ satisfying $P_1 \subset W_{1j}$. We see that $V_j(P_1) = W_{2j}$ if $P_1 \in \mathscr{P}_1$; $V_j(P_1) = \emptyset$ otherwise. Thus

$$V_j \cdot K = \Big(\bigcup_{P_1 \in \mathscr{P}_1} (P_1 \times V_j(P_1))\Big) \cdot \Big(\bigcup_{P_1 \in \mathscr{P}_1} (P_1 \times K(P_1))\Big) \\ = \bigcup_{P_1 \in \mathscr{P}_{1j}} (P_1 \times (W_{2j} \cdot K(P_1))). \tag{8.2.2}$$

Now $0 < \mu(V_j) < \infty$ because of our hypotheses and (8.2.1), therefore $0 < \mu_1(W_{1j}) < \infty$ and $0 < \mu_2(W_{2j}) < \infty$, $j = 1, 2, \ldots, m$. Thus for each set $P_1 \in \mathscr{P}_{1j}$ and each set W_{2j}, $j = 1, 2, \ldots, m$, the expression $\mu_2(W_{2j} \cdot K(P_1))/\mu_2(W_{2j})$ is well defined. For convenience, we abbreviate it as $\alpha_{2j}(P_1)$. From (8.2.1), (8.2.2), and the fact that

$$\mu(P_1 \times (W_{2j} \cdot K(P_1))) = \mu_1(P_1) \cdot \mu_2(W_{2j} \cdot K(P_1))$$

whenever $j = 1, 2, \ldots, m$ and $P_1 \in \mathscr{P}_{1j}$, we see that

$$0 < \alpha < \mu(V_j \cdot K)/\mu(V_j) = \sum_{P_1 \in \mathscr{P}_{1j}} \mu_1(P_1)\alpha_{2j}(P_1)/\mu_1(W_{1j}). \tag{8.2.3}$$

Next, we consider any number $\gamma, 0 < \gamma < \alpha$, and let S_{1j} denote the union of those sets P_1 occurring in the sum in (8.2.3) for which the corresponding numbers $\alpha_{2j}(P_1)$ exceed γ; we let $\mathscr{P}'_{1j}$ denote the family of the sets in question. Keeping in mind that for all sets $P_1 \in (\mathscr{P}_{1j} - \mathscr{P}'_{1j})$ we have $P_1 \subset (W_{1j} - S_{1j})$ and that for those sets we must have $\alpha_{2j}(P_1) \leqslant \gamma$, we see with the help of (8.2.3) that

$$\begin{aligned} \alpha\mu_1(W_{1j}) &< \sum_{P_1 \in \mathscr{P}'_{1j}} \mu_1(P_1)\alpha_{2j}(P_1) + \sum_{P_1 \in (\mathscr{P}_{1j} - \mathscr{P}'_{1j})} \mu_1(P_1)\alpha_{2j}(P_1) \\ &< \sum_{P_1 \in \mathscr{P}'_{1j}} \mu_1(P_1) + \gamma \sum_{P_1 \in (\mathscr{P}_{1j} - \mathscr{P}'_{1j})} \mu_1(P_1) \\ &= \mu_1(S_{1j}) + \gamma\mu_1(W_{1j} - S_{1j}), \quad j = 1, 2, \ldots, m; \end{aligned}$$

this yields

$$\alpha'\mu_1(W_{1j}) < \mu_1(S_{1j}), \quad j = 1, 2, \ldots, m, \tag{8.2.4}$$

where $\alpha' = (\alpha - \gamma)/(1 - \gamma)$, $0 < \alpha' < 1$. We observe that $S_{1j} \neq \emptyset$ and so $\mathscr{P}'_{1j} \neq \emptyset$.

We now let $S_j = S_{1j} \times W_{2j}$, $j = 1, 2, \ldots, m$, and set $S = \bigcup_{j=1}^{m} S_j$. We see that $S_{1j} = S_j(P_2) \cdot W_{1j}$ whenever $P_2 \in \mathscr{P}_2$ and $P_2 \subset W_{2j}$, $j = 1, 2, \ldots, m$. We may thus write (8.2.4) in the form

$$\alpha'\mu_1(W_{1j}) < \mu_1(W_{1j} \cdot S_j(P_2)) \leqslant \mu_1(W_{1j} \cdot S(P_2))$$

for any set $P_2 \in \mathscr{P}_2$, $P_2 \subset W_{2j}$. Since $S(P_2) \in \mathscr{B}_1$ and $\mathscr{B}_1$ is a subset of the smallest Boolean algebra containing $\mathscr{U}_1$, then for any fixed set $P_2 \in \mathscr{P}_2$ we may infer from the sharp weak halo property of $\mathscr{U}_1$ that

$$\mu_1(\textstyle\bigcup' W_{1j}) \leqslant \beta_1(\alpha')\mu_1(S(P_2)), \tag{8.2.5}$$

where $\beta_1(\alpha')$ is the number associated with α' in connection with the sharp weak halo property of $\mathscr{U}_1$, and the prime on the union symbol indicates that the union is taken over all indices j such that $P_2 \subset W_{2j}$.

Given any fixed set $P_1 \in \mathscr{P}_1$, it follows that for any j such that $P_1 \subset S_{1j}$ we have $P_1 \in \mathscr{P}_{1j}$, and so from the definition of S_{1j},

$$\gamma \mu_2(W_{2j}) < \mu(W_{2j} \cdot K(P_1)).$$

Because $K(P_1)$ is a finite union of members of $\mathscr{U}_2$ and $\mathscr{U}_2$ has the property $(WH)''$, we may infer that

$$\mu_2\left(\bigcup{}'' W_{2j}\right) \leqslant \beta_2(\gamma)\mu_2(K(P_1)), \tag{8.2.6}$$

where the double prime over the union symbol indicates that the union is to be taken over all indices j for which $P_1 \subset S_{1j}$. Since $S(P_1)$ is by definition the union of all such sets W_{2j}, then we may write (8.2.6) in the form

$$\mu_2(S(P_1)) \leqslant \beta_2(\gamma) \cdot \mu_2(K(P_1)). \tag{8.2.7}$$

We now set $H = \bigcup_{j=1}^{m} V_j$, $W'' = \bigcup_{j=1}^{m} W_{2j}$, and let $\mathscr{P}_2'$ denote the subfamily of $\mathscr{P}_2$ whose members are subsets of W''. Then

$$H = \bigcup_{P_2 \in \mathscr{P}_2} H(P_2) \times P_2 = \bigcup_{P_2 \in \mathscr{P}_2} H(P_2) \times P_2,$$

where the terms in the union are pairwise disjoint. We note that for any $P_2 \in \mathscr{P}_2'$, $H(P_2) = \bigcup{}' W_{2j}$, whence from (8.2.5) we have, for each $P_2 \in \mathscr{P}_2$,

$$\mu_1(H(P_2)) \leqslant \beta_1(\alpha')\mu_1(S(P_2)),$$

and so, multiplying both sides of this relation by $\mu_2(P_2)$ and summing over $\mathscr{P}_2'$, we obtain

$$\begin{aligned} \mu(H) &= \sum_{P_2 \in \mathscr{P}_2'} \mu(H(P_2) \times P_2) = \sum_{P_2 \in \mathscr{P}_2'} \mu_1(H(P_2)) \cdot \mu_2(P_2) \\ &\leqslant \beta_1(\alpha') \sum_{P_2 \in \mathscr{P}_2} \mu_1(S(P_2)) \cdot \mu_2(P_2) \\ &= \beta_1(\alpha') \sum_{P_2 \in \mathscr{P}_2} \mu(S(P_2) \times P_2) = \beta_1(\alpha')\mu(S). \end{aligned} \tag{8.2.8}$$

Next, we let $S' = \bigcup_{j=1}^{m} S_{1j}$ and let $\mathscr{P}_1'$ denote the subfamily of $\mathscr{P}_1$ whose members are subsets of S'. We see that

$$S = \bigcup_{P_1 \in \mathscr{P}_1} P_1 \times S(P_1) = \bigcup_{P_1 \in \mathscr{P}_1'} P_1 \times S(P_1),$$

$$K = \bigcup_{P_1 \in \mathscr{P}_1} P_1 \times K(P_1) = \bigcup_{P_1 \in \mathscr{P}_1'} P_1 \times K(P_1),$$

where the terms in the unions are pairwise disjoint. Taking any set $P_1 \in \mathscr{P}_1$, multiplying both sides of (8.2.7) by $\mu_1(P_1)$, and summing over $\mathscr{P}_1'$, we obtain

$$\begin{aligned}\mu(S) &= \sum_{P_1 \in \mathscr{P}_1'} \mu(P_1 \times S(P_1)) = \sum_{P_1 \in \mathscr{P}_1'} \mu_1(P_1)\mu_2(S(P_1)) \\ &\leqslant \beta_2(\gamma) \sum_{P_1 \in \mathscr{P}_1'} \mu_1(P_1)\mu_2(K(P_1)) \\ &= \beta_2(\gamma) \sum_{P_1 \in \mathscr{P}_1'} \mu(P_1 \times K(P_1)) = \beta_2(\gamma)\mu(K).\end{aligned} \tag{8.2.9}$$

Putting this together with (8.2.8) we arrive at the result

$$\mu(H) \leqslant \beta_1(\alpha') \cdot \beta_2(\gamma) \cdot \mu(K).$$

Now H was the union of any finite subfamily of those sets making up $S(\alpha, \mu, K)$. Since this last set is an essential union, $\mu(S(\alpha, \mu, K))$ can be approximated arbitrarily closely by suitable finite collections of those sets comprising $S(\alpha, \mu, K)$, hence we conclude that

$$\mu(S(\alpha, \mu, K)) \leqslant \beta_1(\alpha')\beta_2(\gamma)\mu(K).$$

This relation is valid for any γ, $0 < \gamma < \alpha < 1$ and $\alpha' = (\alpha - \gamma)/(1-\gamma)$. If one takes $\gamma = \alpha/2$, then $\alpha' = \alpha/(2-\alpha)$. We may define β so that $\beta(\alpha) = \beta_1(\alpha')\beta_2(\gamma) = \beta_1(\alpha/(2-\alpha))\beta_2(\alpha/2)$, and it is now clear that $\mathscr{U}$ has the property $(WH)''$ with β as its associated function.

By iteration of this process, it follows that for the product of n families $\mathscr{U}_i$, each with an associated halo function β_i, $i = 1, 2, \ldots, n$, a suitable halo function β is given by

$$\beta(\alpha) = \beta_1\left(\frac{\alpha}{2^{n-1}}\right) \cdot \beta_2\left(\frac{\alpha}{2^{n-1} - \alpha}\right) \cdot \beta_3\left(\frac{\alpha}{2^{n-2} - \alpha}\right) \cdots \cdot \beta_n\left(\frac{\alpha}{2-\alpha}\right),$$
$$0 < \alpha < 1.$$

Remarks. (1) If the halo functions β_i, $i = 1, 2, \ldots, n$, are non-increasing (for instance, when they are minimal halo functions), a suitable product halo function β may be defined by $\beta(\alpha) = \beta_1(\frac{\alpha}{2}) \cdot \beta_2(\frac{\alpha}{2}) \ldots \beta_n(\frac{\alpha}{2})$.

(2) If $[\mathscr{U}_i, \delta_i]$, $i = 1, 2, \ldots, n$ are D-bases, then we can form a product D-basis $[\mathscr{U}, \delta]$ by taking for $\mathscr{U}$ all sets of the form $U_1 \times U_2 \times \ldots \times U_n$, where $U_i \in \mathscr{U}_i$, $i = 1, 2, \ldots, n$, and by defining $\delta = \delta_1 \cdot \delta_2 \ldots \delta_n$. Accordingly, if each basis $[\mathscr{U}_i, \delta_i]$ possesses the weak halo property of Def. 7.1, then each family $\mathscr{U}_i$ possesses the corresponding property of Def. 8.1, whence so also does $\mathscr{U}$; and so, finally $[\mathscr{U}, \delta]$ has the weak halo property of Def. 7.1.

Chapter V

The Interval Basis. The Theorem of Jessen-Marcinkiewicz-Zygmund

In this chapter, we take $R = \mathbf{R}^n$; $\mu = \mu_n$ is n-dimensional Borel measure, δ is n-dimensional Euclidean distance, and $\mathscr{I}$ is the family of closed nondegenerate intervals in $\mathbf{R}^n$, that is, the family of sets I of the form $I = E[x : x = (x_1, x_2, \ldots, x_n); \alpha_i \leqslant x_i \leqslant \beta_i, i = 1, 2, \ldots, n], \alpha_i < \beta_i$ for $i = 1, 2, \ldots, n$.

For the basis $\mathfrak{J}$ whose constituents are precisely the family $\mathscr{I}$, with δ as the index of uniform contraction and $E = R$, we shall prove that $\mathfrak{J}$ possesses the property $(WH)''$; that is, there exists a finite-valued function γ defined for all α, $0 < \alpha < 1$, such that for every finite union K of closed intervals of $\mathbf{R}^n$, the inequality $\mu(S(\alpha, K)) \leqslant \gamma(\alpha)\mu(K)$ holds, where $S(\alpha, K)$ is the weak halo corresponding to K (IV. 7). We propose first to obtain a rough estimate for γ, and then refine it. This refinement is closely related to the main theorem of this chapter, namely, the theorem of JESSEN-MARCINKIEWICZ-ZYGMUND [*27*].

The upper and lower derivates with respect to $\mathfrak{J}$, of a function defined on $\mathscr{I}$, are commonly known as its *strong* upper and lower derivates, respectively; if its $\mathfrak{J}$-derivative exists at a point, it is said to be *strongly* derivable at the point. We shall employ this terminology.

1. The interval basis as a weak derivation basis.

1.1. Lemma. *If ψ is any extended real-valued function defined on $\mathscr{I}$, then the (extreme) $\mathfrak{J}$-derivates of ψ are $\mu^\star$-measurable.*

Proof. We let $P = [D^\star\psi > \beta]$, where β is any (finite) real number. For any positive integers h and k, we let $P_{h,k}$ denote the union of all (closed) intervals I for which $\delta(I) < k^{-1}$ and $\psi(I)/\mu(I) > \beta + h^{-1}$. It is clear that $P = \bigcup_h \bigcap_k P_{h,k}$.

For any positive integers h and k, each point $x \in P_{h,k}$ is contained in some interval $I \in \mathscr{I}$, and it is clear that for each $\varepsilon > 0$ we can find a closed hypercube K such that $x \in K \subset I$, $\delta(I) < \varepsilon$. Accordingly, the family of all hypercubes $K \subset P_{h,k}$ constitutes an $\mathfrak{J}$-fine covering of $P_{h,k}$. Invoking Th. IV. 2.2, it follows that there exists a countable disjoint family of these hypercubes whose union equals $P_{h,k}$ (mod $\mathscr{N}^\star$). Thus $P_{h,k}$ is $\mu^\star$-measurable, and so is P. Hence $D^\star\psi$ is $\mu^\star$-measurable. An analogous proof establishes the $\mu^\star$-measurability of $D_\star\psi$.

1.2. Theorem [*6*, 239]. *For any bounded μ-measurable subset M of $\mathbf{R}^n$ we have*

$$\mu(S(\alpha, M)) \leqslant \left(\frac{2}{\alpha}\right)^n \mu(M). \tag{1.2.1}$$

Proof. We consider first the case $n = 1$. We take any finite set of closed intervals $I_1, I_2, \ldots, I_k$ in $\mathbf{R}^1$ for which $\mu(I_j \cdot M) > \alpha\mu(I_j), j = 1, 2, \ldots, k$. If it happens that three of these intervals have a point in common, then it is clear that one of them must be included in the union of the other two. Thus the set $S = \bigcup_{j=1}^{k} I_j$ may be represented as the union of a subcollection of these intervals, say $\{I_{j_r}\}$, $r = 1, 2, \ldots, s$, such that each point of S lies in at most two members of the subcollection. Hence

$$\mu(S) = \mu\left(\bigcup_{r=1}^{s} I_{j_r}\right) \leqslant \sum_{r=1}^{s} \mu(I_{j_r}) < \frac{1}{\alpha} \sum_{r=1}^{s} \mu(M \cdot I_{j_r}) \tag{1.2.2}$$

$$\leqslant \frac{2}{\alpha} \mu\left(\bigcup_{r=1}^{s} (M \cdot I_{j_r})\right) \leqslant \frac{2}{\alpha} \mu(M).$$

Therefore (1.2.1) holds if $n = 1$; in fact, the family $\mathscr{I}$ in $\mathbf{R}^1$ enjoys the sharp weak halo property.

For an arbitrary value of n, we note that $\mathscr{I} = \prod_{i=1}^{n} \mathscr{I}_i$, where $\mathscr{I}_i$ denotes the family of closed intervals in $\mathbf{R}^1$ for $i = 1, 2, \ldots, n$. We may now apply the Remarks following Th. IV. 8.2, to see that $\mathfrak{J}$ possesses the property $(WH)''$ with the halo function γ defined by $\gamma(\alpha) = (2/\alpha)^n$, $0 < \alpha < 1$.

Since the basis $\mathfrak{J} = [\mathscr{I}, \delta]$ clearly enjoys the property (L) (that is, $\mathfrak{J}$ satisfies conditions (S.V. 1) and (S.V. 2) of Def. II. 6.1), thanks to Th. IV. 7.1, of Ch. IV., we are able to assert immediately the following conclusion.

1.3. Corollary. $\mathfrak{J} = [\mathscr{I}, \delta]$ possesses the density property (III. 1) and so derives the μ-integrals of bounded functions.

1.4. Theorem [*6*, 251]. *We assume that $\mathscr{U}$ is a family of bounded open sets in $R = \mathbf{R}^n$, with the property that $\mathscr{H}(G) \subset \mathscr{U}$ whenever $G \in \mathscr{U}$ (cf. the opening remarks of* Ch. IV. 5 *for the definition of $\mathscr{H}(G)$); μ is the Borel measure in $\mathbf{R}^n$.*

Then a necessary and sufficient condition that $[\mathscr{U}, \delta]$ derive every (finite) μ-integral is that there exist a number C depending only on $\mathscr{U}$, such that for any finite disjoint collection of bounded measurable sets $M_1, M_2, \ldots, M_m$ and any positive (finite) numbers $p_1, p_2, \ldots, p_m$, we have

$$\mu(S) \leqslant C \sum_{\nu=1}^{m} p_\nu \mu(M_\nu) \quad \textit{(Condition (C))}, \tag{1.4.1}$$

where S denotes the union of those sets $U \in \mathscr{U}$ satisfying the inequality

$$\sum_{\nu=1}^{m} p_\nu \mu(M_\nu \cdot U) > \mu(U). \tag{1.4.2}$$

Proof. We take up first the sufficiency of Condition (C). Choosing $m = 1$, $p_1 = \alpha^{-1}$, $M_1 = M$, we infer that the union S of those $\mathscr{U}$-sets U satisfying $\mu(M \cdot U) > \alpha \mu(U)$ is of μ-measure less than $C\alpha^{-1}\mu(M)$; thus condition (WH) (IV. 5) holds and $[\mathscr{U}, \delta]$ is a weak derivation basis. We note in passing that the validity of Condition (C) for any finite number of ordered pairs (p_ν, M_ν), $\nu = 1,2,\dots,m$, ensures its validity for enumerably many such pairs.

To prove the derivability μ-almost everywhere of the indefinite integral $\varphi(M) = \int_M f d\mu$ *to* the integrable function f, we may evidently restrict our considerations to the case $f \geqslant 0$.

For $\nu = 1,2,\dots$, we let

$$M_\nu = [\nu - 1 \leqslant f < \nu],$$

and let f_ν denote the function coinciding with f on the set $[f < \nu]$ and taking the value zero on $R_\nu = [f \geqslant \nu]$. We also let N_ν denote the subset of $\bigcup_{i=1}^{\nu} M_i$ where the function $\varphi_\nu(M) = \int_M f_\nu d\mu$ is not derivable to the value of the function $f_\nu = f$. According to Th. II. 2.7 and Lemma IV. 4.2.3.1 (which asserts the μ-measurability of $[\mathscr{U}, \delta]$ derivates), the Lipschitzian function φ_ν is derivable to the value of f_ν μ-almost everywhere; hence N_ν is a μ-nullset.

Because f is integrable, $\sum_{\nu=2}^{\infty} \nu \mu(M_\nu)$ is finite. For $n = 1,2,3,\dots$ and $\varepsilon > 0$ we denote by $S_{n,\varepsilon}$ the union of those sets $U \in \mathscr{U}$ for which

$$\sum_{\nu > n} \nu \mu(M_\nu \cdot U) > \varepsilon \mu(U).$$

For a fixed value of ε, the sequence of sets $(S_{n,\varepsilon})$ is non-increasing as n increases and, because of (1.4.1), we see that

$$\mu\left(\bigcap_{n=1}^{\infty} S_{n,\varepsilon}\right) = 0.$$

We form the union N' of these μ-nullsets for $\varepsilon = 2^{-1}, 3^{-1}, \dots$, and define $N = N' \cup \left(\bigcup_{\nu=1}^{\infty} N_\nu\right)$, which is again a μ-nullset.

Next, we take any point $x \notin N$. For any fixed $\varepsilon > 0$, we choose a positive integer $n_0 > f(x)$, and also sufficiently large so that $x \notin S_{n_0,\varepsilon}$. We have

$$\frac{\varphi(U_k)}{\mu(U_k)} = \sum_{\nu=1}^{n_0} \frac{\varphi(U_k \cdot M_\nu)}{\mu(U_k)} + \sum_{\nu=n_0+1}^{\infty} \frac{\varphi(U_k \cdot M_\nu)}{\mu(U_k)}$$

where (U_k) is any x-contracting sequence of sets.

Now, for each positive integer k,

$$\sum_{\nu=n_0+1}^{\infty} \frac{\varphi(U_k \cdot M_\nu)}{\mu(U_k)} \leqslant \sum_{\nu=n_0+1}^{\infty} \frac{\nu\mu(U_k \cdot M_\nu)}{\mu(U_k)} \leqslant \varepsilon ,$$

and

$$\sum_{\nu=1}^{n_0} \frac{\varphi(U_k \cdot M_\nu)}{\mu(U_k)} = \frac{\varphi_{n_0}(U_k)}{\mu(U_k)} .$$

Since $\lim_k (\varphi_{n_0}(U_k)/\mu(U_k)) = f_{n_0}(x) = f(x)$, then

$$\liminf_k \frac{\varphi(U_k)}{\mu(U_k)} \quad \text{and} \quad \limsup_k \frac{\varphi(U_k)}{\mu(U_k)}$$

both differ from $f(x)$ by not more than ε. Since ε is arbitrary, then φ has a $[\mathscr{U}, \delta]$-derivative at x coinciding with $f(x)$. Since x is an arbitrary point of $R-N$, the derivative exists and coincides with $f(x)$ μ-almost everywhere. This completes the proof of the sufficiency of Condition (C).

To prove the necessity of Condition (C), we assume its negation and so construct a μ-integrable function f whose μ-integral ψ is not μ-almost everywhere derivable to the value of f. It will be seen that the demonstration is similar to the second part of the proof of Th. IV. 5.2.

Assuming the invalidity of Condition (C), then for each positive integer m there exists a finite sequence of pairwise disjoint, bounded measurable sets $M_1^m, M_2^m, \ldots, M_{n_m}^m$ and corresponding positive numbers $p_1^m, p_2^m, \ldots, p_{n_m}^m$ such that the union S^m of those sets $U \in \mathscr{U}$ satisfying the condition

$$\sum_{i=1}^{n_m} p_i^m \mu(M_i \cdot U) > \mu(U)$$

fulfills the relation

$$\mu(S^m) > 4^m \sum_{i=1}^{n_m} p_i^m \mu(M_i^m) .$$

Recalling that $\mathscr{U}$ contains all homothetic images and translates of its members, it can be shown by an elementary though tedious argument that no loss of generality occurs by assuming that $p_i^m \geqslant 1$ for each pair of positive integers m and i, $1 \leqslant i \leqslant n_m$, $m = 1, 2, \ldots$.

For each such pair m and i, we define

$$h(x; S^m) = p_i^m \quad \text{if} \quad x \in S^m \cdot M_i^m$$
$$h(x; S^m) = 0 \quad \text{for all other} \quad x \in R ;$$

then

$$\int_{S^m} h(x; S^m)\, d\mu(x) = \sum_{i=1}^{n_m} p_i^m \mu(S^m \cdot M_i^m) < \frac{1}{4^m} \mu(S^m) , \qquad (1.4.3)$$

and each point of S^m lies in a set $U \in \mathcal{U}$ for which

$$\int_U h(x; S^m) d\mu(x) > \mu(U). \tag{1.4.4}$$

According to Lemma IV. 5.1, taking Q_0 as the set G, for each positive integer m, there exists in the unit cube Q_0 a sequence (S_j^m) of sets belonging to $\mathscr{H}(S^m)$, each of diameter less than m^{-1}, such that

$$\mu(Q_0) = \mu\Big(\bigcup_j S_j^m\Big) \leqslant \sum_j \mu(S_j^m) < 2\mu(Q_0).$$

We denote by T_j^m the homothety or translation mapping S^m onto S_j^m, by $h(x; S_j^m)$ the function corresponding to $h(x; S^m)$ by T_j^m, and by $f^m(x)$, the sum $\sum_j h(x; S_j^m)$. Then, by (1.4.3), we have

$$\int_{Q_0} f^m(x)\, d\mu(x) \leqslant \sum_j \int_{S_j^m} h(x; S_j^m)\, d\mu(x) < \frac{1}{4} \sum_j \mu(S_j^m) < \frac{2}{4^m},$$

for $m = 1, 2, \ldots$. We define $f = \sum_{m=1}^{\infty} f^m$ and note that f is integrable, since

$$\int_{Q_0} f(x) d\mu(x) < 2 \sum_{m=1}^{\infty} \frac{1}{4^m} = \frac{2}{3}. \tag{1.4.5}$$

Since $p_i^m \geqslant 1$ for each i and m, $1 \leqslant i \leqslant n_m$ and $m = 1, 2, \ldots$, then $[f > 0] = [f \geqslant 1]$, so that from (1.4.5) follows $\mu([f > 0]) \leqslant 2/3$. For any given positive integer m, all the points of Q_0, except for those in a certain μ-nullset, lie in some set S_j^m, and so belong to a set $U' \in \mathcal{U}$ that is a homothetic image or translate of set U satisfying (1.4.4), whence

$$\int_{U'} f(x) d\mu(x) \geqslant \int_{U'} h(x; S_j^m)\, d\mu(x) > \mu(U').$$

Accordingly, at μ-almost all points of Q_0, the upper $[\mathcal{U}, \delta]$-derivate of the μ-integral of f is not less than 1. On the other hand, f is zero on a subset of Q_0 of μ-measure at least 1/3. Thus $[\mathcal{U}, \delta]$ does not derive the μ-integral of f, and the necessity of Condition (C) is confirmed.

Remarks. 1. If we denote by ψ the indefinite μ-integral of the function $f = \sum_{\nu=1}^{m} p_\nu c_{M_\nu}$, where c_{M_ν} denotes the characteristic function of M_ν, then the inequality (1.4.2) becomes

$$\psi(U) > \mu(U) \tag{1.4.6}$$

and (1.4.1) becomes

$$\mu(S) \leqslant C\, \|f\|, \tag{1.4.7}$$

where $\|f\|$ denotes the semi-norm in $\mathfrak{L}^1$ of f. Thus the definition of $S = S_\psi$ is meaningful for any μ-integral $\psi \geqslant 0$, and so is Condition (C). If we write

$$S(\alpha, \psi, M) = \bigcup_{\eta > 0} S(\alpha, \psi, \eta, M)$$

(IV. 6), then

$$S(\alpha,\psi,M) = S_{(\psi_M/\alpha)},$$

where $\psi_M(B) = \psi(M \cdot B)$ whenever $B \in \mathscr{M}$.

2. An alternative proof of the sufficiency may be achieved by first proving that Condition (C) holds for any μ-integral and then applying Th. IV. 6.2. For then one has

$$S(\alpha,\varphi,M) \subset S_{(\varphi_M/\alpha)}; \quad \mu(S(\alpha,\varphi,M)) \leqslant \frac{C\,\|\varphi_M\|}{\alpha} = \frac{C}{\alpha}\|\varphi_M\|. \quad (1.4.8)$$

From (1.4.8), we may infer that for any $\alpha > 0$ and any bounded, non-increasing sequence (M_j) of $\mathscr{M}$-sets with $\lim_j \mu(M_j) = 0$, we have $\mu(S(\alpha,\varphi,M_j)) \leqslant C\varphi(M_j)/\alpha$, hence $\mu\left(\lim_j S(\alpha,\varphi,M_j)\right) = 0$. Th. IV. 6.2 now yields the desired result. Notice that the metric plays no role in the proof.

1.5. The open interval basis $[\mathscr{J},\delta]$. The constituents are the open intervals, δ is Euclidean diameter. To each interval $I \in \mathscr{I}$ there corresponds the interior $T(I) = I^0 \in \mathscr{J}$. Conversely, to each J in $\mathscr{J}$ there corresponds its closure $T^{-1}(J) = \bar{J} \in \mathscr{I}$. The derivates with respect to $[\mathscr{J},\delta]$ are μ-measurable (Lemma IV. 4.2.3.1). Th. 1.2 is valid for $[\mathscr{J},\delta]$. To each x-contracting sequence (J_n) with respect to $[\mathscr{J},\delta]$ there corresponds the x-contracting sequence $(T^{-1}(J_n))$ with respect to $[\mathscr{I},\delta]$, but the converse is not true, as may be seen by choosing a sequence of closed intervals (I_n) such that x lies on the boundary of each interval I_n. That is, $[\mathscr{I},\delta]$ has more deriving sequences that $[\mathscr{J},\delta]$. Th. 1.4 is applicable to $[\mathscr{J},\delta]$.

We note that for any μ-integral ψ, $\psi(I) = \psi(I^0)$ and $\psi(J) = \psi(\bar{J})$. Hence if $[\mathscr{J},\delta]$ fails to derive any μ-integral, then $[\mathscr{I},\delta]$ fails to derive it also.

1.5.1. Proposition. The interval bases $[\mathscr{J},\delta]$ and $[\mathscr{I},\delta]$ both fail to derive some μ-integrals.

Proof. If W is a square (open or closed) in the space $\mathbf{R}^2$, then it is not difficult to see that the weak halo $S(\alpha, W)$, for $\alpha = m^{-1}$, is of μ-measure $(1 + 4m\,\mathrm{Log}\,m)\mu(W)$. Consider the finite family of μ-measurable sets consisting of the set $M_1 = W$, the corresponding number $p_1 = m$, and nothing else. Then $\mu(S(\alpha, W)) = \mu(S) = (1 + 4m\,\mathrm{Log}\,m)\mu(W)$. Since the factor $\mathrm{Log}\,m$ is not bounded, it is clear that Condition (C) cannot hold. Consequently, there exists some μ-integral φ that $[\mathscr{J},\delta]$ fails to derive, and so $[\mathscr{I},\delta]$ fails to derive φ also.

2. Theorem of Jessen-Marcinkiewicz-Zygmund (abbreviated J.-M.-Z.). This theorem asserts that the interval basis $\mathfrak{J} = [\mathscr{I},\delta]$ in $\mathbf{R}^n$ derives the Lebesgue integral of each Lebesgue measurable function f for which the function $|f|\,(\mathrm{Log}^+|f|)^{m-1}$ is Lebesgue integrable over the open cube

$$Q_0 = E[x : x = (x_1, x_2, \ldots, x_n), \quad 0 < x_i < 1, \quad i = 1, 2, \ldots, n],$$

and the $\mathfrak{J}$-derivative coincides with f except on a set of Lebesgue measure zero.

We shall prove this theorem with the term "Borel" substituted for "Lebesgue" everywhere above. As we shall see shortly, the extreme $[\mathscr{I}, \delta]$- and $[\mathscr{J}, \delta]$-derivates of integrals coincide and are Borel measurable, regardless of whether the integrals are Lebesgue or Borel. It will thus be seen quite readily that the argument we give can be carried out in the Lebesgue case.

Let f be any Lebesgue (resp.) Borel measurable function defined on $\mathbf{R}$, whose Lebesgue (resp.) Borel integral over any interval K (open or closed) is finite. Let $\psi(K)$ denote the value of this integral. We let μ denote the n-dimensional Borel measure as before.

Let x be any point of $\mathbf{R}$ and (J_m) any sequence of open intervals contracting to x, for which $\lim_m (\psi(J_m)/\mu(J_m)) = l$ ("*nombre derivé*" according to DENJOY). The closed sequence $(\bar{J}_m)$ contracts to x and, since $\psi(J_m) = \psi(\bar{J}_m)$, satisfies $\lim_m (\psi(\bar{J}_m)/\mu(\bar{J}_m)) = l$. On the other hand, if (I_m) is any x-contracting sequence of closed intervals with $\lim_m (\psi(I_m)/\mu(I_m)) = l$, then, because of the absolute continuity of ψ (with respect to Lebesgue or Borel measure, as the case may be), we can obtain a sequence of x-contracting open intervals (J_m) with $\lim_m (\psi(J_m)/\mu(J_m)) = l$ by replacing each closed interval I_m by a slightly larger open interval, $J_m \supset I_m$. It follows that the upper $[\mathscr{I}, \delta]$- and $[\mathscr{J}, \delta]$-derivates of ψ coincide, as do also the lower $[\mathscr{I}, \delta]$- and $[\mathscr{J}, \delta]$-derivates. From Lemma IV. 4.2.3.1, it now follows that the extreme $[\mathscr{I}, \delta]$-derivates of ψ are Borel (i.e., μ-) measurable.

Before proving our theorem, we need some preparatory lemmas. In this connection, we often find it convenient to denote a point $p = (x_1, x_2, \ldots, x_n)$ in $\mathbf{R}^n$ by (x, y), where $x = x_1$, $y = (x_2, \ldots, x_n)$. We also denote the open unit interval $]0, 1[$ in $\mathbf{R}^1$ by X, the corresponding unit interval $]0, 1[$ in $\mathbf{R}^{n-1}$ by Y, so that $Q_0 = X \times Y$. We agree to denote r-dimensional Borel measure by μ_r, whenever this distinction is necessary for clarity of argument.

2.1. Proposition [*44*, 147]. Let f denote a μ_n-integrable (Borel measurable) function defined on the open unit cube Q_0, and let H denote the set of those values of y for which $f(x, y)$ is μ_1-integrable in x over the open interval X. By FUBINI's theorem, $\mu_n(H) = 1$. We also define

$$\begin{aligned} f^\beta(x, y) &= \sup \frac{1}{v-u} \int_u^v f(t, y)\, dt \quad \text{if} \quad y \in H, \\ f^\beta(x, y) &= 0 \quad \text{if} \quad y \in (Y - H), \end{aligned} \tag{2.1.1$'$}$$

and

$$
\begin{aligned}
f^\star(x,y) &= \limsup \frac{1}{v-u}\int_u^v f(t,y)\,dt \quad \text{if} \quad y \in H,\\
f^\star(x,y) &= 0 \quad \text{if} \quad y \in (Y-H),
\end{aligned}
\tag{2.1.1}''
$$

where the supremum in (2.1.1)′ is taken over all u, v such that $0 < u < x < v < 1$ and the limit superior in (2.1.1)″ is taken with respect to u and v subject to these same restrictions and converging to x.

Then f^β and $f^\star$ are μ_n-measurable.

Remark. For $y \in H$, f^β is the greatest mean value of f on the linear intervals parallel to the x-axis and containing $p = (x, y)$, and $f^\star$ is the linear upper derivate of f with respect to x at the point p. Evidently $f^\star \leqslant f^\beta$ on Q_0.

Proof. We denote by f_c the extension of f to the infinite strip $S = E[p : p = (x,y); y \in Y]$ by defining $f_c = f$ on Q_0, and $f_c(x,y) = 0$ for $x \notin X$, $y \in Y$.

We begin by showing that for any two positive numbers a and b, the function $g_{a,b}$ defined by

$$
\begin{aligned}
g_{a,b}(x,y) &= \int_{x-a}^{x+b} f_c(t,y)\,dt \quad \text{if} \quad y \in H\\
g_{a,b}(x,y) &= 0 \qquad \text{if} \quad y \in (Y-H)
\end{aligned}
\tag{2.1.2}
$$

is μ_n-measurable.

To this end, we invoke well known theorems to represent f_c as the $\mathfrak{L}^1(\mu_n)$-limit of n-dimensional step functions $s_1, s_2, \ldots, s_m, \ldots$ on S, each vanishing outside Q_0. Then

$$
\begin{aligned}
&\lim_m \iint_S |f_c(x,y) - s_m(x,y)|\, d\mu_n(x,y)\\
&= \lim_m \iint_{Q_0} |f(x,y) - s_m(x,y)|\, d\mu_n(x,y) = 0\,.
\end{aligned}
$$

For any fixed step function s defined on Q_0, there exists a subdivision of X with

$$x_0 = 0 < x_1 \ldots < x_r = 1\,,$$

and a subdivision of Y into a finite number of non-overlapping $(n-1)$-dimensional intervals $I_1, I_2, \ldots, I_q$ such that for any interval $]x_k, x_{k+1}[$ and any interval I_j, if $x \in]x_k, x_{k+1}[$, then the function

$$g(x,y) = \int_{x-a}^{x+b} s(t,y)\,dt$$

is independent of y for $y \in I_j$. It is also clear that for any fixed $y \in Y$, g is continuous in x. It follows that g is μ_n-measurable.

Since $f(x,y)$ is μ_1-integrable over X for each $y \in H$, then so is $|f(x,y)-s_m(x,y)|$ for $m=1,2,\ldots$. Setting

$$k_m(y) = \int_X |f(t,y)-s_m(t,y)|\,dt$$

for each such y and m, it follows from FUBINI's theorem that

$$\lim_m \int_Y k_m(y)\,d\mu_{n-1}(y) = 0,$$

that is, the sequence (k_m) converges to zero in the space $\mathfrak{L}^1(\mu_{n-1})$. Since any sequence of functions converging in the mean has a subsequence converging almost everywhere, we may, by replacing the sequence (k_m) by a suitable subsequence if necessary, assume without loss of generality that there exists a set $H^\star \subset H$ with $\mu_{n-1}(H^\star) = 1$, such that $\lim_m k_m(y) = 0$ whenever $y \in H^\star$; that is,

$$\lim_m \int_X |f(t,y)-s_m(t,y)|\,dt = 0$$

whenever $y \in H^\star$.

We define

$$g_m(x,y) = \int_{x-a}^{x+b} s_m(t,y)\,dt$$

for each positive integer m, each real x, and each $y \in H^\star$. It follows that for each such m, x, and y we have

$$\begin{aligned}|g_{a,b}(x,y)-g_m(x,y)| &\leqslant \int_{x-a}^{x+b} |f_c(t,y)-s_m(t,y)|\,dt \\ &\leqslant \int_0^1 |f(t,y)-s_m(t,y)|\,dt,\end{aligned}$$

whence $g_{a,b}(x,y) = \lim_m g_m(x,y)$, and so $g_{a,b}$ is the limit of the μ_n-measurable sequence (g_m) μ_n-almost everywhere in S; consequently $g_{a,b}$ is itself μ_n-measurable.

Finally, with the same notation as above, if u_p and u_q are any two rationals such that $0 < u_p < x < u_q < 1$, then the functions defined by

$$\sup \frac{g_{u_p,u_q}(x,y)}{u_q-u_p} \quad \text{if} \quad y \in H, \quad 0 \quad \text{if} \quad y \in (Y-H),$$

and

$$\limsup_{u_p \to x, u_q \to x} \frac{g_{u_p,u_q}(x,y)}{u_q-u_p} \quad \text{if} \quad y \in H, \quad 0 \quad \text{if} \quad y \in (Y-H)$$

are evidently μ_n-measurable. However, these are, respectively, precisely the functions f^β *and* $f^\star$.

2.2. Proposition (Lemma of HARDY-LITTLEWOOD) [44, 145]. If h is a non-negative, μ_1-measurable function on $]0,1[$, and $h \operatorname{Log}^+ h$ is μ_1-integrable over $]0,1[$, then so is h^β, where, for $0 < x < 1$,

$$h^\beta(x) = \sup \frac{1}{v-u} \int_u^v h(t)\,dt\,, \qquad 0 < u < x < v < 1\,,$$

the supremum being taken over all u, v subject to the stated conditions. Furthermore,

$$\int_0^1 h^\beta(x)\,dx \leqslant A \int_0^1 h(x) \operatorname{Log}^+ h(x)\,dx + B\,,$$

where A and B are absolute constants.

Proof. For $0 < x < 1$ we define

$$h_1(x) = \int_0^x h(t)\,dt\,, \qquad J = \int_0^1 h(x) \operatorname{Log}^+ h(x)\,dx\,, \qquad K = \int_0^1 \frac{h_1(x)}{x}\,dx\,.$$

We wish first to establish an upper bound on K in terms of J, which is finite by hypothesis. To this end we introduce the known inequality

$$uv \leqslant u \operatorname{Log} u + e^{v-1}\,,$$

valid for all $u > 0$ and all real v [26, 189]. Setting $u = h(x)$, $v = \frac{1}{2} \operatorname{Log} \frac{1}{x}$, $0 < x < 1$, we obtain

$$\frac{1}{2} h(x) \operatorname{Log} \frac{1}{x} \leqslant h(x) \operatorname{Log} h(x) + \frac{1}{e\sqrt{x}} \leqslant h(x) \operatorname{Log}^+ h(x) + \frac{1}{e\sqrt{x}}\,.$$

Integrating this last inequality yields

$$0 \leqslant \int_0^1 h(x) \operatorname{Log} \frac{1}{x}\,dx \leqslant 2 \int_0^1 h(x) \operatorname{Log}^+ h(x)\,dx + \frac{4}{e} = 2J + \frac{4}{e}\,. \tag{2.2.1}$$

Thus the left-hand side of (2.2.1) is finite. Now for any ε, $0 < \varepsilon < 1$, we have, from integration by parts,

$$\int_\varepsilon^1 h(x) \operatorname{Log} \frac{1}{x}\,dx = \int_\varepsilon^1 \frac{h_1(x)}{x}\,dx - h_1(\varepsilon) \operatorname{Log} \frac{1}{\varepsilon}\,. \tag{2.2.2}$$

Clearly,

$$0 < h_1(\varepsilon) \operatorname{Log} \frac{1}{\varepsilon} = \operatorname{Log} \frac{1}{\varepsilon} \int_0^\varepsilon h(x)\,dx \leqslant \int_0^\varepsilon h(x) \operatorname{Log} \frac{1}{x}\,dx$$
$$\leqslant \int_0^1 h(x) \operatorname{Log} \frac{1}{x}\,dx < \infty\,;$$

hence

$$\lim_{\varepsilon \to 0+} h_1(\varepsilon) \operatorname{Log} \frac{1}{\varepsilon} = 0 .$$

Thus (2.2.2) yields

$$\int_0^1 h(x) \operatorname{Log} \frac{1}{x} \, dx = \int_0^1 \frac{h_1(x)}{x} \, dx ,$$

and using this relation in (2.2.1) gives

$$K \leqslant 2J + 4/e = A \cdot J + B , \tag{2.2.3}$$

where A and B are absolute constants not depending on the function h.

We now introduce the function $\bar{h}$, the non-increasing rearrangement of h [*12*, 276–278; *44*, 143]. As is well known, h and $\bar{h}$ are equimeasurable, so that if we let

$$\bar{h}_1(x) = \int_0^x \bar{h}(t) \, dt , \quad \bar{J} = \int_0^1 \bar{h}(x) \operatorname{Log}^+ \bar{h}(x) \, dx , \quad \text{and} \quad \bar{K} = \int_0^1 \frac{\bar{h}_1(x)}{x} \, dx$$

then $\bar{J} = J$, and the formula corresponding to (2.2.3) becomes

$$\bar{K} \leqslant A \cdot \bar{J} + B = A \cdot J + B . \tag{2.2.4}$$

According to a special case of a well-known inequality [*12*, 291]

$$\int_0^1 h^\beta(x) \, dx \leqslant \int_0^1 (\bar{h})^\beta(x) \, dx . \tag{2.2.5}$$

Using simply the fact that $\bar{h}$ is a non-increasing function, it is easy to verify that

$$\bar{h}^\beta(x) = \frac{1}{x} \int_0^x \bar{h}(t) \, dt = \frac{\bar{h}_1(x)}{x} ; \tag{2.2.6}$$

thus, from (2.2.4), (2.2.5), and (2.2.6) we obtain

$$\int_0^1 h^\beta(x) \, dx \leqslant \int_0^1 \frac{\bar{h}_1(x)}{x} \, dx = \bar{K} \leqslant A \cdot J + B ,$$

as required.

2.3. Proposition. If the non-negative function f defined over the open unit square

$$Q_0 = E[(x, y) : 0 < x < 1, \ 0 < y < 1]$$

is μ_2-measurable and if $f \operatorname{Log}^+ f$ is μ_2-integrable over Q_0, then the upper $\mathfrak{I}$-derivate $D^\star \varphi$ of the indefinite integral φ of f is μ_2-integrable and

$$\iint_{Q_0} D^\star \varphi(x,y)\,d\mu_2(x,y) \leqslant A \iint_{Q_0} f(x,y)\,\mathrm{Log}^+ f(x,y)\,d\mu_2(x,y) + B\,, \tag{2.3.1}$$

where A and B denote the constants of Prop. 2.2.

Proof. We consider the function f^β defined in Prop. 2.1, for $n = 2$. According to this proposition, f^β is μ_2-integrable. We denote by $Y^\star$ the set of those values of y for which $f(x,y)$ $\mathrm{Log}^+ f(x,y)$ is μ_1-integrable in x; then $\mu_1(Y^\star) = 1$. For the same values, $f(x,y)$ is μ_1-integrable in x and, by Prop. 2.2.

$$\int_0^1 f^\beta(x,y)\,dx \leqslant A \int_0^1 f(x,y)\,\mathrm{Log}^+ f(x,y)\,dx + B \tag{2.3.2}$$

for each such x. Thus, integrating (2.3.2) with respect to y, we obtain

$$\iint_{Q_0} f^\beta(x,y)\,d\mu_2(x,y) \leqslant A \iint_{Q_0} f(x,y)\,\mathrm{Log}^+ f(x,y)\,d\mu_2(x,y) + B\,. \tag{2.3.3}$$

In similar fashion, there exists a set $X^\star$ with $\mu_1(X^\star) = 1$ such that $f^\beta(x,y)$ is μ_1-integrable as a function of y whenever $x \in X^*$. According to the derivation theorem for one-dimensional integrals, there exists for each x in $X^\star$ a set $Y(x)$, with $\mu_1(Y(x)) = 1$, such that

$$\limsup_{\substack{w<y<z \\ w\to y,\, z\to y}} \frac{1}{z-w} \int_w^z f^\beta(x,t)\,dt = f^\beta(x,y) \tag{2.3.4}$$

for each $y \in Y(x)$. Let P be the set of points $p = (x,y)$ where this equality holds. Since, by Prop. 2.1, both sides of this relation define μ_2-measurable functions, then the set $Q_0 - P$ is μ_2-measurable and its μ_2-measure may be evaluated with the help of Fubini's theorem. Since $\mu_1(Y(x)) = 1$ for μ_1-almost all x, $0 < x < 1$, then it follows that $\mu_2(Q_0 - P) = 0$; that is, P is μ_2-measurable and $\mu_2(P) = 1$.

On the other hand, for $0 < u < x < v < 1$, $0 < w < y < z < 1$, we have

$$\begin{aligned} \frac{1}{(z-w)(v-u)} \int_w^z \int_u^v f(s,t)\,ds\,dt &= \frac{1}{(z-w)} \int_w^z dt \left(\frac{1}{v-u} \int_u^v f(s,t)\,ds \right) \\ &\leqslant \frac{1}{z-w} \int_w^z f^\beta(x,t)\,dt\,. \end{aligned} \tag{2.3.5}$$

We now take the limit superior in (2.3.5) as u and v converge to x, w and z converge to y, and so obtain

$$D^\star \varphi(x,y) \leqslant f^\beta(x,y) \quad \text{for} \quad (x,y) \in P\,.$$

Hence, by (2.3.3),

$$\iint_{Q_0} D^\star \varphi(x,y)\,d\mu_2(x,y) \leqslant A \iint_{Q_0} f(x,y)\,\mathrm{Log}^+ f(x,y)\,d\mu_2(x,y) + B\,.$$

2.4. Theorem. *The J.-M.-Z. theorem is valid for $n = 2$, (recall the statement at beginning of § 2).*

Proof. We suppose $f \geqslant 0$. Applying (2.3.1) to the function λf, where λ is a positive constant, we obtain

$$\int_{Q_0} D^{\star}\varphi(p)\,d\mu_2(p) \leqslant A \int_{Q_0} f(p)\,\mathrm{Log}^+ \lambda f(p)\,d\mu_2(p) + B/\lambda .$$

Given $\varepsilon > 0$, we choose λ_0 so that $0 < B/\lambda_0 < \varepsilon/2$ and also $\lambda_0 > 1$. Putting $\lambda = \lambda_0$ in this last inequality, we obtain

$$\int_{Q_0} D^{\star}\varphi(p)\,d\mu_2(p) \leqslant A \int_{Q_0} f(p)\,\mathrm{Log}^+ \lambda_0 f(p)\,d\mu_2(p) + \varepsilon/2 . \qquad (2.4.1)$$

We define $f_m(p) = f(p)$ if $0 \leqslant f(p) \leqslant m$, $f_m(p) = 0$ if $f(p) > m$, and $r_m(p) = f(p) - f_m(p)$; and we denote by φ, φ_m, and ρ_m the indefinite integrals of f, f_m and r_m, respectively, for $m = 1, 2, \ldots$.

Taking into account the μ_2-integrability of f and $f\,\mathrm{Log}^+ \lambda_0 f$, we choose m_0 so that

$$\int_{Q_0} r_{m_0}(p)\,d\mu_2(p) < \varepsilon \qquad (2.4.2)$$

and

$$\int_{Q_0} r_{m_0}(p)\,\mathrm{Log}^+ \lambda_0 r_{m_0}(p)\,d\mu_2(p) < \varepsilon/2A . \qquad (2.4.3)$$

Applying the inequality (2.4.1) with f replaced by r_{m_0}, we obtain, with the help of (2.4.3)

$$\int_{Q_0} D^{\star}\rho_{m_0}(p)\,d\mu_2(p) < \varepsilon/2 + \varepsilon/2 = \varepsilon . \qquad (2.4.4)$$

Combining (2.4.4) with (2.4.2), we see that the set $E(\varepsilon)$ consisting of those points p where either $r_{m_0}(p) > \sqrt{\varepsilon}$ or $D^{\star}\rho_{m_0}(p) > \sqrt{\varepsilon}$ is of μ_2-measure less than $2\sqrt{\varepsilon}$.

For any two-dimensional interval $I \subset Q_0$ containing the point p_0, we have

$$\left|\frac{1}{\mu_2(I)} \int_I f(p)\,d\mu_2(p) - f(p_0)\right| \qquad (2.4.5)$$

$$\leqslant \left|\frac{1}{\mu_2(I)} \int_I f_{m_0}(p)\,d\mu_2(p) - f_{m_0}(p)\right| + \left|\frac{1}{\mu_2(I)} \int_I r_{m_0}(p)\,d\mu_2(p) - r_{m_0}(p)\right| .$$

We take the limit superior of both sides of (2.4.5) for $p_0 \notin E(\varepsilon)$ and I contracting to p_0. The first term on the right side tends to zero, provided p_0 is outside a certain set of μ_2-measure zero, since $\mathfrak{J}$ is a weak derivation basis and f_{m_0} is bounded. The second term on the right side is bounded above by $D^{\star}\rho_{m_0}(p) + r_{m_0}(p)$, which in turn does not exceed $2\sqrt{\varepsilon}$. Hence, for almost all $p_0 \notin E(\varepsilon)$,

$$\limsup_{\delta(I)\to 0+} \left| \frac{1}{\mu_2(I)} \int_I f(p)\,d\mu_2(p) - f(p_0) \right| \leqslant 2\sqrt{\varepsilon}.$$

Since ε is arbitrary, we conclude that the limit of the left side of this last expression exists and is zero; that is, $D\varphi(p_0)$ exists and coincides with $f(p_0)$. Since $\lim_{\varepsilon\to 0+} \mu_2(E(\varepsilon)) = 0$, it follows that $D\varphi(p_0) = f(p_0)$ holds μ_2-almost everywhere in Q_0.

2.5. Proposition. If h is a non-negative function on the interval $]0,1[$ in $\mathbf{R}^1$, h^β is the function defined in Prop. 2.2, and $h(\mathrm{Log}^+ h)^r$ is integrable over $]0,1[$, then so is $h^\beta(\mathrm{Log}^+ h^\beta)^{r-1}$; and

$$\int_0^1 h^\beta(x)(\mathrm{Log}^+ h^\beta(x))^{r-1}\,dx \leqslant A_r \int_0^1 h(x)(\mathrm{Log}^+ h(x))^r\,dx + B_r \quad (2.5.1)$$

where A_r and B_r depend only on r.

Proof. Let $\bar{h}$ denote the rearrangement of h in non-increasing order [*12*, 291; *44*, 143]. We also let $k(x) = x(\mathrm{Log}^+ x)^{r-1}$; k is a non-decreasing convex function and thus, according to JENSEN's theorem [*26*, 202],

$$k\left(\frac{1}{x}\int_0^x \bar{h}(s)\,ds\right) \leqslant \frac{1}{x}\int_0^x k(\bar{h}(s))\,ds$$

holds for each x, $0 < x < 1$. Consequently, by virtue of a lemma of HARDY and LITTLEWOOD [*12*, p. 291], we see that

$$\int_0^1 k(h^\beta(x))\,dx \leqslant \int_0^1 k\left(\frac{1}{x}\int_0^x \bar{h}(s)\,ds\right)dx \leqslant \int_0^1 \left(\frac{1}{x}\int_0^x k(\bar{h}(s))\,ds\right)dx. \quad (2.5.2)$$

Since $k(\bar{h})$ is a non-increasing, non-negative, integrable function, it follows that $k(\bar{h}) = \overline{k(h)}$ on $]0,1[$ and also that

$$\overline{k(h)}^\beta(x) = \frac{1}{x}\int_0^x \overline{k(h)}(s)\,ds, \qquad 0 < x < 1.$$

Consequently, (2.5.2) becomes

$$\int_0^1 k(h^\beta(x))\,dx \leqslant \int_0^1 \left(\frac{1}{x}\int_0^x \overline{k(h(s))}\,ds\right)dx = \int_0^1 (\overline{k(h)})^\beta(x)\,dx, \quad (2.5.3)$$

which, because of the equi-measurability of $\overline{k(h)}$ and $k(h)$ and Prop. 2.2, leads to

$$\int_0^1 k(h^\beta(x))\,dx \leqslant A\int_0^1 \overline{k(h(x))}\,\mathrm{Log}^+\,\overline{k(h(x))}\,dx + B$$
$$= A\int_0^1 k(h(x))(\mathrm{Log}^+(k(h(x))))\,dx + B.$$

Recalling the definition of k above, this becomes explicitly

$$\int_0^1 k(h^\beta(x))dx \leqslant A\int_0^1 h(x)(\mathrm{Log}^+ h(x))^{r-1}\,\mathrm{Log}^+(h(x)(\mathrm{Log}^+ h(x))^{r-1})dx + B$$

$$= A\int_0^1 h(x)\,(\mathrm{Log}^+ h(x))^{r-1}(\mathrm{Log}^+ h(x) + (r-1)\mathrm{Log}^+(\mathrm{Log}^+ h(x)))\,dx + B. \tag{2.5.4}$$

Since $\mathrm{Log}^+(\mathrm{Log}^+ h(x)) \leqslant \mathrm{Log}^+ h(x)$, we finally obtain from (2.5.4) the relation

$$\int_0^1 k(h^\beta(x))\,dx \leqslant Ar\int_0^1 h(x)\,(\mathrm{Log}^+ h(x))^r\,dx + B. \tag{2.5.5}$$

Thus the inequality (2.5.1) is established with $A_r = Ar$, $B_r = B$.

2.6. Lemma. *We suppose that*

(1) *The theorem of J.-M.-Z. is valid for* $\mathbf{R}^{n-1}$;

(2) f *is a non-negative* μ_n*-measurable function defined on the open unit cube*

$$Q_0 = E\,[p\colon p = (x_1, x_2, \ldots, x_n);\quad 0 < x_i < 1,\quad i = 1, 2, \ldots, n] \quad \text{in } \mathbf{R}^n;$$

(3) $f(\mathrm{Log}^+ f)^{n-1}$ *is* μ_n*-integrable over* Q_0.

Then the upper $\mathfrak{J}$*-derivate* $D^\star\varphi$ *of the indefinite integral* φ *of* f *is* μ_n*-integrable and*

$$\int_{Q_0} D^\star\varphi(p)\,d\mu_n(p) \leqslant A_{n-1}\int_{Q_0} f(p)\,(\mathrm{Log}^+ f(p))^{n-1}\,d\mu_n(p) + (B_{n-1} + e), \tag{2.6.1}$$

where $A_{n-1} = A\cdot(n-1)$ *and* $B_{n-1} = B$ *are the constants of Prop.* 2.5.

Proof. This is entirely analogous to the proof of Prop. 2.3.

We put $x = x_1$, $y = (x_2, \ldots, x_n)$, $p = (x, y)$, $f(x_1, x_2, \ldots, x_n) = f(x, y)$, and let $f^\beta(x, y)$ have the meaning assigned in (2.1.1). We denote by $Y^\star$ the set of those values of y for which $f(x, y)\,(\mathrm{Log}^+ f(x, y))^{n-1}$ is μ_1-integrable in x; clearly $\mu_{n-1}(Y^\star) = 1$. For these values of y, Prop. 2.5 asserts that $f^\beta(x, y)\,(\mathrm{Log}^+ f^\beta(x, y))^{n-2}$ is μ_1-integrable in x, and

$$\int_0^1 f^\beta(x, y)\,(\mathrm{Log}^+ f^\beta(x, y))^{n-2}\,dx \tag{2.6.2}$$

$$\leqslant A_{n-1}\int_0^1 f(x, y)\,(\mathrm{Log}^+ f(x, y))^{n-1}\,dx + B_{n-1}.$$

Integration of (2.6.2) with respect to y yields

$$\iint_{Q_0} f^\beta(x, y)\,(\mathrm{Log}^+ f^\beta(x, y))^{n-2}\,d\mu_n(x, y) \tag{2.6.3}$$

$$\leqslant A_{n-1}\iint_{Q_0} f(x, y)\,(\mathrm{Log}^+ f(x, y))^{n-1}\,d\mu_n(x, y) + B_{n-1}.$$

We consider next an arbitrary $(n-1)$-dimensional interval J lying in the n-dimensional hyperplane $x = x_1$, and containing the point y, and we take

$$\limsup_{\delta(J)\to 0+} \frac{1}{\mu_{n-1}(J)} \int_J f^\beta(x,t)\, d\mu_{n-1}(t). \tag{2.6.4}$$

By virtue of (2.6.3), there exists a certain set $X^\star$ with $\mu_1(X^\star) = 1$, such that the function $f^\beta(x,y)\,\mathrm{Log}^+ f^\beta(x,y))^{n-2}$ is μ_{n-1}-integrable in y for each $x \in X^\star$. Thus, applying our assumption of the validity of the J.-M.-Z. theorem to the function f^β (recall (2.6.3)), it follows that for each $x \in X^\star$ there exists a set $Y(x)$ with $\mu_{n-1}(Y(x)) = 1$, such that (2.6.4) equals $f^\beta(x,y)$ for each $y \in Y(x)$. Exactly as in the proof of the corresponding part of Prop. 2.3, it follows that the set P of points in $\mathbf{R}^n$ where the equality holds is μ_n-measurable and $\mu_n(P) = 1$.

On the other hand, for $0 < u < x < v < 1$, we have

$$\frac{1}{(v-u)\mu_{n-1}(J)} \int_J \int_u^v f(s,t)\, ds\, d\mu_{n-1}(t) \tag{2.6.5}$$

$$= \frac{1}{\mu_{n-1}(J)} \int_J d\mu_{n-1}(t) \left(\frac{1}{v-u} \int_u^v f(s,t)\, ds \right) \leqslant \frac{1}{\mu_{n-1}(J)} \int_J f^\beta(x,t)\, d\mu_{n-1}(t).$$

Thus, taking the limit superior of both sides of (2.6.5) as u and v converge to x and J converges to y, we obtain

$$D^\star \varphi(x,y) \leqslant f^\beta(x,y) \tag{2.6.6}$$

whenever $(x,y) \in P$. We let Q_1 denote the subset of Q_0 on which $f^\beta \geqslant e$ and define $Q_2 = Q_0 - Q_1$. Then, using (2.6.6) and (2.6.3), we obtain

$$\begin{aligned}
\iint_{Q_0} D^\star \varphi(x,y)\, d\mu_n(x,y) &\leqslant \iint_{Q_0} f^\beta(x,y)\, d\mu_n(x,y) \\
&= \iint_{Q_1} f^\beta(x,y)\, d\mu_n(x,y) + \iint_{Q_2} f^\beta(x,y)\, d\mu_n(x,y) \\
&\leqslant \iint_{Q_0} f^\beta(x,y)\, (\mathrm{Log}^+ f^\beta(x,y))^{n-2}\, d\mu_n(x,y) + e \\
&\leqslant A_{n-1} \iint_{Q_0} f(x,y)\, (\mathrm{Log}^+ f(x,y))^{n-1}\, d\mu_n(x,y) + (B_{n-1} + e).
\end{aligned}$$

2.7. Theorem. *The J.-M.-Z. theorem is true in $\mathbf{R}^n$ for any positive integer n.*

Proof. By Th. 2.4, the J.-M.-Z. theorem holds for $n = 2$. We suppose now that it is valid for the positive integer $n-1 \geqslant 2$. Then, by Lemma 2.6, we have

$$\int_{Q_0} D^\star \varphi(p)\, d\mu_n(p) \leqslant A_{n-1} \int_{Q_0} f(p)\, (\mathrm{Log}^+ f(p))^{n-1}\, d\mu_n(p) + B'_{n-1}, \tag{2.7.1}$$

where $B'_{n-1} = B_{n-1} + e$. Except for the names of the constants, the only difference between (2.7.1) and (2.3.1) is the exponent $n-1$ appearing explicitly on the right side of (2.7.1). The relation (2.3.1) played the key role in the proof of Th. 2.4, and that role was essentially unrelated to the dimensionality of the integrals as well as to the power of $\mathrm{Log}^+ f$ occurring in (2.3.1). Accordingly, if (2.7.1) is substituted for (2.3.1), then the proof of Th. 2.4 carries over virtually verbatim to the present case. Thus, by induction, we establish the truth of our theorem.

In a recent paper [50], A. ZYGMUND has added what he terms "a postscript" to [6] by improving certain specialized results appearing in the earlier paper.

2.8. Theorem. *If f is defined on $\mathbf{R}^n$ and is μ_n-measurable, and if $|f|(\mathrm{Log}^+|f|)^{n-1}$ is integrable on every bounded $\mathcal{M}$-set, then the indefinite integral φ of f is $\mathfrak{J}$-derivable and $D\varphi(p) = f(p)$ μ_n-almost everywhere.*

Proof. Th. 2.7 would clearly be true if Q_0 were replaced by any bounded open cube. Since any open set in $\mathbf{R}^n$ can be represented as the union of countably many disjoint open cubes and a μ_n-nullset, then Th. 2.7 may be applied to the restriction of f to each of these cubes to obtain the desired result.

3. Properties of the halo function as consequences of derivation properties.

In this section, R is Euclidean space $\mathbf{R}^n$, μ is the corresponding Borel measure, and δ is the Euclidean metric in $\mathbf{R}^n$. As before, Q_0 is the unit cube with principal vertices at $(0,0,\ldots,0)$ and $(1,1,\ldots,1)$.

We let ϖ denote a function non-decreasing on $]1,\infty[$, not identically constant there, with $\varpi^+(1) = \lim_{t\to 1+} \varpi(t) > 0$. Although we are not particularly interested in the values of ϖ for $t \in [0,1]$, we find it convenient to define ϖ for that interval and, in fact, to be constantly zero there. We let $\mathfrak{L}_\varpi$ denote the class of functions defined over R, vanishing outside of Q_0, and with $\varpi(|f|)$ integrable over Q_0.

3.1. Theorem. *If a Busemann-Feller basis (Ch. I. 3.6) $[\mathcal{U}, \delta]$ derives the μ_n-integrals of all non-negative $\mathfrak{L}_\varpi$-functions, then there exists a number C, not depending on α or M, for which*

$$\mu_n(S(\alpha, M)) \leqslant C\varpi(1/\alpha)\mu_n(M), \tag{3.1.1}$$

holds whenever $M \in \mathcal{M}$, M is bounded, and $0 < \alpha < 1$.

Proof. (Recall the necessity proofs in Ths. IV. 5.2, and V. 1.4.) We assume that the theorem is false, and we select a sequence (C_m) of positive numbers such that

$$\sum_{m=1}^{\infty} \frac{1}{C_m} < \frac{1}{2}\varpi^+(1). \tag{3.1.2}$$

Then, for each positive integer m, we may determine a bounded set $M_m \in \mathcal{M}$ of positive μ_n-measure and a number α_m, such that

$$\mu_n(S(\alpha_m, M_m)) > C_m \varpi(1/\alpha_m)\mu_n(M_m), \quad 0 < \alpha_m < 1. \tag{3.1.3}$$

For $m = 1,2,\ldots$, we define $S_m = S(\alpha_m, M_m)$; and for each such value of m we choose a sequence of sets (S_m^k), each of diameter less than m^{-1}

and homothetic to S_m, with their union covering Q_0 except for at most a μ_n-nullset, and satisfying the relation

$$\sum_k \mu_n(S_m^k) < 2\mu_n(Q_0) = 2\,. \tag{3.1.4}$$

The possibility of such a choice is ensured by Lemma IV. 5.1.

Now we let M_m^k denote a set obtained from M_m by the same homothetical transformation that yielded S_m^k from S_m, for $m, k = 1, 2, \dots$. From (3.1.3) it follows that

$$\mu_n(S_m^k) > C_m \varpi(1/\alpha_m)\mu_n(M_m^k)\,. \tag{3.1.5}$$

Next, we define

$$\begin{aligned} f_m(p) &= 1/\alpha_m \quad \text{if } p \in Q_0 \cdot \Big(\bigcup_k M_m^k\Big) \\ f_m(p) &= 0 \quad \text{for all other points } p \in \mathbf{R}^n, \end{aligned} \tag{3.1.6}$$

and

$$f(p) = \sup_{m \geq 1} f_m(p)\,. \tag{3.1.7}$$

It is clear that either $f(p) = 0$ or there is some positive integer m such that $f(p) \geq f_m(p) = 1/\alpha_m > 1$, whence $\varpi(f(p)) \geq \varpi(f_m(p)) \geq \varpi^+(1) > 0$. From these facts, it follows readily that for all $p \in \mathbf{R}^n$, we have

$$\varpi(f(p)) \leq \sum_{m=1}^{\infty} \varpi(f_m(p))\,.$$

Hence, with the help of (3.1.2), (3.1.3), and (3.1.5), we see that

$$\begin{aligned} \int_{\mathbf{R}^n} \varpi(f(p))\, d\mu_n(p) &\leq \sum_{m=1}^{\infty} \int_{Q_0} \varpi(f_m(p))\, d\mu_n(p) \\ &\leq \sum_{m=1}^{\infty} \Big(\sum_k \varpi(1/\alpha_m)\mu_n(M_m^k)\Big) \leq \sum_{m=1}^{\infty} \sum_k \frac{1}{C_m}\mu_n(S_m^k) \\ &\leq 2\sum_{m=1}^{\infty} \frac{1}{C_m} < \varpi^+(1)\,; \end{aligned} \tag{3.1.8}$$

f belongs to $\mathfrak{L}_\varpi$.

Also, for each positive integer m and μ-almost all points $p \in Q_0$, there exists a set S_m^k such that $p \in S_m^k$. Consequently, by the definition of S_m^k, there exists a set $U_m \in \mathscr{U}$ such that $p \in U_m$ and

$$\mu_n(U_m \cdot M_m^k) > \alpha_m \mu_n(U_m)\,.$$

Since $U_m \subset S_m^k$ and the diameter of S_m^k is less than m^{-1}, the same is true of the diameter of U_m. Since $f(p) \geq 1/\alpha_m$ whenever $p \in \bigcup_k M_m^k$, then

$$\frac{1}{\mu_n(U_m)} \int_{U_m} f(q)\, d\mu_n(q) \geq \frac{1}{\mu(U_m)} \cdot \frac{1}{\alpha_m} \mu_n(U_m \cdot M_m^k) > 1\,. \tag{3.1.9}$$

Thus, under the assumption that the derivation theorem holds for f, (3.1.9) leads to the conclusion that $f(p) \geqslant 1$ for μ_n-almost all $p \in Q_0$. According to our earlier observation, we must then have $\varpi(f(p)) \geqslant \varpi^+(1)$ μ-almost everywhere in Q_0; therefore

$$\int_{Q_0} \varpi(f(p))\, d\mu_n(p) \geqslant \varpi^+(1),$$

in contradiction with (3.1.8). This proves the theorem.

Remarks. (1) The integrability of $\varpi(|f|)$ over Q_0 depends only on the behaviour of $\varpi(t)$ for large values of t.

(2) According to the definition, the weak halo is empty for $\alpha \geqslant 1$, hence its measure is zero. This is why we are not especially concerned with the values taken by ϖ on $[0,1]$.

3.2. Theorem. *If, for every non-negative integrable function f in $\mathfrak{L}_\varpi$, the integral ψ of f is $\mathfrak{I}$-derivable to f μ_n-almost everywhere, then*

$$\varpi(t) > Ct(\mathrm{Log}^+ t)^{n-1} \tag{3.2.1}$$

whenever $t > 1$, for a suitable constant $C > 0$.

Proof. We apply Th. 3.1 to the case $\mathscr{U} = \mathscr{I}$. We consider first the case $n = 2$. For the open unit square Q_0, the weak halo $S(\alpha, Q_0)$ is of μ_2-measure $(4/\alpha)\,\mathrm{Log}\,(1/\alpha) + 1$. (cf. Prop. 1.5.1). Therefore, $(1/\alpha\,\mathrm{Log}\,(1/\alpha) < C'\varpi(1/\alpha)$ for $0 < \alpha < 1$, so that $\varpi(t) > Ct(\mathrm{Log}\,t)$.

For $n = 3$, the weak halo $S(\alpha, Q_0)$ turns out to have μ_3-measure $\frac{4}{\alpha}(\mathrm{Log}\,\frac{1}{\alpha})^2 + \frac{4}{\alpha}(\mathrm{Log}\,\frac{1}{\alpha}) + \frac{2}{\alpha} - 1$, therefore $(1/\alpha)(\mathrm{Log}\,(1/\alpha))^2 < C\varpi(1/\alpha)$ for $\alpha < 1$, and $\varpi(t) > Ct(\mathrm{Log}\,t)^2$. It may be noted that we are not interested in the exact value of $\mu_n(S(\alpha, Q_0))$, but only in the order of magnitude.

In the general case, it can be shown that the term of highest order encountered in computing the value of the halo of a hypercube is $k(1/\alpha)(\mathrm{Log}\,(1/\alpha))^{n-1}$; therefore

$$(1/\alpha)(\mathrm{Log}\,(1/\alpha))^{n-1} < C'\varpi(1/\alpha),$$

whenever $0 < \alpha < 1$, whence $\varpi(t) > Ct(\mathrm{Log}\,t)^{n-1}$.

Consequences of Th. 3.2. (1) Combining Th. 3.2 with the theorem of J.-M.-Z., we see that if the integrals of all non-negative functions $f \in \mathfrak{L}_\varpi$ are $\mathfrak{I}$-derivable to f almost everywhere, then the functions $|f|(\mathrm{Log}^+|f|)^{n-1}$ are also integrable.

(2) Let ε be a bounded, positive-valued function on $1 < t < \infty$ such that $\lim_{t \to +\infty} \varepsilon(t) = 0$ and

$$\varpi(t) = t(\mathrm{Log}\,t)^{n-1}\varepsilon(t)$$

is non-decreasing. Then there exists a function $f \in \mathfrak{L}_\varpi$ whose integral is not $\mathfrak{I}$-derivable to f almost everywhere. Otherwise, (3.2.1) would hold, which is impossible.

3.3. Theorem. *For the basis* $[\mathcal{I}, \delta]$ *we define the minimal halo function* γ_m (IV., V.) *by*

$$\gamma_m(\alpha) = \sup \frac{\mu_n(S(\alpha, M))}{\mu_n(M)}, \tag{3.3.1}$$

where the supremum is taken among all bounded $\mathcal{M}$*-sets of positive measure in* $\mathbf{R}^n$.

Then

$$0 < \liminf_{\alpha \to 0+} \frac{\gamma_m(\alpha)}{\frac{1}{\alpha}(\operatorname{Log}\frac{1}{\alpha})^{n-1}} \leqslant \limsup_{\alpha \to 0+} \frac{\gamma_m(\alpha)}{\frac{1}{\alpha}(\operatorname{Log}\frac{1}{\alpha})^{n-1}} < \infty, \tag{3.3.2}$$

so that $(1/\alpha)(\operatorname{Log}(1/\alpha))^{n-1}$ *is the exact order of* $\gamma_m(\alpha)$ *as* α *tends to zero.*

Proof. According to the theorem of J.-M.-Z. and Th. 3.1, we have,

$$\mu_n(S(\alpha, M)) \leqslant C(1/\alpha)(\operatorname{Log}(1/\alpha))^{n-1}\mu_n(M)$$

for all bounded measurable sets of positive measure, whence

$$\frac{\gamma_m(\alpha)}{\frac{1}{\alpha}(\operatorname{Log}\frac{1}{\alpha})^{n-1}} \leqslant C. \tag{3.3.3}$$

On the other hand, if we take for M the hypercube Q_0, we obtain

$$\gamma_m(\alpha) \geqslant \kappa \cdot \tfrac{1}{\alpha} \cdot (\operatorname{Log}\tfrac{1}{\alpha})^{n-1},$$

where κ is a positive constant; thus

$$\frac{\gamma_m(\alpha)}{\frac{1}{\alpha}(\operatorname{Log}\frac{1}{\alpha})^{n-1}} \geqslant \kappa. \tag{3.3.4}$$

Since the numbers C and κ appearing in (3.3.3) and (3.3.4) are absolute constants, the validity of (3.3.2) is now apparent.

3.4. Question. Ths. IV. 5.2 and V. 1.4 asserted the equivalence of derivation and halo properties. Th. 3.1 deduces a property of the halo function from the property of derivability of the $\mathfrak{L}_\varpi$ functions. The question naturally arises whether the converse of Th. 3.1 is true: if $[\mathcal{U}, \delta]$ is a Busemann-Feller basis in $\mathbf{R}^n$, ϖ a non-decreasing function defined on $]1, \infty[$ satisfying $\varpi^+(1) > 0$, and $C\varpi(1/\alpha)$ is a halo function of $[\mathcal{U}, \delta]$, then must the basis derive the integrals of $\mathfrak{L}_\varpi$ functions to their respective integrands almost everywhere?

In particular, for the case $\mathcal{U} = \mathcal{I}$, the results of Th. 3.3 for $n = 2$ can be proved in an elementary manner [*5*, 245–246]. It would be interesting to deduce, at least for $n = 2$, the theorem of J.-M.-Z.

In [*22*], it is shown in a quite general setting that the existence of a halo condition implies the derivability of certain classes of set functions, but the results there obtained supply only a partial answer to the question raised above concerning Busemann-Feller bases.

4. Saks' counterexample [*42*, 238−242]. We wish to show that the theorem of J.-M.-Z. is the best possible result, in a sense. We shall state the theorem we propose to prove and then give its proof in stages as consequences of certain constructions that will be carried out. It is worth remarking that the consequence (2) of Th. 3.2 enunciates a special case of the theorem now to be proved.

4.0. Theorem. *Given an arbitrary positive function* σ *defined on* $[0, \infty[$ *such that* $\liminf_{t \to +\infty} \sigma(t) = 0$, *there exists a non-negative measurable function* f *defined on the unit cube* $Q_0 =]0,0,\ldots,0; 1,1,\ldots,1[$ *of* $\mathbf{R}^n$, *such that* $\sigma(f)f(\operatorname{Log}^+ f)^{n-1}$ *is integrable over* Q_0, *and the upper* $\mathfrak{I}$*-derivate* $D^\star \varphi(x) = \infty$ *at* μ_n*-almost all points of* Q_0, *where* φ *is the indefinite* μ_n*-integral of* f.

Proof. We shall first construct for $n = 2$ a function f and then show that f has the desired properties; afterwards, we shall indicate necessary modifications in the construction for arbitrary n.

4.1. Bohr's construction [*7*, 689−691]. We let $S =] a, c; b, d [$ denote an open interval in the plane, take an arbitrary number $\tau > 1$, and let N denote the greatest positive integer not exceeding τ.

We consider the set of open subintervals of S,

$$I_j^1 = \left] a,c; a + \frac{(b-a)}{N}j, c + \frac{d-c}{j} \right[\tag{4.1.1}$$

for $j = 1, 2, \ldots, N$. We let U^1 and I^1 denote, respectively, the union and intersection of the intervals defined by (4.1.1). It is easily seen that

$$\mu_2(I_j^1) = N\mu_2(I^1)$$

for $j = 1, 2, \ldots, N$; and

$$\mu_2(U^1) = N\left(1 + \frac{1}{2} + \ldots + \frac{1}{N}\right)\mu_2(U^1) \geqslant N \operatorname{Log} N \mu_2(I^1).$$

We may now clearly subdivide $S - U^1$ into a finite number of disjoint open intervals whose union is $S - U^1$ except for a μ_2-nullset. To each of these we apply the operation just carried out on S, and then repeat this procedure sufficiently many times so that the μ_2-measure (area) of the remainder set is less that $\mu_2(S)/(N + 1)^2$. Finally, we divide this remainder set into a finite number of disjoint open intervals $J^1, \ldots, J^r$ covering all of it except for a μ_2-nullset. Thus, we obtain a subdivision of S into a set of open intervals

$$I_1^1, \ldots, I_N^1, I_1^2, \ldots, I_N^2, \ldots, I_1^s, \ldots, I_N^s, J^1, \ldots, J^r.$$

We denote by U^i and I^i, respectively, the union and the intersection of the intervals I^i_j, $j = 1,2,\dots,N$, $i = 1,2,\dots,s$. For each such i and j we have then

$$\mu_2(I^i_j) = N\mu_2(I^i), \tag{4.1.2}$$

$$\mu_2(U^i) \geqslant N \operatorname{Log} N \mu_2(I^i), \quad \text{and} \tag{4.1.3}$$

$$\mu_2(J^1 + \cdots + J^r) \leqslant \mu_2(S)/(N+1)^2. \tag{4.1.4}$$

Now we let w be a function assuming the value τ on the union of the intervals I^i and J^k, $i = 1,2,\dots,s$, $k = 1,2,\dots,r$, and vanishing elsewhere on S. By virtue of (4.1.3),

$$\sum_{i=1}^{s} \mu_2(I^i) \leqslant \frac{4}{(N+1)\operatorname{Log}(N+1)} \sum_{i=1}^{s} \mu_2(U^i) \leqslant \frac{4\mu_2(S)}{\tau \operatorname{Log} \tau}.$$

We adopt the convention that for $t = 0$, $t \operatorname{Log} t = \lim_{\varepsilon \to 0+} \varepsilon \operatorname{Log} \varepsilon = 0$, and likewise $t \operatorname{Log}^+ t = 0$. Then, by (4.1.4),

$$\int_S w(p) \operatorname{Log} w(p) d\mu_2(p) = \tau \operatorname{Log} \tau \left(\sum_{i=1}^{s} \mu_2(I^i) + \sum_{k=1}^{r} \mu_2(J^k) \right) \leqslant 5\mu_2(S).$$

On the other hand, from (4.1.2) and the definition of w, it follows that

$$\int_{I^i_j} w(p) d\mu_2(p) = \tau \mu_2(I^i_j) \geqslant \mu_2(I^i_j);$$

$$\int_{J^k} w(p) d\mu_2(p) = \tau \mu_2(J^k).$$

We may summarize the above discussion in the following way: Given an open interval $S \subset Q_0$ and any number $\tau > 1$, there exists a function $w_{S,\tau}$ defined on S, vanishing outside of S, and enjoying the properties:

(A_1) $w_{S,\tau}$ takes on the values 0 and τ, the value τ being assumed on a finite union of open intervals, the value 0 everywhere else;

(A_2) $$\int_S w_{S,\tau}(p) \operatorname{Log}(w_{S,\tau}(p)) d\mu_2(p) \leqslant 5\mu_2(S);$$

(A_3) each point $p_0 \in S$, except for at most a μ_2-nullset, belongs to an open interval $I_0 \subset S$ such that

$$\int_{I_0} w_{S,\tau}(p) d\mu_2(p) \geqslant \mu_2(I_0).$$

4.2. Construction of the function f. We let (t_l), $l = 1, 2, \dots$ denote an increasing sequence of real numbers such that $\lim_{l \to \infty} t_l = \infty$ and $\lim_{l \to \infty} \sigma(t_l) = 0$, which exists since $\liminf_{t \to \infty} \sigma(t) = 0$. We denote by ε a continuous, non-increasing function on $[0, \infty[$ that coincides with σ at the points of the sequence (t_l). We can obviously assume that the values of ε do not exceed 1. We shall define a sequence of non-negative functions (w_m), $m = 0, 1, 2, \dots$, so as to satisfy the following conditions:

(C_1) w_m takes on a finite number of values, but at least two; say, for instance

$$0 = \alpha_0^{(m)} < \alpha_1^{(m)} < \cdots < \alpha_{n_m}^{(m)}$$

each value being taken over a set that is representable as the union of a finite number of open intervals, to within a set of μ_2-measure zero;

$$(C_2) \quad \varepsilon(\alpha_1^{(m)}) < 2^{-m}, \quad \alpha_1^{(m)} > 2^m, \quad \text{and} \quad \alpha_1^{(m)} > \sum_{\nu=0}^{m-1} \alpha_{n_\nu}^{(\nu)}$$

for each $m \geqslant 0$;

$$(C_3) \qquad \frac{w_0(p)}{[\varepsilon(\alpha_1^{(0)})]^{1/2}} + \frac{w_1(p)}{[\varepsilon(\alpha_1^{(1)})]^{1/2}} + \cdots + \frac{w_m(p)}{[\varepsilon(\alpha_1^{(m)})]^{1/2}}$$

admits only values belonging to the set $\{t_l\}$ and the value zero;

$$(C_4) \qquad \int_{Q_0} w_m(p)\, d\mu_2(p) \leqslant \int_{Q_0} w_m(p) \operatorname{Log} w_m(p)\, d\mu_2(p) \leqslant 5;$$

(C_5) for μ_2-almost all points p_0 in Q_0, there exists an open interval $I_0 \subset Q_0$ such that

$$p_0 \in I_0, \quad \delta(I_0) < m^{-1}, \quad \text{and} \quad \int_{I_0} w_m(p)\, d\mu_2(p) \geqslant \mu_2(I_0).$$

We shall construct the desired sequence recursively. We take $w_0 = w_{Q_0,\alpha}$, where α is a constant greater than 1 such that $\alpha/[\varepsilon(\alpha)]^{1/2}$ is one of the values of the sequence (t_l), $l = 1, 2, \ldots$. We put $\alpha_0^{(0)} = 0$, $\alpha_1^{(0)} = \alpha$, $n_0 = 2$. If we agree that the summation occurring in (C_2) refers to an empty set of values, and consequently is zero, when $m = 0$, then conditions $(C_1)-(C_5)$ are satisfied for $m = 0$. Suppose now that the functions w_m are already defined and satisfy conditions $(C_1)-(C_5)$ for $m = 0, 1, \ldots, h-1$. We have to construct a function w_h so that $(C_1)-(C_5)$ hold for $m = 0, 1, \ldots, h-1, h$.

We partition Q_0 into a finite number of open intervals I_q^{h-1}, $q = 1, 2, \ldots, n_h$, each of diameter less than $1/h$, so that the sum

$$\sum_{m=0}^{h-1} \frac{w_m(p)}{[\varepsilon(\alpha_1^{(m)})]^{1/2}}$$

has a constant value $\beta_q^{(h-1)}$ throughout I_q^{h-1}. Its values on the remainder of Q_0, which comprises a set of μ_2-measure zero, are of no interest. This may obviously be accomplished by virtue of condition (C_1), which holds for $m = 0, 1, 2, \ldots, h-1$ because of our inductive assumption.

We now choose a number $\alpha_1^{(h)}$ sufficiently large to satisfy condition (C_2) for $m = h$. Since

$$\frac{t}{[\varepsilon(t)]^{1/2}}$$

is continuous and tends to $+\infty$ as t tends to $+\infty$, it is clear that the value of $\alpha_1^{(h)}$ may be determined so that additionally

$$\beta_1^{(h-1)} + \frac{\alpha_1^{(h)}}{[\varepsilon(\alpha_1^{(h)})]^{1/2}}$$

belongs to the set $\{t_l\}$. The numbers $\alpha_2^{(h)}, \ldots, \alpha_{n_h}^{(h)}$ are then defined successively so that

$$\alpha_1^{(h)} < \alpha_2^{(h)} < \cdots < \alpha_{n_h}^{(h)}$$

and so that the values

$$\beta_2^{(h-1)} + \frac{\alpha_2^{(h)}}{[\varepsilon(\alpha_1^{(h)})]^{1/2}}, \ldots, \beta_{n_h}^{(h-1)} + \frac{\alpha_{n_h}^{(h)}}{[\varepsilon(\alpha_1^{(h)})]^{1/2}}$$

all belong to the set $\{t_l\}$.

Next, we set

$$w_h = w_{I_q^{h-1}, \alpha_q^{(h)}}$$

on I_q^{h-1}, $q = 1, 2, \ldots, n_h$, where $w_{I_q^{h-1}, \alpha_q^{(h)}}$ is a function satisfying the conditions $(A_1)-(A_3)$ of 4.1, with $S = I_q^{h-1}$ and $\tau = \alpha_q^{(h)}$. It is readily seen that w_h and $\alpha_1^{(h)}, \alpha_2^{(h)}, \ldots, \alpha_{n_h}^{(h)}$ satisfy all the conditions $(C_1)-(C_5)$ for $m = h$.

Finally, we set

$$f(p) = \sum_{m=0}^{\infty} \frac{w_m(p)}{[\varepsilon(\alpha_1^{(m)})]^{1/2}} \quad \text{for} \quad p \in Q_0, \quad f(p) = 0, \quad p \in (\mathbf{R}^2 - Q_0). \tag{4.2.1}$$

4.3. *f possesses the asserted property for $n = 2$.*

(1) We shall prove first that the function $\sigma(f) \cdot f \cdot \mathrm{Log}^+ f$ is μ_2-integrable. To this end, we denote by $f_m(p)$ the m-partial sum of the right-hand side of (4.2.1), and we define

$$R_m = E[p: f_m(p) > f_{m-1}(p)], \quad m = 1, 2, \ldots.$$

Clearly $p \in R_m$ iff $w_m(p) > 0$, which holds in turn by (C_1) iff $w_m(p) \geqslant \alpha_1^{(m)}$. Hence

$$R_m = E[p: w_m(p) > 0] = E[p: w_m(p) \geqslant \alpha_1^{(m)}], \quad m = 1, 2, \ldots \tag{4.3.1}$$

Thus, if $p \in R_m$, we have from (C_2)

$$f_{m-1}(p) \leqslant \sum_{\nu=0}^{m-1} \frac{\alpha_{n_\nu}^{(\nu)}}{[\varepsilon(\alpha_1^{(m)})]^{1/2}} < \frac{\alpha_1^{(m)}}{[\varepsilon(\alpha_1^{(m)})]^{1/2}} \leqslant \frac{w_m(p)}{[\varepsilon(\alpha_1^{(m)})]^{1/2}}.$$

Consequently,

$$f_m(p) = f_{m-1}(p) + \frac{w_m(p)}{[\varepsilon(\alpha_1^{(m)})]^{1/2}} < \frac{2\, w_m(p)}{[\varepsilon(\alpha_1^{(m)})]^{1/2}}, p \in R_m.$$

Hence, because of (C_4), we may write

$$\int_{R_m} f_m(p) \, \mathrm{Log}^+ (f_m(p)) \, d\mu_2(p)$$

$$\leqslant \frac{2}{[\varepsilon(\alpha_1^{(m)})]^{1/2}}\left[\int_{R_m} w_m(p)\,\mathrm{Log}^+(w_m(p))\,d\mu_2(p)\right.$$

$$\left. + \left(\mathrm{Log}^+ \frac{2}{[\varepsilon(\alpha_1^{(m)})]^{1/2}}\right)\int_{R_m} w_m(p)\,d\mu_2(p)\right]$$

$$\leqslant \frac{10}{[\varepsilon(\alpha_1^{(m)})]^{1/2}}\left[1 + \mathrm{Log}^+ \frac{2}{[\varepsilon(\alpha_1^{(m)})]^{1/2}}\right], \qquad m = 1, 2, \ldots. \tag{4.3.2}$$

According to (C_3), if $f_m(p) > 0$, then $f_m(p)$ is a number belonging to the set $\{t_l\}$, so that $\sigma(f_m(p)) = \varepsilon(f_m(p))$. This relation is true, in particular, if $p \in R_m$, whence $w_m(p) \geqslant \alpha_1^{(m)}$. Since the values taken by ε are all less than 1 and $f_m(p) \geqslant w_m(p)/[\varepsilon(\alpha_1^{(m)})]^{1/2}$, then $f_m(p) > \alpha_1^{(m)}$, whenever $p \in R_m$; thus, because ε is non-increasing, $\varepsilon(f_m(p)) \leqslant \varepsilon(\alpha_1^{(m)})$. Therefore, from (4.3.2) it follows that

$$\int_{R_m} \sigma(f_m(p)) \cdot f_m(p) \cdot \mathrm{Log}^+(f_m(p))\,d\mu_2(p) \leqslant \varepsilon_m, \tag{4.3.3}$$

where ε_m represents

$$10\,[\varepsilon(\alpha_1^{(m)})]^{1/2}\left[1 + \mathrm{Log}^+ \frac{2}{[\varepsilon(\alpha_1^{(m)})]^{1/2}}\right], \qquad m = 1, 2, \ldots.$$

By virtue of the first of the relations (C_2) we have

$$\sum_{m=1}^{\infty} \varepsilon_m < \infty. \tag{4.3.4}$$

Because of (C_1) and (C_2), each term in the sum representing f is either 0 or exceeds 1. Hence $f(p) = \infty$ iff an infinite number of terms in the sum are positive, i.e., iff $w_m(p) > 0$ for an infinite number of values of m, which, by (4.3.1), means $p \in \limsup_m R_m$. By (C_2) and (C_4), $\mu_2(R_m) \leqslant 5/\alpha_1^{(m)} < 5 \cdot 2^{-m}$; thus $\sum_{m=1}^{\infty} \mu_2(R_m) < \infty$, and so $\mu_2(\limsup_m R_m) = 0$. Hence the set on which f is infinite is of μ_2-measure zero.

We let $E_m = E[p : f(p) = f_m(p) > f_{m-1}(p)]$ for $m = 1, 2, \ldots$. Clearly $E_m \subset R_m$ for each such m. We also let E denote the subset of Q_0 on which either $f(p) = 0$ or $f(p) = \infty$. From above, we see that $0 < f(p) < \infty$ implies that only finitely many values of $w_m(p)$, $m = 0, 1, 2, \ldots$, are positive, whence there exists some m such that $f(p) = f_m(p) > f_{m-1}(p)$, i.e., $p \in \bigcup_{m=1}^{\infty} E_m$. Thus $Q_0 = E \cup \left(\bigcup_{m=1}^{\infty} E_m\right)$ and, recalling (4.3.3) and (4.3.4), we obtain

$$\int_{Q_0} \sigma(f(p)) f(p)\,\mathrm{Log}^+(f(p))\,d\mu_2(p)$$

$$\leqslant \sum_{m=1}^{\infty} \int_{E_m} \sigma(f_m(p)) f_m(p)\,\mathrm{Log}^+(f_m(p))\,d\mu_2(p) < \infty.$$

(2) We shall now prove that $D^\star \varphi(p) = \infty$ at μ_2-almost all points p_0 in Q_0. From (C_5) we see that for each positive integer m and μ_2-almost all points p_0 in Q_0, there corresponds an open interval $I_0 \subset Q_0$ such that $p_0 \in I_0$, $\delta(I_0) < m^{-1}$, and

$$\int_{I_0} f(p) \, d\mu_2(p) \geqslant \frac{1}{[\varepsilon(\alpha_1^{(m)})]^{1/2}} \int_{I_0} w_m(p) \, d\mu_2(p) \geqslant \frac{\mu_2(I_0)}{[\varepsilon(\alpha_1^{(m)})]^{1/2}} .$$

Since $\lim_{m \to \infty} \varepsilon(\alpha_1^{(m)}) = 0$, this means that the upper $\Im$-derivate of the indefinite integral φ of f equals $+\infty$ at μ_2-almost points of Q_0.

Remark. It appears that the result just proved can be sharpened slightly upon taking account of the special nature of the exceptional sets of μ_2-measure zero arising in the construction described above. For instance, we see that *each* point $p_0 \in S$ lies in the closure $\bar{I}_0$ of the interval I_0 described in (A_3), so that (A_3) could be expressed without exception on p_0, upon replacing I_0 by $\bar{I}_0$ in its assertion. Similarly, the exceptional sets arising in constructing each function w_m, $m = 0, 1, \ldots$ are linear boundaries of certain intervals, so that (C_5) could be expressed for each point $p_0 \in Q_0$ upon replacing I_0 by $\bar{I}_0$. Then (2) above would follow for each point $p_0 \in Q_0$.

One could start the construction using closed sets rather than open intervals; the overlap on the boundaries would cause no essential difficulty since the sets in question are of μ_2-measure zero.

4.4. Modification of the proof for arbitrary values of n. In BOHR's construction, we started from a set (4.1.1) of intervals with a common corner on the curve $(x - a)(y - b) = (b - a)(d - c)/N = \mu_2(S)/N$. Thus the total area $\mu_2(U^1)$ of these intervals was approximately equal to

$$\frac{\mu_2(S)}{N} \int_{(b-a)/N}^{b-a} \frac{dx}{x} = \frac{\mathrm{Log}\, N}{N} \mu_2(S) .$$

If, instead of an interval in the plane, we were to consider an interval in the space $\mathbf{R}^n$, namely

$$S = \,]a_1, a_2, \ldots, a_n; \quad b_1, b_2, \ldots, b_n[,$$

then we should start BOHR's construction from a set of n-dimensional intervals with a common corner $(a_1, a_2, \ldots, a_n)$ and their opposite corners belonging to the $(n-1)$-dimensional hypersurface $(x_1 - a_1)(x_2 - a_2) \ldots (x_n - a_n) = \mu_n(S)/N^{n-1}$. The measure of the union of these intervals may easily be computed (cf. the proof of Th. 3.2); it is approximately equal to $((\mathrm{Log}\, N)/N)^{n-1} \mu_n(S)$, while their intersection is an interval of measure $\mu_n(S)/N^{n-1}$. Accordingly, in passing to the space $\mathbf{R}^n$, we have merely to replace $N \,\mathrm{Log}\, N$ in (4.1.3) by $N (\mathrm{Log}\, N)^{n-1}$ (with a suitable coefficient),

$(N+1)^2$ in (4.1.4) by $(N+1)^n$, and μ_2 by μ_n. This explains the role of the $(n-1)st$ power of $\mathrm{Log}^+|f|$ in the general enunciation of Th. 4.0.

5. The parallelepipedon basis [6, 243–247). The intervals considered in the preceding section were defined as (open) parallelepipeda whose sides were parallel to the coordinates axes, and thus of fixed direction. We now denote by $\mathscr{P}$ the family of all (open) rectangular parallelepipeda in $\mathbf{R}^n$. We shall establish the following result.

5.0. Theorem. *For the basis $[\mathscr{P}, \delta]$ of rectangular parallelepipeda, the Density Theorem does not hold.*

Proof. We shall give an explicit proof for the case $n = 2$. The construction for arbitrary values of n is essentially the same, but the notation becomes more complicated.

According to Th. IV. 5.3, it is sufficient to produce a set U^0 of finite measure and a number α_0 between 0 and 1 for which the weak halo $S(\alpha_0, U^0)$ is of infinite μ_2-measure.

5.1. Principles of the method. Let $T = D(a\,b\,c)$ be a triangular domain (open connected set) with vertices a, b, c. We determine points a', b', so that

$$\overrightarrow{ab'} = 2\overrightarrow{ab}, \quad \overrightarrow{ac'} = 2\overrightarrow{ac}.$$

Thus, every point q of $T' = D(ab'c')$ is inside a rectangle R (rectangular parallelipipedon for $n = 2$) that depends on q, such that

$$\mu_2(R \cdot T) > \mu_2(R)/2;$$

consequently,

$$T' \subset S\ (1/2, T).$$

The construction to follow exhibits an enumerable family of triangular domains $T_n = D(a_n b_n c_n)$ whose union has finite μ_2-measure, but the union of the corresponding triangles $T_n' = D(a_n b_n' c_n')$ has infinite μ_2-measure.

5.2. Pattern of the elementary construction. Basic relations. The rectangular x- and y-axes being fixed, we take the side bc on the x-axis, put $\delta(bc) = \alpha > 0$, and let η denote the ordinate of a. The points b' and c' are the intersections of the lines ab and ac, respectively, with the line $y = -\eta$. Take η' between 0 and η and denote by s, q, b'' (respectively, p, r, c'') the points of intersection of the line ab (respectively, ac) with the lines $y = \eta + \eta'$, $y = \eta - \eta'$, and $y = -\eta - \eta'$. The parallel lines pq and rs intersect the segments bc and $b''c''$ at points u, v and u'', v'' respectively.

We have $\delta(q, r) = \delta(u, v) = (\eta'/\eta)\alpha$ and

$$\delta(b, v) + \delta(u, c) = \alpha + (\eta'/\eta)\alpha = \tfrac{\alpha}{\eta}(\eta + \eta'). \tag{5.2.1}$$

We set $D_1 = D(p\,u\,c)$, $D_2 = D(s\,v\,b)$, and $D = D(a\,b\,c)$.

We observe that

$$\mu_2(D_1 \cup D_2) = \mu_2(D) + \mu_2(D(s\,a\,r)) + \mu_2(D(p\,a\,q))\,; \tag{5.2.2}$$

$$\mu_2(D_1 \cup D_2) - \mu_2(D) = \frac{{\eta'}^2}{\eta}\,\alpha\,.$$

Additionally,

$$D(a\,b''\,c'') \subset D(p\,u''\,c'') \cup D(s\,v''\,b'')\,.$$

The area of the trapezium domain $D(b'\,c'\,b''\,c'')$ exceeds $2\alpha\eta'$. It is essential for the sequel that its area increases linearly with η', while $\mu_2(D_1 \cup D_2) - \mu_2(D)$ is proportional to η'^2.

5.3. Construction of the sequence. We start with the configuration of 5.2, and set $D = D_1^1 = D(a\,b\,c) = U^1$, $\alpha = \alpha^1$, and $\eta = \eta^1$. The upper indices over the numbers α^1 and η^1 must not be mistaken for exponents, and the same will apply to the numbers α_k^m and η^m, to be defined shortly.

We also set $D_1 = D_1^2$, $D_2 = D_2^2$, $D_1^2 \cup D_2^2 = U^2$, $\eta' + \eta^1 = \eta^2$, $b' = b^1, c' = c^1, b'' = b^2, c'' = c^2$. We take $\eta' = \eta^1/2$, whence $\eta^2 = (3/2)\eta^1$.

We proceed by recursion. We assume that the triangular domains D_k^m, $m = 1,2,\ldots,h$, $k = 1,2,\ldots,2^{m-1}$ are already defined. Each has a side of length α_k^m on the x-axis. The corresponding vertices for $k = 1,2,\ldots,2^{m-1}$ lie on a line parallel to the x-axis, namely $y = \eta^m$. We put $U^m = \bigcup_{k=1}^{2^{m-1}} D_k^m$.

Additionally, we assume that the following equations, already established for $m = 1$ and $m = 2$, are valid for $m = 1,2,\ldots,h$:

(a) $\eta^m = \eta^1(1 + 1/2 + \cdots + (1/m))$;

(b) $\alpha_1^m + \alpha_2^m + \cdots + \alpha_{2^{m-1}}^m = \alpha^1\left(\dfrac{\eta^m}{\eta^1}\right)$;

(c) $\mu_2(U^m) = \mu_2(D_1^m \cup \cdots \cup D_{2^{m-1}}^m) = \alpha^1\eta^1\left(1 + \dfrac{1}{2^2} + \cdots + \dfrac{1}{m^2}\right) - \dfrac{\alpha^1\eta^1}{2}$;

(d) $D(a\,b^m\,c^m) \subset S(1/2, U^m)$, where b^m and c^m are the points of intersection of the lines $a\,b$ and $a\,c$, respectively, with the line $y = -\eta^m$.

The $(h+1)$ st step is accomplished as follows: We apply the construction described in (5.2) to each triangle D_k^h for $\eta' = \eta^1/(h+1)$. To D_1 and D_2 there will then correspond D_{2k-1}^{h+1} and D_{2k}^{h+1} with sides on the x-axis of length α_{2k-1}^{h+1} and α_{2k}^{h+1} and with the common height

$$\eta^h + \eta^1/(h+1) = \eta^1\left(1 + \frac{1}{2} + \cdots + \frac{1}{h+1}\right).$$

We have thus realized condition (a) for $m = h + 1$. We now set out to prove that (b), (c), and (d) also hold for $m = h + 1$.

According to (5.2.1)

$$\alpha_{2k-1}^{h+1} + \alpha_{2k}^{h+1} = \frac{\alpha_k^h}{\eta^h} \cdot \eta^{h+1};$$

therefore, using (b) with $m = h$, we have

$$\sum_{k=1}^{2^h} \alpha_k^{h+1} = \frac{\eta^{h+1}}{\eta^h} \sum_{k=1}^{2^{h-1}} \alpha_k^h = \alpha^1 \frac{\eta^{h+1}}{\eta^1}.$$

This is condition (b) for $m = h + 1$.

According to (5.2.2)

$$\mu_2(D_{2k-1}^{h+1} \cup D_{2k}^{h+1}) - \mu_2(D_k^h) = \left(\frac{\eta^1}{h+1}\right)^2 \frac{\alpha_k^h}{\eta^h}.$$

We set $U^{h+1} = \bigcup_{k=1}^{2^h} D_k^{h+1}$. We have

$$\mu_2(U^{h+1}) = \mu_2\left(\bigcup_{k=1}^{2^h} D_k^{h+1}\right) = \mu_2\left(\bigcup_{k=1}^{2^{h-1}} D_k^h\right) + \mu_2\left(\bigcup_{k=1}^{2^{h-1}} ((D_{2k-1}^{h+1} \cup D_{2k}^{h+1}) - D_k^h)\right)$$

$$= \alpha^1\eta^1\left(1 + \frac{1}{2^2} + \cdots + \frac{1}{h^2}\right) - \frac{\alpha^1\eta^1}{2} + \left(\frac{\eta^1}{h+1}\right)^2 \left(\frac{1}{\eta^h}\right)\left(\sum_{k=1}^{2^{h-1}} \alpha_k^h\right)$$

$$= \alpha^1\eta^1\left(1 + \frac{1}{2^2} + \cdots + \frac{1}{h^2}\right) - \frac{\alpha^1\eta^1}{2} + \left(\frac{\eta^1}{h+1}\right)^2 \left(\frac{1}{\eta^h}\right)\left(\frac{\alpha^1\eta^1}{\eta^1}\right)$$

$$= \alpha^1\eta^1\left(1 + \frac{1}{2^2} + \cdots + \frac{1}{h^2} + \frac{1}{(h+1)^2}\right) - \frac{\alpha^1\eta^1}{2}$$

which is condition (c) for $m = h + 1$.

Finally, we denote by b^{h+1} and c^{h+1} the intersection points of the line $y = -\eta^{h+1}$ with the lines ab and ac, respectively. Applying the procedure of (5.2) to each domain D_k^{h+1}, $k = 1, 2, \ldots, 2^h$, we see upon forming the union that

$$D(ab^{h+1}c^{h+1}) \subset \bigcup_{k=1}^{2^h} S(1/2, D_k^{h+1}),$$

whence

$$D(ab^{h+1}c^{h+1}) \subset S(1/2, U^{h+1}).$$

This is condition (d) for $m = h + 1$.

5.4. Construction of U°. The sequence of sets (U^m) is non-decreasing, and we put $U^0 = \lim_m U^m$. From (d) in (5.3), we have

$$\mu_2(U^0) = \lim_m \mu_2(U^m) = \alpha^1\eta^1 \sum_{m=1}^{\infty} \frac{1}{m^2} - \frac{\alpha^1\eta^1}{2} < \infty.$$

On the other hand,

$$S(1/2, U^m) \subset S(1/2, U^0) \quad \text{and} \quad D(ab^m c^m) \subset S(1/2, U^m)$$

both hold for each positive integer m. Hence $S(1/2, U^0)$ includes the angular domain bounded by the lines ab and ac, which is a set of infinite measure, and therefore so is $S(1/2, U^0)$.

6. Saks' "rarity" theorem [*42*, 257–261]. We shall prove that, in a sense to be made precise later, those functions f whose integrals φ have an upper $\mathfrak{I}$-derivate that is not everywhere infinite are the exceptions rather than the rule.

We take $n = 2$, $Q_0 =]0,0;1,1[$. The functions to be considered in this section vanish outside Q_0.

6.1. Lemma. *For any positive integer ν_0 there exists a non-negative and μ_2-measurable function f_{ν_0} that satisfies the following conditions:*

$$\int_{\bar{Q}_0} f_{\nu_0}(p)\, d\mu_2(p) \leqslant 2\nu_0^{-1}\,; \tag{6.1.1}$$

to each point $p_0 = (x_0, y_0)$ in $\bar{Q}_0$ there corresponds an interval I_0 centered at p_0 such that $\delta(I_0) \leqslant 4\nu_0^{-1}$ and

$$\int_{I_0} f_{\nu_0}(p)\, d\mu_2(p) \geqslant 4^{-1}\nu_0\,\mu_2(I_0)\,. \tag{6.1.2}$$

Proof. We divide the square $\bar{Q}_0$ into ν_0^2 equal, non-overlapping, closed squares $\bar{Q}_\nu$, $\nu = 1,2,\dots,\nu_0^2$. By virtue of Bohr's construction in (4.1), for every positive integer N, each square $\bar{Q}_\nu$ may be considered as a union of a finite number of closed intervals

$$I^1_{\nu,1}, \dots, I^1_{\nu,N}, I^2_{\nu,1}, \dots, I^2_{\nu,N}, \dots, I^s_{\nu,1}, \dots, I^s_{\nu,N}, J^1_\nu, \dots, J^r_\nu$$

such that, for $\quad i = 1,2,\dots,s \quad$ and $\quad j = 1,2,\dots,r$,

$$\mu_2(I^i_{\nu,1}) = \mu_2(I^i_{\nu,2}) = \cdots = \mu_2(I^i_{\nu,N}) = \frac{1}{(1 + 1/2 + \cdots + 1/N)}\,\mu_2(U^i_\nu)\,, \tag{6.1.3}$$

where $U^i_\nu = \bigcup_{j=1}^{N} I^i_{\nu,j}$, and

$$\sum_{k=1}^{r} \sum_{\nu=1}^{\nu_0^2} \mu_2(J^k_\nu) \leqslant 1/\nu_0^2\,; \tag{6.1.4}$$

$$\mu_2(I^i_\nu) > 0\,, \quad \text{where} \quad I^i_\nu = \bigcap_{j=1}^{N} I^i_{\nu,j}\,; \tag{6.1.5}$$

the sets U^i_ν and J^k_ν are non-overlapping. (6.1.6)

We take N sufficiently large so that

$$1 + \frac{1}{2} + \cdots + \frac{1}{N} \geqslant \nu_0^2\,;$$

then (6.1.3) becomes

$$\mu_2(I^i_{\nu,j}) \leqslant \frac{1}{\nu_0^2}\mu_2(U^i_\nu). \tag{6.1.7}$$

Let us now define

$$\begin{aligned} f_{\nu_0}(p) &= \frac{1}{\nu_0}\cdot\frac{\mu_2(U^i_\nu)}{\mu_2(I^i_\nu)} \quad \text{whenever} \quad p\in I^i_\nu\,; \\ & i = 1,2,\ldots,s, \quad \nu = 1,2,\ldots,\nu_0^2\,; \\ f_{\nu_0}(p) &= \nu_0 \quad \text{whenever} \quad p\in J^k_\nu, \quad k = 1,2,\ldots,r, \\ & \nu = 1,2,\ldots,\nu_0^2, \quad \text{and} \\ f_{\nu_0}(p) &= 0 \quad \text{elsewhere in } \mathbf{R}^2. \end{aligned} \tag{6.1.8}$$

Since the intervals on which f_{ν_0} is defined are not pairwise disjoint, the above values do not define f_{ν_0} uniquely everywhere in $\mathbf{R}^2$. Wherever ambiguity occurs, we agree to define the value $f_{\nu_0}(p)$ as the supremum of the values assigned in accordance with the above.

In view of (6.1.6) and (6.1.4), we have

$$\begin{aligned} \int_{\bar{Q}_0} f_{\nu_0}(p)\,d\mu_2(p) &\leqslant \nu_0^{-1}\left(\sum_{i=1}^{s}\sum_{\nu=1}^{\nu_0^2}\mu_2(U^i_\nu)\right) + \nu_0\sum_{k=1}^{r}\sum_{\nu=1}^{\nu_0^2}\mu_2(J^k_\nu) \\ &\leqslant \nu_0^{-1} + \nu_0^{-1} = 2\nu_0^{-1}. \end{aligned} \tag{6.1.9}$$

Let p_0 denote an arbitrary point in $\bar{Q}_0$. If p_0 belongs to some interval $I^i_{\nu,j}$, then it follows from (6.1.8) and (6.1.7) that

$$\int_{I^i_{\nu,j}} f_{\nu_0}(p)\,d\mu_2(p) \geqslant \nu_0^{-1}\mu_2(U^i_\nu) \geqslant \nu_0\,\mu_2(I^i_{\nu,j}). \tag{6.1.10}$$

We denote by I_0 the smallest interval with center at p_0 that includes $I^i_{\nu,j}$. Since f_{ν_0} is non-negative and $\delta(I^i_{\nu,j}) < 2\nu_0^{-1}$, we see at once that the interval I_0 satisfies the conditions of (6.1.2). The same obviously holds if p_0 belongs to an interval J^k_ν and if I_0 denotes the smallest interval with center at p_0 and including J^k_ν.

6.2. Theorem. *Let L be the space of functions f defined on* $\mathbf{R}^2$*, vanishing outside* $\bar{Q}_0$*,* μ_2*-measurable and integrable, the norm in L being, as usual,*

$$\|f\| = \int_{R^2}|f(p)|\,d\mu_2(p) = \int_{\bar{Q}_0} f(p)\,d\mu_2(p).$$

Then, except for the functions belonging to a set of Baire first category in L, the integral of each function in L has upper $\mathfrak{J}$*-derivate equal to* $+\infty$ *at each point of* $\bar{Q}_0$*.*

Proof. For any positive integer m, let A_m denote the set of functions f in L with the following property:

(P) There exists a point p_0 in $\bar{Q}_0$ such that for any interval I of center p_0 and diameter not exceeding m^{-1} we have $|\varphi(I)| \leqslant m\mu_2(I)$, where φ denotes the μ_2-integral of f. Since p_0 will depend on m and f we express this dependence by writing $p_0 = p_0(m, f)$.

We assert that A_m is closed. Indeed, let (f_i), $i = 1, 2, \ldots$ denote a sequence of functions in A_m converging to $f^\star$ in L, and let $p_i = p_i(m, f_i)$ denote the points associated with f_i by condition (P) above, for $i = 1, 2, \ldots$. Since $\bar{Q}_0$ is compact, there is a subsequence $p_{i_1}, \ldots, p_{i_j}, \ldots$, of these points converging to a point $p^\star$ of $\bar{Q}_0$. Consider an arbitrary interval $I^\star$ of center $p^\star$ and diameter not exceeding m^{-1}, and denote by $I_j^\star$ the translate of $I^\star$ with center p_{i_j}. Then $\delta(I_j^\star) \leqslant m^{-1}$. We denote by φ_{i_j} and $\varphi^\star$ the integrals of f_{i_j} and $f^\star$, respectively. Since $f_{i_j} \in A_m$, then $\varphi_{i_j}(I^\star) \leqslant m\mu_2(I_j^\star)$, and since the functions f_{i_j} converge to $f^\star$ in L, then the inequality

$$\varphi^\star(I^\star) = \lim_j \varphi_{i_j}(I_j^\star) \leqslant m\mu_2(I_j^\star) = m\mu_2(I^\star)$$

is valid. Hence $f^\star \in A_m$, and A_m is closed.

Next we shall prove that A_m is non-dense. In fact, let $f \in A_m$, and for a given positive number ε, let f_0 be a bounded measurable function vanishing outside $\bar{Q}_0$, for which

$$\|f - f_0\| \leqslant \varepsilon. \tag{6.2.1}$$

We select a function f_{ν_0}, $\nu_0 \geqslant 2\varepsilon^{-1}$, in accordance with Lemma 6.1, and let

$$g_{\nu_0} = f_0 + f_{\nu_0};$$

by (6.2.1) and (6.1.1) we have

$$\|g_{\nu_0} - f\| \leqslant \varepsilon + 2\nu_0^{-1} \leqslant 2\varepsilon. \tag{6.2.2}$$

Let M be an upper bound of $|f_0|$. By virtue of (6.1.2), if $\nu_0 > 4(M + m)$, then to any point p_0 in $\bar{Q}_0$, there corresponds an interval I_0 with center p_0 such that $\delta(I_0) \leqslant 4\nu_0^{-1} < m^{-1}$ and

$$\int_{I_0} |g_{\nu_0}(p)|\, d\mu_2(p) \geqslant (4\nu_0^{-1} - M)\mu_2(I_0) > m\mu_2(I_0).$$

Hence, for values of ν_0 sufficiently large, g_{ν_0} does not belong to A_m; and by (6.2.2), f is the limit in L of a sequence of elements not belonging to A_m. Thus the sets A_m are non-dense, $\bigcup_{m=1}^{\infty} A_m$ is a set of first category, and the complementary set $L - \bigcup_{m=1}^{\infty} A_m$ coincides with that of functions in L whose μ_2-integrals all have their upper $\mathfrak{J}$-derivate equal to $+\infty$ everywhere in $\bar{Q}_0$.

Chapter VI

A. P. Morse's Blankets

We now consider some of A. P. Morse's blankets (I. 3.3). Throughout this chapter R denotes a metric space metrized by δ. At times, R and δ will be specialized. The terms bounded, open, Borel, etc., will be used relative to δ. We denote by $\delta(A)$ the δ-diameter of an arbitrary set $A \subset R$. The spreads of all Morse's blankets are families of bounded Borel sets. He also introduces a Carathéodory measure function (outer measure) φ, finite on bounded sets [cf. *44*, 43]. We may regard our measure μ, defined on the Borel sets and finite on bounded Borel sets, to have been induced by such a function φ, we let $\mu^\star$ denote, as usual, the completion of μ. Since $\mu^\star$ agrees with μ on the bounded Borel subsets of R, we never need to refer explicitly to φ.

1. Nets [*29*, 226].

1.1. Definition. $\mathscr{H}$ is a *net* whenever $\mathscr{H}$ is a countable family of bounded, non-vacuous Borel sets such that

(i) $\mathscr{H} \cdot E[B: H \subset B]$ is a finite family whenever $H \in \mathscr{H}$;

(ii) If B_1 and B_2 belong to $\mathscr{H}$ and $B_1 \cdot B_2 \neq \emptyset$, then either $B_1 \subset B_2$ or $B_2 \subset B_1$.

1.2. Definition. A family of sets $\mathscr{H}$ *enmeshes* a set $A \subset R$ iff for each $x \in A$ and each $\varepsilon > 0$ there exists a set B such that $x \in B$, $B \in \mathscr{H}$, and $\delta(B) < \varepsilon$.

For the remainder of this section, we assume that $\mathscr{H}$ is a net enmeshing some set in R. $\mathscr{H}$ thus determines a basis $[\mathscr{H}, \delta]$ whose spread is $\mathscr{H}$ and whose domain is the largest set enmeshed by $\mathscr{H}$. If x is a point of the domain of $[\mathscr{H}, \delta]$, then the deriving sequences associated with x are all those ordinary sequences of sets $H \in \mathscr{H}$ such that $x \in H$ and whose δ-diameters tend to zero. If A is any subset of the domain of $[\mathscr{H}, \delta]$, then a subfamily $\mathscr{H}' \subset \mathscr{H}$ is an $[\mathscr{H}, \delta]$-fine covering of A iff for each $x \in A$ there is some $[\mathscr{H}, \delta]$-deriving sequence converging to x, all of whose constituents belong to $\mathscr{H}'$.

1.3. Theorem. $[\mathscr{H}, \delta]$ *possesses the strong Vitali property* (II. 6.1) *with respect to any Radon measure.*

Proof. Because Haupt's adaptation property (II. 1.6) holds we see, from Prop. II. 6.3, that only the R.S.V. property needs to be established.

Accordingly, we take an arbitrary set A included in the domain of $[\mathscr{H}, \delta]$ and any $[\mathscr{H}, \delta]$-fine covering $\mathscr{H}'$ of A. Def. 1.1 above ensures that if $B \in \mathscr{H}'$ then there exists a maximal set $H_B \in \mathscr{H}'$ that includes B. If B and B' are any two sets in $\mathscr{H}'$, then either $H_B \cdot H_{B'} = \emptyset$ or else $H_B = H_{B'}$, again by virtue of Def. 1.1. Thus the family $\mathscr{H}'' = E[H: H = H_B$ for some set $B \in \mathscr{H}']$ is a countable disjoint subfamily of $\mathscr{H}'$ that clearly covers A', and the proof is complete.

1.4. Corollary. $[\mathscr{H}, \delta]$ derives each Radon measure ψ.
Proof. Recall Th. II. 4.4.

2. Hives. Hives [*29*, 215] constitute an interesting application of blankets having the strong Vitali property with respect to Borel measure (II. 6.1). Throughout this section, R is taken as Euclidean n-space, δ is the usual metric, and μ is n-dimensional Borel measure, whose completion $\mu^\star$ is classical n-dimensional Lebesgue measure.

2.1. Definition. $\mathscr{H}$ is a *hive* whenever $\mathscr{H}$ is a family of bounded closed convex subsets of R such that

(i) if $B \in \mathscr{H}$, then $\mu(B) > 0$;

(ii) if B_1 and B_2 belong to $\mathscr{H}$, then B_1 includes or is included in some translate of B_2.

Throughout this section we consider a basis $[\mathscr{H}, \delta]$ whose spread is a hive, and whose deriving sequences are defined as in (1.1).

2.2. Theorem. *$[\mathscr{H}, \delta]$ is a special* MORSE *basis (Def.* IV. 2.2.1*).*

Proof. We choose a positive integer q such that $2 < (1 + 2^{-n})^q$ and define Δ on $\mathscr{H}$ so that $\Delta(B) = (\mu(B))^q$ whenever $B \in \mathscr{H}$.

For any convex set $B \subset R$ we define

$$B^\nabla = E[z: z = 2x - y \quad \text{for some } x \text{ and some } y \text{ in } B].$$

It follows directly that B^∇ is convex and coincides with

$$E[t: t = x + (y - z) \quad \text{for some } x, \text{ some } y, \text{ and some } z \text{ in } B].$$

It is easily verified that $t \in B^\nabla$ iff there exists a translate B' of B such that $t \in B'$ and $B \cdot B' \neq \emptyset$.

We consider any two sets A and B in $\mathscr{H}$ with $A \cdot B \neq \emptyset$ and $\Delta(A) \leqslant 2\Delta(B)$. We shall show that $A \subset B^{\nabla\nabla}$.

If some translate B' of B includes A, then $B \cdot B' \neq \emptyset$ and $B' \subset B^\nabla$, whence $A \subset B^\nabla \subset B^{\nabla\nabla}$. The remaining possibility is that some translate A' of A includes B. In this case, since $A \cdot B \neq \emptyset$, we have $A \cdot A' \neq \emptyset$, so that $A \subset A'^\nabla$. The desired conclusion $A \subset B^{\nabla\nabla}$ will then follow if we can show that $A' \subset B^\nabla$.

We establish this by contradiction, and thus we suppose that $x_0 \in (A' - B^\nabla \cdot A')$. We let

$$B' = E[z: z = (x_0 + y)/2 \quad \text{for some} \quad y \in B].$$

If $z \in B \cdot B'$, then $z = (x_0 + y)/2$ and $z = y'$ for suitable members y and y' of B, whence $x_0 = 2y' - y \in B^\nabla$, a contradiction. Thus $B \cdot B' = \emptyset$.

We let a be a vector such that each point $t \in A'$ is of the form $t = z + a$, where $z \in A$. If z is an arbitrary point of B', then $z = (x_0 + y)/2$ for some $y \in B$. Since $B \subset A'$, there exists $v \in A$ such that $y = v + a$; since $x_0 \in (A' - B^{\nabla} \cdot A')$, there exists $u \in A$ such that $x_0 = u + a$. Thus $z = (x_0 + y)/2 = a + (u + v)/2$, and so $z \in A'$. Hence $B' \subset A'$; therefore $B \cup B' \subset A$. From the definition of B' and the fact that $B \cdot B' = \emptyset$, we have

$$(1 + 2^{-n})\mu(B) = \mu(B) + \mu(B') = \mu(B \cup B') \leqslant \mu(A') = \mu(A).$$

Recalling our definition of q we obtain

$$2\Delta(B) = 2(\mu(B))^q \leqslant [(1 + 2^{-n})\mu(B)]^q \leqslant (\mu(A))^q = \Delta(A).$$

This contradiction shows that $A' \subset B^{\nabla}$.

Hence $A \subset B^{\nabla\nabla}$ in all cases, and consequently $H(\Delta, 2, B) \subset B^{\nabla\nabla}$ whenever $B \in \mathscr{H}$. If B happens to have central symmetry, it is easily checked that $\mu(B^{\nabla}) = 3^n \mu(\mathrm{B})$, whence $\mu(B^{\nabla\nabla}) = 3^{2n}\mu(B)$. Otherwise, it can be shown that $\mu(B^{\nabla}) \leqslant (n + 2)^n \mu(B)$ [*4*, p. 53, formula (1)], whence $\mu(B^{\nabla\nabla}) \leqslant (n + 2)^{2n}\mu(B)$ and $[\mathscr{H}, \delta]$ is a special MORSE basis.

2.3. Corollary. $[\mathscr{H}, \delta]$ has the strong Vitali property with respect to μ.

Proof. Recall Th. IV. 22, and Prop. II. 6.3.

The theorem just proved shows, in particular, that a satisfactory derivation theory holds with respect to Borel measure μ for a large class of rectangles that are long, thin, and oriented in every direction, despite the limitation expressed by Th. V. 5.1.

3. Fundamental covering theorems [*30*, 418–442]. The theorems of this section do not require that R be a metric space, although its later applications do.

3.1. Definition. Ω is a *covering relation* iff Ω is a relation and $x \in A$ whenever $(x, A) \in \Omega$.

3.2. Definition. Ω is *Δ-restrained* iff Δ is a non-negative real-valued function and Ω is a covering relation such that $y \notin A$ and $\Delta(y, B) \leqslant 2\Delta(x, A)$ whenever (x, A) and (y, B) are different members of Ω with $x \in B$.

3.3. Theorem. *If Ω is a covering relation with domain Q and Δ is a non-negative function bounded on Ω, then Ω includes a Δ-restrained subrelation whose range covers Q.*

Proof. We let $\Omega_1 = \Omega$, set

$$\Delta_1 = \sup_{(x, A) \in \Omega} \Delta(x, A) < \infty,$$

and select $(y, B) \in \Omega_1$, so that $\Delta(y_1, B_1) \geqslant (\Delta_1/2)$. Then for any member (x, A) of Ω_1 we have $\Delta(x, A) \leqslant \Delta_1 \leqslant 2\Delta(y_1, B)$. If $Q \subset B_1$, then the set $\{(y_1, B_1)\}$ is the required Δ-restrained subrelation of Ω. Otherwise, we may continue this procedure inductively. We suppose that λ is an ordinal number and that for all ordinals $\iota < \lambda$ the (possibly transfinite) sequence $((y_\iota, B_\iota))$ of members of Ω has been defined so that the set of its values constitutes a Δ-restrained subrelation of Ω, with the additional properties that $(y_\iota, B_\iota) \in \Omega_\iota$ and $\Delta(y_\iota, B_\iota) \geqslant (\Delta_\iota/2)$ for each $\iota < \lambda$, where Ω_ι is the subset of Ω for which $(x, A) \in \Omega_\iota$ iff $x \in \left(Q - Q \cdot \left(\bigcup_{\iota' < \iota} B_{\iota'}\right)\right)$, and

$$\Delta_\iota = \sup_{(x, A) \in \Omega_\iota} \Delta(x, A).$$

If $Q \subset \bigcup_{\iota < \lambda} B_\iota$, then the set $\{(y_\iota, B_\iota)\}$, $\iota < \lambda$, is the desired Δ-restrained subrelation. If not, then we define Ω_λ as that subset of Ω for which $(x, A) \in \Omega_\lambda$ iff $x \in \left(Q - Q \cdot \left(\bigcup_{\iota < \lambda} B_\iota\right)\right)$, define

$$\Delta_\lambda = \sup_{(x, A) \in \Omega_\lambda} \Delta(x, A),$$

and select $(y_\lambda, B_\lambda) \in \Omega_\lambda$ so that $\Delta(y_\lambda, B_\lambda) \geqslant (\Delta_\lambda/2)$.

We want to prove that the set consisting of the ordered pairs (y_ι, B_ι) obtained by the inductive procedure just described is a Δ-restrained subset of Ω. First, it is clear that the family of the sets B_ι covers Q. Because for any ordinal λ, $y_\lambda \in \left(Q - Q \cdot \left(\bigcup_{\iota < \lambda} B_\iota\right)\right)$, it follows that $y_\lambda \notin B_\iota$ if $\iota < \lambda$. Let us consider any two different pairs (y_ι, B_ι) and (y_λ, B_λ), and suppose that $y_\iota \in B_\lambda$. Then $\iota < \lambda$ and $y_\lambda \notin B_\iota$. Now $\Delta(y_\iota, B_\iota) \geqslant (\Delta_\iota/2)$, and since $(y_\lambda, B_\lambda) \in \Omega_\iota$, then

$$\Delta(y_\lambda, B_\lambda) \leqslant \Delta_\iota \leqslant 2\Delta(y_\iota, B_\iota).$$

This completes the proof.

3.4. Definition. Ω is *diametrically restrained* iff Ω is a covering relation, each member of the range of Ω is a subset of a metric space R, such that $\delta(B) \leqslant 2\delta(A)$ and $y \notin A$ whenever (x, A) and (y, B) are different members of Ω and $x \in B$.

The following is an obvious special case of Th. 3.3.

3.5. Corollary. If Ω is a covering relation with domain Q whose range is a family of subsets of R with uniformly bounded diameters, then Ω includes a diametrically restrained subrelation whose range covers Q.

3.6. Theorem. *If Δ is a bounded non-negative function whose domain is a non-vacuous family of sets $\mathscr{F}$ and $\alpha > 1$, then $\mathscr{F}$ includes a disjoint subfamily $\mathscr{G}$ such that for each set $B \in \mathscr{F}$ there exists a set $C \in \mathscr{G}$ with $B \cdot C \neq \emptyset$, $\Delta(B) \leqslant \alpha\Delta(C)$.*

Proof. This involves defining a possibly transfinite sequence of sets in a manner quite similar to that used in the proof of the first part of Th. IV. 2.2. Accordingly, we leave verification to the reader.

4. Star blankets. A. S. Besicovitch [*3*, 103–110] proved that the circular blanket in the plane has the strong Vitali property with respect to Radon measures. A. P. Morse extended this result to the more general class of star blankets in n-dimensional space [*30*][1]. We shall prove his theorem after some preparatory work. Throughout this section we assume that R is n-dimensional Euclidean space with the usual metric δ and μ is Borel measure in R. Addition, subtraction, and multiplication of points in R by a scalar are denoted in the usual manner. The origin or null element of R is denoted by 0. For any points x, y in R we define $\|x-y\| = \delta(x,y)$.

4.1. Definition. If $B \subset R$ and $\kappa \in B$, then the *internal radius of B at x* is defined as the supremum of those numbers r for which the closed ball $S(x,r)$, with center at x and radius r, is included in B.

4.2. Definition. The *hub* of a set $B \subset R$ is the set of those points $y \in B$ such that $((1-t)x + ty) \in B$ whenever $x \in B$ and $0 \leqslant t \leqslant 1$. We denote this set by hub (B).

It is easily seen that $B = \text{hub}(B)$ iff B is convex.

4.3 Definition. If $B \subset R$ and $x \in \text{hub}(B)$, then we define the *hub radius of B at x* as the internal radius of hub (B) at x, which we agree to abbreviate as $h(x,B)$.

4.4. Definition. $\mathfrak{B}$ is a *star blanket* iff $\mathfrak{B}$ is a blanket with closed constituents B and for each point x in its domain,

$$\operatorname*{Lim\,sup}_{x \in B \;\; \delta(B) \to 0+} \frac{\delta(B)}{h(x,B)} < \infty .$$

4.5. Definition. For any set A, $\nu(A)$ denotes the number of points in A if A is finite and $\nu(A) = \infty$ if A is infinite.

4.6. Definition. For any set A, $\mathscr{C}(A)$ is defined as the family of all sets X that satisfy $\delta(A) \leqslant 2\delta(X)$ and $X \cdot A \neq \emptyset$.

4.7. Theorem. *If* $1 \leqslant \lambda < \infty$, Ω *is a diametrically restrained covering relation with range* $\mathscr{R}$, $A \in \mathscr{R}$, *and*

$$\delta(X) < \lambda h(z,X) \quad \textit{whenever} \quad (z,X) \in \Omega, \quad \textit{then} \quad \nu(\mathscr{R} \cdot \mathscr{C}(A)) < (9\lambda)^{3n}.$$

[1] If $n = 1$ and in the setting of two functions F and α of bounded variation on an interval [a, b] we find a version of the universal derivation theorem for the central derivative in Daniell [Trans. Amer. Math. Soc. **19**, 353–362 (1918); p. 353]. The general case (for $n=1$) was treated by R. L. Jeffery [Trans. Amer. Math. Soc. **34**, 645–675 (1934); p. 656, Th. VI]. We quote also R. L. Jeffery [Canad. J. Math. **10**, 617–626 (1958)] and H. W. Ellis, and R. L. Jeffery [Canad. J. Math. **19**, 225–241 (1967)]. In the latter papers a Vitali property is used. We mention particularly the "Vitali -property" in the last paper, p. 226.

Proof. From Def. 3.2, it follows that if $(z, X) \in \Omega$ and $(y, X) \in \Omega$, then $z = y$. Thus, for each set $X \subset R$ there is a unique point z associated with X such that $(z, X) \in \Omega$. Let us agree to denote this point by the corresponding lower case letter; thus $(x, X) \in \Omega$, $(a, A) \in \Omega$, etc. We also agree to let

$$x^\star = x-a;\ K_X = S(x,(4\lambda)^{-1}\,\delta(X)) \text{ (cf. Def. 4.1 above)};$$
$$\kappa = \mu(S(0,1));$$
$$\mathscr{P}_1 = \mathscr{R}\cdot E[X: \|x^\star\| \leqslant 8\lambda\delta(A)]\cdot\mathscr{C}(A)];$$
$$\mathscr{P}_2 = \mathscr{R}\cdot E[X: \|x^\star\| > 8\lambda\delta(A)]\cdot\mathscr{C}(A)].$$

The following relations are essentially immediate consequences of the above definitions and our hypotheses:

If $0 < r < \infty$, $z \in R$, and $B \in \mathscr{R}$, then (4.7.1)
$\mu(S(z,r)) = \kappa r^n$, $\delta(B) < \lambda h(b,B)$; $K_B \subset S(b,\lambda^{-1}\delta(B)) \subset B$;

If B and C are different members of $\mathscr{R}$ such that $b \in C$, then $c \notin B$ and $\lambda^{-1}\delta(B) < \|b-c\| \leqslant \delta(C) \leqslant 2\delta(B)$. If $B = C$, we surely have $\lambda^{-1}\delta(B) \leqslant \delta(C) \leqslant 2\delta(B)$; (4.7.2)

If B and C are different members of $\mathscr{R}$, then $(2\lambda)^{-1}\delta(B) < \|b-c\|$; $K_B \cdot K_C = \emptyset$. (These relations can be established by contradiction with the help of (4.7.1) and (4.7.2).) (4.7.3)

We complete the proof in five steps.

Step I: If $B \in \mathscr{P}_2$, $C \in \mathscr{P}_2$, $8\lambda\delta(A) < \|b^\star\| \leqslant \|c^\star\|$, and $\big\| \|b^\star\|^{-1} b^\star - \|c^\star\|^{-1} c^\star \big\| \leqslant (8\lambda)^{-1}$, then $b \in C$.

Proof. We let $s = \|c^\star\| \cdot \|b^\star\|^{-1}$, $t = s^{-1}$, note that $1 \leqslant s < \infty$ and $0 < t \leqslant 1$, observe that $C \cdot A \neq \emptyset$ since $C \in \mathscr{C}(A) \subset \mathscr{P}_2$, accordingly choose $x \in C \cdot A$, let $y = (1-s)x + sb$, and check that $b = (1-t)x + ty$. If we can now show that $\|y-c\| < h(c,C)$, then $y \in \operatorname{hub}(C)$ and, since $x \in C$, we can infer finally that $b \in C$. We proceed with this in mind. We have

$$\begin{aligned} y-c &= (1-s)x + sb - c = (1-s)(x-a) + s(b-a) - (c-a) \\ &= (1-s)(x-a) + sb^\star - c^\star \\ &= (1-s)(x-a) + \|c^\star\| \cdot (\|b^\star\|^{-1}\cdot b^\star - \|c^\star\|^{-1}\cdot c^\star). \end{aligned}$$

Using the hypotheses of Step I and the fact that $x \in A$, we obtain

$$\begin{aligned} \|y-c\| &\leqslant (s-1)\|x-a\| + \|c^\star\|\cdot(8\lambda)^{-1} \\ &\leqslant s\delta(A) + \|c^\star\|\cdot(8\lambda)^{-1} < (8\lambda)^{-1}(s\|b^\star\| + \|c^\star\|) \\ &= (8\lambda)^{-1}(\|c^\star\| + \|c^\star\|) = (4\lambda)^{-1}\|c^\star\|. \end{aligned} \tag{4.7.4}$$

Since $x \in A \cdot C$ and $C \in \mathscr{C}(A)$, then $\delta(A) \leqslant 2\delta(C)$ and so

$$\begin{aligned} \|c^\star\| &= \|c-a\| \leqslant \|c-x\| + \|x-a\| \leqslant \delta(C) + \delta(A) \\ &\leqslant 3\delta(C) < 3\lambda h(c,C). \end{aligned} \tag{4.7.5}$$

Putting (4.7.4) and (4.7.5) together yields $\|y-c\| < h(c,C)$ as we required.

Step II: If $B \subset R$ and $\delta(B) \leqslant (8\lambda)^{-1}$, then

$$\nu(\mathscr{P}_2 \cdot E[C: \|c^\star\|^{-1} c^\star \in B]) \leqslant 8^n \lambda^{2n}.$$

Proof. It is sufficient to prove that $\nu(\mathscr{H}) \leqslant 8^n \lambda^{2n}$ whenever $\mathscr{H}$ is a finite subfamily of $\mathscr{P}_2 \cdot E[C: \|c^\star\|^{-1} \cdot c^\star \in B]$. We may assume that $\mathscr{H} \neq \emptyset$ and, since $\mathscr{H}$ is finite, we may determine $B_0 \in \mathscr{H}$ so that $8\lambda\delta(A) < \|b_0^\star\| \leqslant \|c^\star\|$ whenever $C \in \mathscr{H}$. If $C \in \mathscr{H}$, then we infer from the definition of $\mathscr{H}$ that $\| \|c^\star\|^{-1} \cdot c^\star - \|b_0^\star\|^{-1} \cdot b_0^\star \| \leqslant \delta(B) \leqslant (8\lambda)^{-1}$, and so, by Step I, $b_0 \in C$. Hence by (4.7.2), $\lambda^{-1}\delta(B_0) \leqslant \delta(C) \leqslant 2\delta(B_0)$. Combining this with (4.7.1) yields

$$K_C \subset C \subset S(b_0, 2\delta(B_0)); \quad \mu(K_C) = \kappa((4\lambda)^{-1}\delta(C))^n \geqslant \kappa((4\lambda^2)^{-1}\delta(B_0))^n \tag{4.7.6}$$

whenever $C \in \mathscr{H}$. From (4.7.6), and (4.7.3), we obtain

$$\kappa((4\lambda^2)^{-1}\delta(B_0))^n \nu(\mathscr{H}) \leqslant \sum_{C \in \mathscr{H}} \mu(K_C) \leqslant \mu(S(b_0, 2\delta(B_0)) = \kappa(2\delta(B_0))^n.$$

This yields $\nu(\mathscr{H}) \leqslant 8^n \lambda^{2n}$.

Step III: $\nu(\mathscr{P}_2) \leqslant (512)^n \lambda^{3n}$.

Proof. We set $q = (32\lambda)^{-1}$ and let $\mathscr{F}$ be the family of all balls of the form $S(z, q)$, where $z \in S(0,1)$. We use Th. 3.6, with $\Delta = \delta$, to find a disjoint family $\mathscr{K} \subset \mathscr{F}$ such that for each set $C \in \mathscr{F}$ there is some set $B \in \mathscr{K}$ with $B \cdot C \neq \emptyset$.

For any set $B \in \mathscr{K}$, we let

$$\tilde{B} = E[x: \|x - y\| \leqslant q \quad \text{for some} \quad y \in B],$$

and note that $B \subset S(0,2)$, $\delta(\tilde{B}) = 4q = (8\lambda)^{-1}$, and $\mu(B) = \kappa q^n$ whenever $B \in \mathscr{K}$. It is easily seen that $S(0,2) \subset \bigcup_{B \in \mathscr{K}} \tilde{B}$. Since $\mathscr{K}$ is disjoint we, therefore, have

$$\kappa q^n \nu(\mathscr{K}) = \sum_{B \in \mathscr{K}} \mu(B) \leqslant \mu(S(0,2)) = \kappa \cdot 2^n,$$

whence $\nu(\mathscr{K}) \leqslant 2^n \cdot q^{-n} = 64^n \lambda^n$. Also,

$$\mathscr{P}_2 = \mathscr{P}_2 \cdot E[C: \|c^\star\|^{-1} \cdot c^\star \in S(0,1)] \subset \bigcup_{B \in \mathscr{K}} \mathscr{P}_2 \cdot E[C: \|c^\star\|^{-1} c^\star \in \tilde{B}].$$

Thus, using Step II, we obtain

$$\begin{aligned}\nu(\mathscr{P}_2) &\leqslant \sum_{B \in \mathscr{K}} \nu(\mathscr{P}_2 \cdot E[C: \|c^\star\|^{-1} \cdot c^\star \in \tilde{B}]) \leqslant \sum_{B \in \mathscr{K}} 8^n \lambda^{2n} \\ &= 8^n \lambda^{2n} \nu(\mathscr{K}) \leqslant (512)^n \lambda^{3n}.\end{aligned}$$

Step IV. $\nu(\mathscr{P}_1) \leqslant (128)^n \lambda^{2n}$.

Proof. If $A, C \in \mathscr{P}_1$, $A \neq C$, and $y \in K_C$, then from (4.7.3) we see that

$$\|y-a\| \leqslant \|y-c\| + \|c-a\| \leqslant (4\lambda)^{-1}\delta(C) + \|c-a\| < 2\|c-a\| = 2\|c^\star\| \leqslant 16\lambda\delta(A).$$

Thus

$$K_C \subset S(a, 16\lambda\,\delta(A)). \tag{4.7.7}$$

If $A = C$, (4.7.7) is obviously valid. Since $\mathscr{P}_1 \subset \mathscr{C}(A)$, then $\delta(A) \leqslant 2\delta(C)$ and $\mu(K_C) = ((4\lambda)^{-1}\delta(C))^n \cdot \kappa \geqslant ((8\lambda)^{-1}\delta(A))^n \cdot \kappa$ whenever $C \in \mathscr{P}_1$. Thus, with the help of (4.7.7) we have

$$\kappa((8\lambda)^{-1}\delta(A))^n v(\mathscr{P}_1) = \sum_{C\in\mathscr{P}_1} \kappa(8\lambda)^{-1}\delta(A))^n \leqslant \sum_{C\in\mathscr{P}_1} \mu(K_C)$$
$$\leqslant \mu(S(a, 16\lambda\delta(A)) = \kappa(16\lambda\delta(A))^n.$$

From this it follows that $v(\mathscr{P}_1) \leqslant (128)^n\lambda^{2n}$.

Step V: $v(\mathscr{R}\cdot\mathscr{C}(A)) < (9\lambda)^{3n}$.

Proof. From Steps III and IV, we see that $v(\mathscr{R}\cdot\mathscr{C}(A)) = v(\mathscr{P}_1 + \mathscr{P}_2) = v(\mathscr{P}_1) + v(\mathscr{P}_2) < (729)^n\lambda^{3n} = (9\lambda)^{3n}$.

4.8. Lemma. *If λ is a positive integer, $\eta = (9\lambda)^{3n}$, Ω is a diametrically restrained covering relation with range $\mathscr{R}$, $\sup_{B\in\mathscr{R}}\delta(B) < \infty$, and $\delta(B) < \lambda h(z,B)$ holds for each $(z,B)\in\Omega$, then there exists a sequence of countable disjoint families $\mathscr{Q}_i$, $i = 1,2,\ldots,\eta$, such that $\mathscr{R} = \bigcup_{i=1}^{\eta}\mathscr{Q}_i$.*

Proof. Using Th. 3.6, we define $\mathscr{Q}_1$ as a disjoint subfamily of $\mathscr{R}$ such that for each set $B\in\mathscr{R}$ there exists $C\in\mathscr{Q}_1$ with $C\cdot B \neq \emptyset$, $\delta(B) \leqslant \delta(C)$. We proceed inductively. Assuming that $\mathscr{Q}_1,\ldots,\mathscr{Q}_{j-1}$, $2 \leqslant j \leqslant \eta$ have been defined, we use Th. 3.6 to obtain a disjoint subfamily $\mathscr{Q}_j$ of $\mathscr{R} - \bigcup_{i=1}^{j-1}\mathscr{Q}_i$ such that for each set $B \in \left(\mathscr{R} - \bigcup_{i=1}^{j-1}\mathscr{Q}_i\right)$ there exists $C\in\mathscr{Q}_j$ with $C\cdot B \neq \emptyset$, $\delta(B) \leqslant 2\delta(C)$.

Clearly $\mathscr{Q}_i\cdot\mathscr{Q}_j = \emptyset$ if $i \neq j$, $1 \leqslant i, j \leqslant \eta$. Since the members of $\mathscr{R}$ all have interior points and so are of positive n-dimensional Borel measure, then each family $\mathscr{Q}_i$, $i = 1,2,\ldots,\eta$ is countable.

To show that $\mathscr{R} = \bigcup_{i=r}^{\eta}\mathscr{Q}_i$, we assume the contrary and suppose that $B_0 \in \left(\mathscr{R} - \bigcup_{i=1}^{\eta}\mathscr{Q}_i\right)$. Then for $j = 1,2,\ldots,\eta$ we have $B_0 \in \left(\mathscr{R} - \bigcup_{i=1}^{j}\mathscr{Q}_i\right)$ and there exists $C\in\mathscr{Q}_j$ with $B_0\cdot C \neq \emptyset$, $\delta(B_0) \leqslant 2\delta(C)$, whence $\mathscr{Q}_j\cdot\mathscr{C}(B_0) \neq \emptyset$ and $v(\mathscr{Q}_j\cdot\mathscr{C}(B_0)) \geqslant 1$, in accordance with Defs. 4.5 and 4.6. Hence

$$v(\mathscr{R}\cdot\mathscr{C}(B_0)) \geqslant \sum_{j=1}^{\eta} v(\mathscr{Q}_j\cdot\mathscr{C}(B_0)) \geqslant \eta = (9\lambda)^{3n},$$

contradicting Th. 4.7. Therefore $\mathscr{R} = \bigcup_{i=1}^{\eta}\mathscr{Q}_i$.

4.9. For the remainder of this chapter, we let ψ denote an arbitrary Radon measure defined and finite on the bounded Borel subsets of R^n. In analogy with the notational conventions of (I.1), we let $\psi^\star$ denote the completion of ψ, $\bar{\psi}$ the outer measure associated with ψ, and $\mathcal{N}_\psi^\star$ the family of the $\psi^\star$-nullsets.

4.9.1. Lemma. *Let λ be a positive integer and let Ω be a covering relation with domain $X \subset R$ and range $\mathscr{V}$, subject to the following conditions:*

(i) $\bar{\psi}(X) < \infty$;

(ii) *$\mathscr{V}$ is a family of closed sets in R and* $\sup_{B \in \mathscr{V}} \delta(B) < \infty$;

(iii) $\delta(B) < \lambda h(z, B)$ *for each* (z, B) *in* Ω;

(iv) *for each* $\varepsilon > 0$ *and each* $x \in X$ *there exists a set* $(x, B) \in \Omega$ *with* $\delta(B) < \varepsilon$.

Then, given any finite disjoint collection of sets $B_1, B_2, \ldots, B_k$ in $\mathscr{V}$, there exists a countable disjoint subfamily of $\mathscr{V}$ that includes the given sets and covers X (mod $\mathcal{N}_\psi^\star$).

Proof. We let $F_1 = \bigcup_{i=1}^{k} B_i$, $X_1 = X - X \cdot F_1$. If $\bar{\psi}(X_1) = 0$, then the family consisting of the sets $B_1, B_2, \ldots, B_k$ fulfills our requirements. If not, we let Ω_1 denote the subrelation of Ω with range $\mathscr{V}_1$ such that $(z, B) \in \Omega_1$ iff $z \in X_1$ and $B \cdot F_1 = \emptyset$. Since F_1 is closed, it follows from (iv) above that the domain of Ω is X_1 and $\mathscr{V}_1$ covers X_1.

According to Cor. 3.5, there exists a diametrically restrained subrelation Ω_1' of Ω_1 whose range $\mathscr{V}_1'$ covers X_1. By Lemma 4.8, there is a finite sequence of countable disjoint families $\mathscr{Q}_i'$, $i = 1, 2, \ldots, \eta = (9\lambda)^{3n}$, such that $\mathscr{V}_1' = \bigcup_{i=1}^{\eta} \mathscr{Q}_i'$. We let $Q_i' = \bigcup \mathscr{Q}_i'$, $i = 1, 2, \ldots, \eta$. Then $X_1 = \bigcup_{i=1}^{\eta} Q_i' \cdot X_1$ and so $\bar{\psi}(X_1) \leqslant \sum_{i=1}^{\eta} \bar{\psi}(Q_i' \cdot X_1)$. Consequently, $\bar{\psi}(X_1 \cdot Q_{i_0}) \geqslant (1/\eta)\bar{\psi}(X_1)$ for some i_0, $1 \leqslant i_0 \leqslant \eta$, and thus there is a finite subfamily $\mathscr{Q}_1$ of $\mathscr{Q}_{i_0}$ whose union T_1 satisfies

$$\bar{\psi}(T_1 \cdot X_1) \geqslant (1/2\eta)\bar{\psi}(X_1).$$

We keep in mind that $\mathscr{Q}_1$ is disjoint and $T_1 \cdot F_1 = \emptyset$.

We now let $X_2 = X_1 - X_1 \cdot T_1$ and, since T_1 is closed, if $\bar{\psi}(X_2) > 0$, we may repeat the process just carried out to find a finite disjoint subfamily $\mathscr{Q}_2$ of $\mathscr{V}$ such that $T_2 = \bigcup \mathscr{Q}_2$, T_2 is closed, $T_1 \cdot T_2 = \emptyset$, and $\bar{\psi}(X_2 \cdot T_2) \geqslant (1/2\eta)\bar{\psi}(X_2)$. Proceeding thus inductively we obtain a finite or infinite sequence $\mathscr{Q}_1, \mathscr{Q}_2, \ldots, \mathscr{Q}_p, \ldots$, of finite disjoint subfamilies of $\mathscr{V}$ such that for each positive integer m not exceeding the number of terms in the sequence, $T_m = \bigcup \mathscr{Q}_m$, $T_m \cdot T_k = \emptyset$ if $k \neq m$, and

$$\bar{\psi}\left(\left(X_1 - X_1 \cdot \left(\bigcup_{i=1}^{m-1} T_i\right)\right) \cdot T_m\right) \geqslant (1/2\eta)\bar{\psi}\left(X_1 - X_i\left(\bigcup_{i=1}^{m-1} T_i\right)\right).$$

The process stops if, for some value of m, $\bar{\psi}\left(X_1 - X_1 \cdot \left(\bigcup_{i=1}^{m-1} T_i\right)\right) = 0.$ Otherwise, it continues indefinitely; but in that case, we use the procedure employed in the proof of Th. IV. 3.1, to obtain

$$\bar{\psi}\left(X_1 - X_1\left(\bigcup_{i=1}^{\infty} T_i\right)\right) = 0.$$

We let $\mathscr{Q} = \bigcup_i \mathscr{Q}_i$, where the index runs through either a finite set of values or through all positive integers, according to the situation just discussed. In any case, $\mathscr{Q}$ is a countable collection of disjoint members of $\mathscr{V}$, and $\mathscr{Q}$ covers X_1 (mod $\mathscr{N}_{\psi}^{\star}$). We need only adjoin $B_1, B_2, \ldots, B_k$ to $\mathscr{Q}$ to obtain a countable disjoint subfamily of $\mathscr{V}$ that covers X (mod $\mathscr{N}_{\psi}^{\star}$) and includes the given sets.

4.9.2. Theorem. *If $\mathfrak{B}$ is a star blanket, then $\mathfrak{B}$ has the sharp strong Vitali property (IV. 1.3) with respect to ψ.*

Proof. We take an arbitrary subset X of E, the domain of $\mathfrak{B}$, with $\bar{\psi}(X) < \infty$, and an arbitrary $\mathfrak{B}$-fine covering $\mathscr{V}$ of X. We let Ω denote the covering relation consisting of all ordered pairs (x, B) such that $x \in X$, $B \in \mathscr{V}$, and $\delta(V) < 1$. The domain of Ω is X and its range $\mathscr{V}' \subset \mathscr{V}$ covers X. According to Def. 4.4, we have, for each $x \in X$,

$$\operatorname*{Lim\,sup}_{(x,B)\in\Omega,\ \delta(B)\to 0+} \frac{\delta(B)}{h(x,B)} < \infty .$$

For any positive integer λ, we let S_λ denote the open ball of center 0, radius λ, and we define

$$X_\lambda = X \cdot S_\lambda \cdot E\left[x\colon \operatorname*{Lim\,sup}_{(x,B)\in\Omega,\ \delta(B)\to 0+} \frac{\delta(B)}{h(x,B)} < \lambda\right].$$

Clearly (X_λ) is an expanding sequence of sets with $X = \lim_\lambda X_\lambda$. For each λ, we let Ω_λ denote that subfamily of Ω such that $(x,B) \in \Omega_\lambda$ iff $x \in X_\lambda$, $B \subset S_\lambda$, and $\delta(B) < \lambda h(x,B)$. It follows that the range $\mathscr{V}_\lambda \subset \mathscr{V}'$ of Ω_λ is a $\mathfrak{B}$-fine covering of X_λ. Thus Ω_λ satisfies the hypothese of Lemma 4.9.1, so that its conclusions are valid.

Hence $\mathscr{V}'$, and also $\mathscr{V}$, satisfies the hypotheses of Lemma IV. 2.3, so that $\mathscr{V}$ has the sharp ψ-(R.S.V.) property with respect to X. Thus $\mathfrak{B}$ has the sharp ψ-(R.S.V.) property, also the sharp ψ-(S.V.) property, thanks to Haupt's adaptation property (II. 1.6).

Bibliography

Part I

1. Alfsen, E. M.: Some coverings of Vitali type. Math. Annalen **159**, 203–216 (1965).
2. Bertolini, F.: Le funzioni misurabili di punto (d'ultrafiltro) e la derivazione delle funzioni d'insieme (di soma) nella teoria algebraica della misura. Ann. Sc. Norm. Sup. Pisa III, **XII**, 163–207 (1958).
3. Besicovitch, A. S.: A general form of the covering principle and relative differentiation of additive functions. Proc. Camb. Philos. Soc. **41**, 103–110 (1945).
4. Bonnesen, T., u. W. Fenchel: Theorie der konvexen Körper. Ergeb. der Math. **13** (1934).
5. Burkill, J. C.: On the differentiability of multiple integrals. Jour. London. Math. Soc. **26**, 244–249 (1951).
6. Busemann, H., u. W. Feller: Zur Differentiation der Lebesgueschen Integrale. Fund. Math. **XXII**, 226–256 (1934).
7. Carathéodory, C.: Vorlesungen über Reelle Funktionen. Leipzig: 1927.
8. Denjoy, A.: Une extension du théorème de Vitali. Amer. Jour. Math. **73**, 314–356 (1951).
9. Dieudonné, J.: Sur un théorème de Jessen. Fund. Math. **XXVII**, 242–248 (1950).
10. Hahn, H., and A. Rosenthal: Set Functions. University of New Mexico (1948).
11. Halmos, P.: Measure Theory. New York: Van Nostrand 1950.
12. Hardy, G. H., J. E. Littlewood, and G. Polya: Inequalities. Cambridge: 1934.
13. Hartnett, W. E., and A. H. Kruse: Differentiation of set functions using Vitali coverings. Trans. Amer. Math. Soc. **96**, 185–209 (1960).
14. Haupt, O.: Zum Beweise des Lebesgueschen Ableitungssatzes. Sitzungsberichte der Bayer. Akademie der Wissenschaften Mathematisch-naturwissenschaftlichen Klasse, No. **14**, 171–174 (1949).
15. –, G. Aumann u. C. Y. Pauc: Differential- und Integralrechnung. 2nd ed., vol. III. Berlin: W. de Gruyter 1955.
16. –, u. C. Y. Pauc: Über die Ableitung absolut additiver Mengenfunktionen. Archiv der Math. **1**, 23–28 (1948).
17. – – Vitalische Systeme in Booleschen σ-Verbänden. Sitzungsberichte der Bayer. Akademie der Wissenschaften. Mathematisch-naturwissenschaftlichen Klasse **14**, 187–207 (1950).
18. – – Propriétés de mesurabilité de bases de dérivation. Port. Math. **13**, 37–54 (1953).
19. Hayes, C.: Differentiation with respect to Φ-pseudo-strong blankets and related problems. Proc. Amer. Math. Soc. **3**, 283–296 (1952).
20. – Differentiation of some classes of set functions. Proc. Camb. Philos. Soc. **48**, 374–382 (1952).
21. – A sufficient condition for the differentiation of certain classes of set functions. Proc. Camb. Philos. Soc. **54**, 346–353 (1958).
22. – A condition of halo type for the differentiation of classes of integrals. Canad. Jour. Math. **18**, 1015–1023 (1966).
23. –, and A. P. Morse: Some properties of annular blankets. Proc. Amer. Math. Soc. **1**, 107–126 (1950).
24. – – Convexical blankets. Proc. Amer. Math. Soc. **1**, 719–730 (1950).
25. –, and C. Y. Pauc: Full individual and class differentiation theorems in their relations to halo and Vitali properties. Canad. Jour. Math. **7**, 221–274 (1955).
26. Hewitt, E., and K. Stromberg: Real and abstract analysis. Berlin-Heidelberg-New York: Springer 1965.
27. Jessen, B., J. Marcinkiewicz, and A. Zygmund: Note on the differentiability of multiple integrals. Fund. Math. **25**, 217–234 (1935).

28. Kenyon, H., and A. P. Morse: Runs. Pac. Jour. Math. **8**, 811 – 824 (1958).
29. Morse, A. P.: A theory of covering and differentiation. Trans. Amer. Math. Soc. **55**, 205 – 235 (1944).
30. – Perfect blankets. Trans. Amer. Math. Soc. **61**, 418 – 442 (1947).
31. –, and J. F. Randolph: The Φ-rectifiable subsets of the plane. Trans. Amer. Math. Soc. **55**, 236 – 305 (1944).
32. Nöbeling, G.: Analytische Topologie. Berlin-Göttingen-Heidelberg: Springer 1954.
33. Papoulis, A.: On the strong differentiability of the indefinite integral. Trans. Amer. Math. Soc. **69**, 130 – 141 (1950).
34. Pauc, C. Y.: Compléments à la théorie de la dérivation de fonctions d'ensemble suivant de Possel et A. P. Morse. Paris: C. R. Acad. Sci. **231**, 1406 – 1408 (1950).
35. – La dérivation dans les reseaux incomplets et les fonctions de Haar. Paris: C. R. Acad. Sci. **232**, 1387 – 1389 (1951).
36. – Les théorèmes fort et faible de Vitali et les conditions d'évanescence de halos. Paris: C. R. Acad. Sci. **232**, 1727 – 1729 (1951).
37. – Contributions à une théorie de la differentiation de fonctions d'intervalles sans hypothèse de Vitali. Paris: C. R. Acad. Sci. **236**, 1937 – 1939 (1953).
38. – Ableitungsbasen, Prätopologie und starker Vitalischer Satz. J. reine und angew. Math. **191**, 69–91 (1953).
39. De Possel, R.: Dérivation abstraite des fonctions d'ensemble. Jour. de Math. pures et appl. **15**, 391 – 409 (1936).
40. – Sur la généralisation de la notion de système dérivant. Paris: C. R. Acad. Sci. **224**, 1137 – 1139 (1947).
41. – Sur les systèmes dérivants et l'extension du théorème de Lebesgue relatif à la dérivation d'une fonction de variation bornée. Paris: C. R. Acad. Sci. **224**, 1197 — 1198 (1947).
42. Saks, S.: Remark on the differentiability of the Lebesgue indefinite integral. Fund. Math. **22**, 257 – 261 (1934).
43. – On the strong derivatives of functions of an interval. Fund. Math. **25**, 235 – 252 (1935).
44. – Theory of the integral. Warsaw 1937.
45. Trjitzinsky, W. J.: Théorie métrique dans les espaces où il y a une mésure. Mémorial des sciences mathématiques CXLIII (1960).
46. Wecken, F.: Abstrakte Integrale und fastperiodische Funktionen. Math. Zeit. **45**, 377 – 404 (1939).
47. Young, L. C.: On area and length. Fund. Math. **35**, 275 – 302 (1948).
48. Younovitch, B.: Sur les systèmes dérivants et l'extension du théorème de Lebesgue relatif à la dérivation d'une fonction de variation bornée. Comptes rendus (Doklady) de l'Acad. des Sci., de l'U.R.S.S. **30**, 112 – 114 (1940).
49. Zygmund, A.: On the differentiability of multiple integrals. Fund. Math. **23**, 143 – 149 (1934).
50. – A note on the differentiability of integrals. Colloquium Mathematicum **XVI**, 199 – 204 (1967).

Bruckner, A. M.: Differentiation of Integrals. To appear soon.

PART II

Martingales and Cell Functions

Chapter I

Theory without an Intervening Measure

We begin by assembling those concepts concerning additive and σ-additive set functions that will be used in the sequel.

1. Additive functions. Let $\mathscr{B}$ be an abstract Boolean σ-algebra with zero element $\bigcirc$ and unit element E. We regard $\mathscr{B}$ as fixed. Following Carathéodory, the elements of $\mathscr{B}$ are sometimes called *somas*. The symbols $\leqslant$, $\vee$, and $\wedge$ will be used to denote order (partial order in the old terminology), supremum, and infimum, respectively, in $\mathscr{B}$, whereas $\subset$, $\cup$, and $\cap$ will be used for the corresponding set relations and operations. We shall sometimes use the shorter notations $\vee \mathscr{K}$ and $\wedge \mathscr{K}$ for $\vee \{K: K \in \mathscr{K}\}$ and $\wedge \{K: K \in \mathscr{K}\}$, respectively. We shall write $A \wedge B$ or $A \cdot B$ for the infimum of A and B in $\mathscr{B}$, $A \vee B$ for the supremum of A and B in $\mathscr{B}$. If $A \in \mathscr{B}$ and $\mathscr{H} \subset \mathscr{B}$, then $A \wedge \mathscr{H}$ denotes the set of elements $\{A \wedge H: H \in \mathscr{H}\}$. This set is called the *trace of $\mathscr{H}$ on A*. We denote the operations greatest lower bound, least upper bound, minimum, and maximum on the extended real line by g.l.b., l.u.b., min., and max. respectively.

If $\mathscr{A}$ is a Boolean subalgebra of $\mathscr{B}$ with unit E, we let $\mathscr{M}_{\mathscr{A}}$ denote the set of applications (functions) φ of $\mathscr{A}$ in the extended real line, finitely additive and satisfying $\varphi(\bigcirc) = 0$. If φ and ψ belong to $\mathscr{M}_{\mathscr{A}}$, then the function $\varphi + \psi$, defined as usual by $(\varphi + \psi)(A) = \varphi(A) + \psi(A)$, provided the right hand side makes sense (thus excluding $\infty - \infty$), is again a member of $\mathscr{M}_{\mathscr{A}}$. Similarly $\lambda\varphi \in \mathscr{M}_{\mathscr{A}}$ if $\varphi \in \mathscr{M}_{\mathscr{A}}$ and λ is a real number. The function $\varphi \in \mathscr{M}_{\mathscr{A}}$ is said to be of *bounded variation (bounded variation from above (below))* if the set of values $\varphi(A)$, $A \in \mathscr{A}$, is bounded (bounded from above (below)). φ is said to be of *semi-bounded variation* if it is of bounded variation from above or from below.

We agree that $\varphi \leqslant \psi$ iff $\varphi(A) \leqslant \psi(A)$ for each $A \in \mathscr{A}$, and we denote the infimum and the supremum in $\mathscr{M}_{\mathscr{A}}$ of a subset $\mathscr{R}$ of $\mathscr{M}_{\mathscr{A}}$ by $\inf \mathscr{R}$ and $\sup \mathscr{R}$, respectively, whenever these functions exist.

1.1. Proposition [2]. The existence of $\sup \mathscr{R}$ ($\inf \mathscr{R}$) in $\mathscr{M}_{\mathscr{A}}$ is ensured by the following conditions:
S^s(S^i) $\mathscr{R}$ contains a function of bounded variation from below (from above).

1.2. Proposition [*2*]. If $\mathscr{R}$ satisfies $\mathbf{S}^s$, then for each $A \in \mathscr{A}$, the relation

$$(\sup \mathscr{R})(A) = \text{l.u.b.}\,(\varphi_1(A_1) + \cdots + \varphi_k(A_k)),$$

where $A = \bigvee_{i=1}^{k} A_i$ denotes any finite disjoint decomposition of the element A into elements A_i, $i = 1, 2, \ldots, k$, of $\mathscr{A}$, and $\varphi_1, \varphi_2, \ldots, \varphi_k$ are any functions in $\mathscr{R}$ such that $\varphi_1(A_1) + \varphi_2(A_2) + \cdots + \varphi_k(A_k)$ is meaningful.

We can say that the above equation is the *individual* definition of $\sup \mathscr{R}$. A corresponding equation holds for $(\inf \mathscr{R})(A)$.

Thus, for each $\varphi \in \mathscr{M}_{\mathscr{A}}$, the functions $\varphi^+ = \sup(\varphi, 0)$, $\varphi^- = \sup(-\varphi, 0)$, and $\varphi^T = \sup(-\varphi, \varphi) = \varphi^+ + \varphi^-$ exist; they are called the *positive*, *negative*, and *total* variations of φ, respectively. Thus φ is of bounded variation iff $\varphi^T(E)$ is finite or, equivalently, $\varphi^+(E)$ and $\varphi^-(E)$ are both finite. We also call the value of $\varphi^T(E)$ the *norm of* φ and denote it by $\|\varphi\|$. We note that $\varphi^T(E) = 0$ iff φ is identically zero.

From Props. 1.1, and 1.2, we obtain

1.3. Proposition. The space $\mathscr{M}_{\mathscr{A}}^T$ of functions of bounded variation in $\mathscr{M}_{\mathscr{A}}$ is a Riesz space, conditionally complete as a lattice ("complétement réticulé" according to Bourbaki.) If φ is of semi-bounded variation, then $\varphi = \varphi^+ - \varphi^-$.

We note that, in general, $\varphi^+(A) \neq (\varphi(A))^+$, the latter being $\max(\varphi(A), 0)$; and, more generally, that $\sup \mathscr{R}(A) \neq \sup\{\varphi(A); \varphi \in \mathscr{R}\}$. However, as a consequence of Prop. 1.2, we have

1.4. Proposition. If $\mathscr{R}$ is directed (filtering to the right) with respect to the natural order $\leqslant$ and satisfies $\mathbf{S}^s$, then for all $A \in \mathscr{A}$,

$$(\sup \mathscr{R})(A) = \sup\{\varphi(A) \colon \varphi \in \mathscr{R}\}$$

or, in Bourbaki's notation,

$$\sup \mathscr{R} = \operatorname{env} \sup\{\varphi \colon \varphi \in \mathscr{R}\}.$$

If, in the formulation of Prop. 1.4, we replace "right" by "left", "$\mathbf{S}^s$" by "$\mathbf{S}^i$", and "sup" by "inf", we obtain the dual of that proposition.

1.5. Corollaries. 1. The inequality $\operatorname{env}\sup \mathscr{R} \leqslant \sup \mathscr{R}$ is always true. The equality $\sup\{\varphi(A) \colon \varphi \in \mathscr{R}\} = (\sup \mathscr{R})(A)$ holds if A is an atom of $\mathscr{A}$.

2. Let $\mathscr{C}$ be a Boolean subalgebra of $\mathscr{A}$ with unit E. Let $\varphi|\mathscr{C}$ denote the restriction to $\mathscr{C}$ of a function φ of $\mathscr{M}_{\mathscr{A}}$; thus $\varphi|\mathscr{C} \in \mathscr{M}_{\mathscr{C}}$. Then, if $\mathscr{R} \subset \mathscr{M}_{\mathscr{A}}$, we must have $\sup\{\varphi|\mathscr{C} \colon \varphi \in \mathscr{R}\} \leqslant (\sup \mathscr{R})|\mathscr{C}$, the symbol sup on the left representing the $\mathscr{M}_{\mathscr{C}}$-supremum, and on the right the $\mathscr{M}_{\mathscr{A}}$-supremum. Equality holds if $\mathscr{R}$ is directed (filtering) to the right.

2. σ-additive functions. By a $\mathscr{K}$*-partition* of an element $B \in \mathscr{B}$, where $\mathscr{K} \subset \mathscr{B}$, we shall mean an enumerable disjoint set $\mathscr{P}$ of elements of $\mathscr{K}$

that are non-zero and with $\vee \mathscr{P} = B$. If φ is an extended real-valued function defined on $\mathscr{P}$, we write $\varphi(\mathscr{P}) = \sum_{K \in \mathscr{P}} \varphi(K)$, provided that at least one of the sums

$$\sum_{K \in \mathscr{P}} (\varphi(K))^+ \quad \text{or} \quad \sum_{K \in \mathscr{P}} (\varphi(K))^- \quad \text{is finite}.$$

A function $\varphi \in \mathscr{M}_{\mathscr{A}}$ is called *σ-additive* iff for all $A \in \mathscr{A}$ and all $\mathscr{A}$-partitions $\mathscr{P}$ of A, the sum $\varphi(\mathscr{P})$ exists and equals $\varphi(A)$. We denote by $\mathscr{M}^{\sigma}_{\mathscr{A}}$ the set of those functions in $\mathscr{M}_{\mathscr{A}}$ that are σ-additive.

2.1. Proposition. (The nullsequence test.) A function φ in $\mathscr{M}_{\mathscr{A}}$ of bounded variation is σ-additive iff φ possesses the following property: if (S_n) is any decreasing nullsequence, then $\lim_n \varphi(S_n) = 0$.

2.2. Corollary. Any function $\varphi \in \mathscr{M}_{\mathscr{A}}$ whose absolute value is majorized (dominated) by a σ-additive function of bounded variation is itself of bounded variation and σ-additive.

2.3. Proposition. [*37*, 320; *4*, 158]. If $\mathscr{R}$ is a subset of $\mathscr{M}^{\sigma}_{\mathscr{A}}$, then condition $\mathbf{S}^s$ (Prop. 1.1) implies the existence of $\sup \mathscr{R}$ in $\mathscr{M}^{\sigma}_{\mathscr{A}}$ and its coincidence with $\sup \mathscr{R}$ in $\mathscr{M}_{\mathscr{A}}$.

2.4. Corollary. The space $\mathscr{M}^{T,\sigma}_{\mathscr{A}}$ consisting of those functions in $\mathscr{M}_{\mathscr{A}}$ that are of bounded variation and σ-additive is a conditionally complete Riesz space (Prop. 1.3), a subspace of $\mathscr{M}^{T}_{\mathscr{A}}$.

Remark. Condition $\mathbf{S}^s$ may hold and $\inf \mathscr{R}$ may exist in both $\mathscr{M}_{\mathscr{A}}$ and in $\mathscr{M}^{\sigma}_{\mathscr{A}}$, without being equal.

2.5. Example. $\mathscr{R}$ can be minorized by a function of $\mathscr{M}_{\mathscr{A}}$ of bounded variation below and $\inf_{\mathscr{M}_{\mathscr{A}}} \neq \inf_{\mathscr{M}^{\sigma}_{\mathscr{A}}}$: Let E be the set of positive integers together with zero, $\mathscr{B}$ the Boolean σ-algebra of all subsets of E, φ_n the measure defined on $\mathscr{B}$ by $\varphi_n(A)$ = number of elements of A that are greater than n if n is positive, and $\varphi_0(A) = 0$. We let $\mathscr{R} = (\varphi_n)$, $n = 1, 2, \ldots$. Evidently $\mathscr{R}$ satisfies the condition $\mathbf{S}^s$, and $\inf \mathscr{M}^{\sigma}_{\mathscr{A}}$ vanishes identically, whereas $\inf \mathscr{R}$ in $\mathscr{M}_{\mathscr{A}}$ vanishes on the finite sets and takes the value $+\infty$ on the infinite ones.

According to the theorem of HAHN [*21*, 17], a σ-additive function defined on a Boolean σ-algebra is of semi-bounded variation and hence admits a Jordan decomposition (Prop. 1.3). Explicitly, φ is of bounded variation from above or from below, depending on whether $\varphi(E) < \infty$ or $\varphi(E) > -\infty$.

Let $\mathscr{C}$ denote the Boolean σ-extension of $\mathscr{A}$; that is, the smallest Boolean σ-algebra containing $\mathscr{A}$. $\mathscr{C}$ is also the Borelian extension of $\mathscr{A}$, meaning the $\delta\sigma$-extension of $\mathscr{A}$. According to a classical theorem [*21*, 80], a positive σ-additive function defined on $\mathscr{A}$ can be extended to a positive σ-additive function defined on $\mathscr{C}$, the extension being unique if the original function is σ-finite. Combining these facts, we infer the following result.

2.6. Proposition. In order that an additive function φ defined on $\mathscr{A}$ may be extended to a σ-additive one defined on the Borelian extension of $\mathscr{A}$, it is necessary and sufficient that φ be σ-additive and of semi-bounded variation. The extension is unique if φ^T is σ-finite.

A function φ in $\mathscr{M}_{\mathscr{A}}$ is said to be *σ-additive above* iff φ^+ is σ-additive; analogously, φ is *σ-additive below* iff φ^- is σ-additive. In accordance with HAHN's decomposition theorem, a σ-additive function is σ-additive above and below. The reverse implication holds thanks to the Jordan decomposition for functions of semi-bounded variation.

A positive (more accurately, non-negative) function $\varphi \in \mathscr{M}_{\mathscr{A}}$ is called *purely simply additive* iff every σ-additive function $\psi \in \mathscr{M}_{\mathscr{A}}$ satisfying $0 \leqslant \psi \leqslant \varphi$ vanishes identically. A function $\varphi \in \mathscr{M}_{\mathscr{A}}$ of variable sign is called *purely simply additive (above, below)* iff $\varphi^T(\varphi^+, \varphi^-)$ is purely simply additive.

2.7. Proposition.[60]. Any function $\varphi \in \mathscr{M}_{\mathscr{A}}$ of bounded variation can be represented as the sum of a σ-additive function φ_c and a purely simply additive function φ_p; the decomposition is unique. If $0 \leqslant \varphi$, then φ_c is the greatest of all σ-additive functions with $0 \leqslant \psi \leqslant \varphi$. We have $(\varphi^+)_c = (\varphi_c)^+$, $(\varphi^+)_p = (\varphi_p)^+$. Analogous results hold if '+' is replaced by '−'. These show the legitimacy of adopting the notation φ_c^+, φ_p^+, etc.

Prop. 2.7 follows readily from Prop. 2.3. In case $0 \leqslant \varphi$, it suffices to define φ_c as the supremum in $\mathscr{M}_{\mathscr{A}}$ of those σ-additive positive functions ψ that are majorized (dominated) by φ. The proof when φ is of variable sign and of bounded variation may be reduced to that of the case $0 \leqslant \varphi$ thanks to the Jordan decomposition (Prop. 1.3).

For a function φ in $\mathscr{M}_{\mathscr{A}}$ of bounded variation, we define the *deficiency of σ-additivity* as $\|\varphi_p\|$. Thus, the σ-additivity of φ is equivalent to the vanishing of the deficiency of σ-additivity.

3. Premartingales, semi-martingales, and martingales. Let Θ denote a non-empty set of "parameters", directed by a transitive relation $\preccurlyeq$, to be regarded as fixed permanently. In other words, Θ is a *set filtering to the right*, or an *increasing filtering set.* If $\rho, \tau \in \Theta$, then $\rho \preccurlyeq \tau$ is read "ρ precedes τ" or, equivalently "τ follows ρ". A subset Δ of Θ is called *terminal* iff there exists in Θ an index ρ such that $\rho \preccurlyeq \tau$ implies $\tau \in \Delta$, *cofinal* iff $\Theta - \Delta$ is not terminal.

The *filtering families (Moore-Smith sequences)* occurring in the sequel have as their set of parameters either Θ or a terminal subset of Θ. We shall write (a_τ) instead of the longer $(a_\tau)_{\tau \in \Theta \preccurlyeq}$; similarly, we shall write $\lim a_\tau$

instead of $\lim_{\tau\in\Theta;\,\ll} a_\tau$. If Θ is the set of natural numbers and $\ll$ is their natural order, then the Moore-Smith sequences are ordinary sequences.

By *stochastic basis* we mean a filtering family (to the right) of Boolean σ-subalgebras $\mathscr{B}_\tau$ of $\mathscr{B}$, with E as their unit.

By *premartingale* of basis $(\mathscr{B}_\tau)$ we mean a filtering family $\boldsymbol{\Phi} = (\varphi_\tau)$, $\tau \in \Theta$, where φ_τ denotes, for each $\tau \in \Theta$, a σ-additive function defined on $\mathscr{B}_\tau$. In what follows, $(\mathscr{B}_\tau)$ must be considered as permanently fixed, so we shall omit the qualifying phrase "of basis $(\mathscr{B}_\tau)$" henceforth.

In Part II, Chapters I, II, and III, we shall consider only *increasing bases*; that is, $\mathscr{B}_\rho \subset \mathscr{B}_\tau$ whenever $\rho \ll \tau$. $\mathscr{A}$ will denote the Boolean subalgebra of $\mathscr{B}$ consisting of the union of the families $(\mathscr{B}_\tau)$, $\tau \in \Theta$. Each element A of $\mathscr{A}$ belongs to $\mathscr{B}_\tau$ for some terminal set of parameters, denoted by Δ_A. The *trace* of the premartingale $\boldsymbol{\Phi} = (\varphi_\tau)$ on $A \in \mathscr{A}$ is defined as the sequence of the restrictions of φ_τ to $A \wedge \mathscr{B}_\tau$ for $\tau \in \Delta_A$.

A premartingale is called a *submartingale* [*15*] (upper *semi-martingale* in [*13*]) iff $A \in \mathscr{B}_\rho$ and $\rho \ll \tau$ together imply $\varphi_\rho(A) \leqslant \varphi_\tau(A)$ or, in other words, iff $\rho \ll \tau$ implies $\varphi_\rho \ll \varphi_\tau | \mathscr{B}_\rho$, where $\varphi_\tau | \mathscr{B}_\rho$ denotes the restriction of φ_τ to $\mathscr{B}_\rho$. The reverse inequality defines a *supermartingale* [*15*] (lower semi-martingale in [*13*]). The conjunction of these inequalities, namely $\varphi_\rho = \varphi_\tau | \mathscr{B}_\rho$ for $\rho \ll \tau$ defines a *martingale.* For a martingale, $\varphi_\tau(E)$ is independent of τ. The trace of a submartingale on an element A of $\mathscr{A}$ is a submartingale.

The sum of two premartingales (φ_τ) and (ψ_τ) is defined as $(\varphi_\tau + \psi_\tau)$ provided $\varphi_\tau + \psi_\tau$ is meaningful for each $\tau \in \Theta$. Similarly, for a real finite number λ, $\lambda(\varphi_\tau) = (\lambda\varphi_\tau)$. The sum of two martingales or two submartingales is a martingale or a submartingale, respectively, provided it exists. If $\boldsymbol{\Phi}$ is a submartingale and $0 \leqslant \lambda$, then $\lambda\boldsymbol{\Phi}$ is also a submartingale. By definition, $(\varphi_\tau) \leqslant (\psi_\tau)$ means that $\varphi_\tau \leqslant \psi_\tau$ for each $\tau \in \Theta$, thus $\varphi_\tau(A) \leqslant \psi_\tau(A)$ for each τ in Θ and each $A \in \mathscr{A}$. In particular, if we denote by $\boldsymbol{\Omega}$ the null martingale whose functions vanish identically, then $\boldsymbol{\Omega} \leqslant \boldsymbol{\Phi}$ iff $0 \leqslant \varphi_\tau$ for each $\tau \in \Theta$.

A premartingale $\boldsymbol{\Phi} = (\varphi_\tau)$ is said to be *terminally uniformally bounded (from above, from below)* more briefly, of *bounded variation (from above, from below)* iff Θ admits a terminal subset Δ such that for $\tau \in \Delta$ and $A \in \mathscr{B}_\tau$, the set of numbers $\{\varphi_\tau(A)\}$ is bounded (from above, from below). If $\boldsymbol{\Phi}$ is a martingale, one can always take $\Delta = \Theta$; the same holds for a submartingale with respect to boundedness from above. When it exists, the sum of two premartingales of bounded variation is also of bounded variation; the sum $\varphi_\tau + \psi_\tau$ certainly exists for τ in a terminal subset. Similar statements hold for products by a scalar. The *norm* $\|\boldsymbol{\Phi}\|$ of a premartingale of bounded variation is defined as $\limsup_\tau \varphi_\tau^T(E)$.

4. Ordered space of martingales of basis ($\mathscr{B}_\tau$). Let $\mathfrak{C}$ be a set of premartingales. For each $\tau_0 \in \Theta$ we denote by $\mathscr{R}_{\tau_0}$ the set of those functions of parameter τ_0 in the premartingales of $\mathfrak{C}$. Consequently $\mathscr{R}_{\tau_0}$, the τ_0-*section of* $\mathfrak{C}$, is a set of σ-additive functions defined on $\mathscr{B}_{\tau_0}$. We state the following hypothesis:

$\mathbf{S}^s_{\mathfrak{C}}$: for each $\tau_0 \in \Theta$ there exists a premartingale $\boldsymbol{\Phi} = (\varphi_\tau)$ of $\mathfrak{C}$ such that $\varphi_{\tau_0}(E) > -\infty$; that is, each family $\mathscr{R}_{\tau_0}$ possesses the property $\mathbf{S}^s$.

In these circumstances, by virtue of Prop. 1.1, we can define

$$\xi_{\tau_0} = \sup \mathscr{R}_{\tau_0} = \sup \{\varphi_{\tau_0} : \boldsymbol{\Phi} \in \mathfrak{C}\}, \tag{4.0.1}$$

where the supremum refers to $\mathscr{M}_{\mathscr{B}_{\tau_0}}$. From Prop. 2.3, it follows that ξ_{τ_0} is σ-additive for each τ_0. Consequently, $\boldsymbol{\Xi} = (\xi_\tau)$ is the smallest of all premartingales $\boldsymbol{\Psi}$ such that $\boldsymbol{\Phi} \leqslant \boldsymbol{\Psi}$ for each $\boldsymbol{\Phi}$ in $\mathfrak{C}$.

4.1. Proposition [*37*, 320]. If $\mathfrak{C}$ is a set of submartingales satisfying $\mathbf{S}^s_{\mathfrak{C}}$, then the premartingale $\boldsymbol{\Xi} = (\xi_\tau)$ defined by (4.0.1) is a submartingale, the least submartingale greater than all $\boldsymbol{\Phi}$ in $\mathfrak{C}$.

Proof. Let τ_0 and ζ_0 in Θ be fixed, with $\tau_0 \ll \zeta_0$. Since $\varphi_{\zeta_0} \leqslant \xi_{\zeta_0}$ whenever $\boldsymbol{\Phi} \in \mathfrak{C}$, then $\varphi_{\zeta_0} | \mathscr{B}_{\tau_0} \leqslant \xi_{\zeta_0} | \mathscr{B}_{\tau_0}$. Furthermore $\varphi_{\tau_0} \leqslant \varphi_{\zeta_0} | \mathscr{B}_{\tau_0}$ by the definition of a submartingale. Hence $\varphi_{\tau_0} \leqslant \xi_{\zeta_0} | \mathscr{B}_{\tau_0}$ whenever $\boldsymbol{\Phi} \in \mathfrak{C}$. Therefore $\xi_{\tau_0} \leqslant \xi_{\zeta_0} | \mathscr{B}_{\tau_0}$, and so $\boldsymbol{\Xi} = (\xi_\tau)$ is a submartingale. From the definition of ξ_τ it follows immediately that $\boldsymbol{\Xi}$ represents the least submartingale greater than all $\boldsymbol{\Phi}$ in $\mathfrak{C}$.

In particular, let $\boldsymbol{\Phi} = (\varphi_\tau)$ be a martingale and $\mathfrak{C} = \{\boldsymbol{\Phi}, \boldsymbol{\Omega}\}$. We see that (φ^+_τ) is the smallest submartingale majorizing $\boldsymbol{\Phi}$ and $\boldsymbol{\Omega}$, (φ^-_τ) is the smallest submartingale majorizing $-\boldsymbol{\Phi}$ and $\boldsymbol{\Omega}$; (φ^T_τ) is the smallest submartingale majorizing $\boldsymbol{\Phi}$ and $-\boldsymbol{\Phi}$.

If the set $\mathfrak{C}$ of premartingales is filtering with respect to the order $\leqslant$ then, for each $\tau_0 \in \Theta$, the τ_0-section $\mathscr{R}_{\tau_0}$ enjoys the same property and we can write

$$\xi_{\tau_0} = \operatorname{env} \sup \{\varphi_{\tau_0} : (\varphi_\tau) \in \mathfrak{C}\}\,.$$

Let $\boldsymbol{\Xi} = (\xi_\tau)$ be a submartingale such that

$$\xi_\rho(E) > -\infty \qquad \text{for some } \rho \text{ in } \Theta. \tag{4.1.1}$$

We define $\xi^m_\tau(A) = \sup_{\tau \ll \delta} \xi_\delta(A) = \lim_{\tau \ll \delta} \xi_\delta(A)$ for all $\tau \in \Theta$ and $A \in \mathscr{B}_\tau$; more briefly expressed,

$$\xi^m_\tau = \operatorname{env} \sup_{\tau \ll \delta} (\xi_\delta | \mathscr{B}_\tau)\,. \tag{4.1.2}$$

Props. 1.4 and 2.3 imply that ξ^m_τ is σ-additive; consequently, we have the following result.

4.2. Proposition. Assuming that the submartingale $\boldsymbol{\Xi} = (\xi_\tau)$ satisfies (4.1.1), then $\boldsymbol{\Xi}^m = (\xi^m_\tau)$ is a martingale, namely, the smallest of all martingales $\boldsymbol{\Psi}$ for which $\boldsymbol{\Xi}^m \leqslant \boldsymbol{\Psi}$; also $\boldsymbol{\Xi}^m$ is of bounded variation from below. $\boldsymbol{\Xi}^m$ is called the *integral of* $\boldsymbol{\Xi}$.

Remark. If we discard (4.1.1) but assume that the functions ξ_τ^m defined by (4.1.2) are σ-additive, then $\boldsymbol{\Xi} = (\xi_\tau^m)$ is a martingale, the smallest of all martingales $\boldsymbol{\Psi}$ for which $\boldsymbol{\Xi} \leqslant \boldsymbol{\Psi}$. If $\boldsymbol{\Xi}^m$ is of bounded variation from above, then $\boldsymbol{\Xi}$ has the same property.

Let $\mathfrak{M}$ be the ordered set of all martingales of basis $(\mathscr{B}_\tau)$. We denote the infimum and supremum in $\mathfrak{M}$ (if they exist) by $\inf_{\mathfrak{M}}$ and $\sup_{\mathfrak{M}}$, respectively. From Props. 4.1, and 4.2, we deduce readily the following result.

4.3. Proposition. Let $\mathfrak{C}$ be a set of martingales satisfying $\mathbf{S}_{\mathfrak{C}}^s$; that is, for each $\tau_0 \in \Theta$, $\varphi_{\tau_0}(E) > -\infty$ for at least one martingale $(\varphi_\tau) \in \mathfrak{C}$. Then $\sup_{\mathfrak{M}} \mathfrak{C}$ exists and is equal to (ξ_τ^m) defined by (4.0.1) and (4.1.2). If $\mathfrak{C}$ is filtering with respect to the order $\leqslant$, then for each $\tau_0 \in \Theta$,

$$\xi_\tau^m = \operatorname{env\,sup} \{\varphi_{\tau_0} : (\varphi_\tau) \in \mathfrak{C}\} .$$

4.4. Corollary. The space of all martingales of bounded variation is a conditionally complete Riesz space.

For each martingale $\boldsymbol{\Phi}$ there exist the martingales

$$\boldsymbol{\Phi}^+ = \sup_{\mathfrak{M}} (\boldsymbol{\Phi}, \boldsymbol{\Omega}), \qquad \boldsymbol{\Phi}^- = \sup_{\mathfrak{M}} (-\boldsymbol{\Phi}, \boldsymbol{\Omega}), \qquad \text{and} \qquad \boldsymbol{\Phi}^T = \sup_{\mathfrak{M}} (-\boldsymbol{\Phi}, \boldsymbol{\Phi}),$$

and we have

$$\begin{aligned} \boldsymbol{\Phi}^+ &= (\varphi_\tau^{+m}), \quad \text{where} \quad \varphi_\tau^{+m} = \lim_{\tau \ll \delta} (\varphi_\delta^+ \mid \mathscr{B}_\tau);\\ \boldsymbol{\Phi}^- &= (\varphi_\tau^{-m}), \quad \text{where} \quad \varphi_\tau^{-m} = \lim_{\tau \ll \delta} (\varphi_\delta^- \mid \mathscr{B}_\tau);\\ \boldsymbol{\Phi}^T &= (\varphi_\tau^{Tm}), \quad \text{where} \quad \varphi_\tau^{Tm} = \lim_{\tau \ll \delta} (\varphi_\delta^T \mid \mathscr{B}_\tau); \end{aligned} \tag{4.4.1}$$

and $\boldsymbol{\Phi}^T = \boldsymbol{\Phi}^+ + \boldsymbol{\Phi}^-$. The martingales $\boldsymbol{\Phi}^+$, $\boldsymbol{\Phi}^-$, and $\boldsymbol{\Phi}^T$ are called, respectively, the *positive, negative,* and *total variations of* $\boldsymbol{\Phi}$. We see that $\boldsymbol{\Phi}$ is of bounded variation (from above, from below) iff $\boldsymbol{\Phi}^T (\boldsymbol{\Phi}^+, \boldsymbol{\Phi}^-)$ is finite, i.e., if $\varphi_\tau^{Tm}(E)\,(\varphi_\tau^{+m}(E), \varphi_\tau^{-m}(E))$ is finite. The norm of a martingale $\boldsymbol{\Phi}$ of bounded variation can be written

$$\|\boldsymbol{\Phi}\| = \lim_\tau \varphi_\tau^T(E) = \varphi_\tau^{Tm}(E)$$

for any $\tau \in \Theta$; $\|\boldsymbol{\Phi}\| = 0$ is equivalent to $\boldsymbol{\Phi} = \boldsymbol{\Omega}$.

From the relations (4.4.1) the Jordan decomposition for martingales follows immediately.

4.5. Proposition. If a martingale $\boldsymbol{\Phi}$ is of bounded variation either from above or from below, then $\boldsymbol{\Phi} = \boldsymbol{\Phi}^+ - \boldsymbol{\Phi}^-$.

5. Integrals of premartingales. Let $\boldsymbol{\Phi} = (\varphi_\tau)$ be a premartingale for which the following condition holds:

$\mathbf{T}_1^s(\boldsymbol{\Phi})$: $\varphi_\rho(E) > -\infty$ for a terminal set of indices ρ.

Let $\tau \in \Theta$ be regarded as fixed. For every $\eta \gg \tau$ we set

$$\psi_\tau^\eta = \sup_{\eta \ll \delta} (\varphi_\delta | \mathscr{B}_\tau). \tag{5.0.1}$$

where the supremum refers to $\mathscr{M}_{\mathscr{B}_\tau}$. This supremum exists by virtue of Prop. 2.3, and coincides with the supremum with respect to $\mathscr{M}_{\mathscr{B}_\tau}^\sigma$.

We shall assume in addition the condition

$\mathbf{T}_s^2(\boldsymbol{\Phi})$: For each $\tau \in \Theta$ there corresponds $\eta \gg \tau$ such that $\psi_\tau^\eta(E) < +\infty$.

We note that $\tau \ll \eta_1 \ll \eta_2$ implies $\psi_\tau^{\eta_1}(E) \geqslant \psi_\tau^{\eta_2}(E)$.

Thanks to the hypothesis $\mathbf{T}_2^s(\boldsymbol{\Phi})$, the infimum relative to $\mathscr{M}_{\mathscr{B}_\tau}^\sigma$ of the functions ψ_τ^η for $\eta \gg \tau$ exists and is the infimum in $\mathscr{M}_{\mathscr{B}_\tau}^\sigma$. We denote it by φ_τ^s. Since the family $(\psi_\tau^\eta : \tau \ll \eta)$ is decreasing for fixed τ, it follows from Props. 1.4, and 2.3, that

$$\varphi_\tau^s = \operatorname*{env\,inf}_{\tau \ll \eta} \left(\sup_{\eta \ll \delta} (\varphi_\delta | \mathscr{B}_\tau) \right), \tag{5.0.2}$$

or, more briefly,

$$\varphi_\tau^s = \limsup_{\tau \ll \delta} (\varphi_\delta | \mathscr{B}_\tau). \tag{5.0.3}$$

We represent the premartingale (φ_τ^s) by $\boldsymbol{\Phi}^s$ and call it the *upper integral of* $\boldsymbol{\Phi}$. The lower integral $\boldsymbol{\Phi}^i$ is defined analogously under the conditions

$$\mathbf{T}_1^i(\boldsymbol{\Phi}) = \mathbf{T}_1^s(-\boldsymbol{\Phi}) \quad \text{and} \quad \mathbf{T}_2^i(\boldsymbol{\Phi}) = \mathbf{T}_2^s(-\boldsymbol{\Phi}).$$

5.1. Proposition. Under the conditions $\mathbf{T}_1^s(\boldsymbol{\Phi})$ and $\mathbf{T}_2^s(\boldsymbol{\Phi})$ (securing the existence of $\boldsymbol{\Phi}^s$), the upper integral $\boldsymbol{\Phi}^s$ is a submartingale and $\varphi_\tau^s(E) < +\infty$ for all $\tau \in \Theta$.

Proof. The inequality $\varphi_\tau^s(E) < +\infty$ is evident. We consider fixed parameters τ, ζ in Θ with $\tau \ll \zeta$. The η-family $(\psi_\tau^\eta : \tau \ll \eta)$ is decreasing, hence

$$\varphi_\tau^s = \inf_{\tau \ll \eta} \psi_\tau^\eta = \inf_{\zeta \ll \eta} \psi_\tau^\eta .$$

From Cor. 1.5 and the inequalities $\tau \ll \zeta \ll \eta$, it follows that

$$\psi_\tau^\eta = \sup_{\eta \ll \delta} (\varphi_\delta | \mathscr{B}_\tau) \leqslant \left(\sup_{\eta \ll \delta} (\varphi_\delta | \mathscr{B}_\zeta) \right) | \mathscr{B}_\tau = \psi_\zeta^\eta | \mathscr{B}_\tau .$$

The η-family $(\psi^\eta{}_\zeta | \mathscr{B}_\tau : \zeta \ll \eta)$ for fixed ζ and τ is also decreasing, which implies, by virtue of the last part of the corollary for the case of an infimum,

$$\inf_{\zeta \ll \eta} (\psi_\zeta^\eta | \mathscr{B}_\tau) = (\inf_{\zeta \ll \eta} \psi_\zeta^\eta) | \mathscr{B}_\tau = \varphi_\zeta^s | \mathscr{B}_\tau .$$

This yields $\inf_{\tau \ll \eta} \psi_\tau^\eta \leqslant \varphi_\zeta^s | \mathscr{B}_\tau$, and so $\varphi_\tau^s \leqslant \varphi_\zeta^s | \mathscr{B}_\tau$. Thus $\boldsymbol{\Phi}^s$ is a submartingale.

5.2. Proposition. A sufficient condition for $\mathbf{T}_1^s(\Phi)$ to hold is that Φ be of bounded variation from below. In this case, $\mathbf{T}_2^s(\Phi)$ implies Φ to be of bounded variation and Φ^s to be of bounded variation from below.

Proof. Suppose that Φ is of bounded variation from below. Then there exists a terminal set Δ of parameters such that $\sup(\varphi_\rho^-(E):\rho\in\Delta) = \kappa < +\infty$. Thus condition $\mathbf{T}_1^s(\Phi)$ holds.

Now suppose that $\mathbf{T}_2^s(\Phi)$ also holds. Choose any element τ, and then select $\eta(\tau)$ so that $\psi_\tau^{\eta(\tau)}(E) < +\infty$, $\tau \ll \eta(\tau)$.

By virtue of (5.0.1), if $\eta(\tau) \ll \delta$ we have

$$\varphi_\delta(E) \leqslant \psi_\tau^{\eta(\tau)}(E).$$

For $\delta\in\Delta$ we have in addition

$$\varphi_\delta^-(E) \leqslant \kappa,$$

whence

$$\varphi_\delta^+(E) = \varphi_\delta(E) + \varphi_\delta^-(E) \leqslant \psi_\tau^{\eta(\tau)}(E) + \kappa.$$

Since this holds for each δ in a terminal set, then Φ is of bounded variation. Finally, for any η in Θ and δ in Δ such that $\tau \ll \eta \ll \delta$, we deduce from (5.0.1) the inequality

$$\varphi_\delta|\mathscr{B}_\tau \leqslant \psi_\tau^\eta|\mathscr{B}_\tau = \psi_\tau^\eta$$

which implies $(\psi_\tau^\eta)^-(E) \leqslant \kappa$.

The η-family $(\psi_\tau^\eta:\tau\ll\eta)$ is decreasing for fixed τ, so that from Props. 1.2 and 1.4 we obtain

$$(\varphi_\tau^s)^-(E) \leqslant \sup_{\tau\ll\eta}(\psi_\tau^\eta)^-(E) \leqslant \kappa,$$

and so Φ^s is of bounded variation from below.

Remarks. (1). If Φ is of bounded variation and enjoys properties $\mathbf{T}_2^s(\Phi)$ and $\mathbf{T}_2^i(\Phi)$, then Φ^s may still not be of bounded variation (§ 10).

(2) If Φ is of bounded variation but does not satisfy $\mathbf{T}_2^s(\Phi)$, then we can define the σ-additive function φ_τ^s as the η-infimum of the functions ψ_τ^η with respect to $\mathscr{M}_{\mathscr{B}_\tau}^\sigma$ since the functions ψ_τ^η are minorized by a function of bounded variation. However, in this case the family (φ_τ^s) may not be a submartingale. For example, let E be the set of positive integers, $\mathscr{B}$ the σ-algebra of all subsets of E, $\Theta = E$, $\ll$ the natural order on E, $\mathscr{B}_1 = \{\emptyset, E\}$, and $\mathscr{B}_\tau = \mathscr{B}$ for $\tau = 2,3,\ldots$. We define $\varphi_1(\emptyset) = 0$, $\varphi_1(E) = 1$. For $\tau > 1$, let φ_τ be the measure defined on $\mathscr{B}_\tau = \mathscr{B}$ by placing a unit mass at τ. The premartingale $\Phi = (\varphi_\tau:\tau = 1,2,\ldots)$ is of bounded variation because $0 \leqslant \varphi_\tau \leqslant 1$. For any η whatever, $\psi_1^\eta(\emptyset) = 0$, $\psi_1^\eta(E) = 1$. For $\tau > 1$, $\psi_\tau^\eta(A) =$ number of elements of A that are at least as great as η. Consequently, $\varphi_1^s(\emptyset) = 0$, $\varphi_1^s(E) = 1$, whereas φ_τ^s vanishes identically as a function in $\mathscr{M}_{\mathscr{B}_\tau}^\sigma$ whenever $\tau > 1$. On the other hand, the infimum in $\mathscr{M}_{\mathscr{B}_\tau}$ of the η-family (ψ_τ^η), for $1 < \tau < \eta$, vanishes on the finite subsets

of E and takes the value $+\infty$ on the infinite sets. This furnishes an example for the possible cleavage of the infima in $\mathcal{M}_{\mathscr{B}_\tau}$ and $\mathcal{M}^\sigma_{\mathscr{B}_\tau}$ if the condition $\mathbf{S}^i$ (Prop. 1.1) is not satisfied; moreover, all functions in this example are positive. To avoid this phenomenon, we assume $\mathbf{T}^s_1(\boldsymbol{\Phi})$, $\mathbf{T}^s_2(\boldsymbol{\Phi})$, $\mathbf{T}^i_1(\boldsymbol{\Phi})$, and $\mathbf{T}^i_2(\boldsymbol{\Phi})$.

5.3. Definition. A premartingale $\boldsymbol{\Phi}$ satisfying the conditions $\mathbf{T}^s_1(\boldsymbol{\Phi})$, $\mathbf{T}^s_2(\boldsymbol{\Phi})$, $\mathbf{T}^i_1(\boldsymbol{\Phi})$, and $\mathbf{T}^i_2(\boldsymbol{\Phi})$ is called *integrable* if $\boldsymbol{\Phi}^s$ and $\boldsymbol{\Phi}^i$ are equal. In this case, $\boldsymbol{\Phi}^s = \boldsymbol{\Phi}^i$ is a martingale taking finite values, called the *integral of* $\boldsymbol{\Phi}$.

5.4. Proposition. Let $\boldsymbol{\Phi} = (\varphi_\tau)$ be a submartingale.

(1) $\mathbf{T}^s_1(\boldsymbol{\Phi})$ is equivalent to the condition (4.1.1), namely, there exists $\rho \in \Theta$ such that $\varphi_\rho(E) > -\infty$. $\mathbf{T}^i_1(\boldsymbol{\Phi})$ is equivalent to $\varphi_\tau(E) < +\infty$ for each $\tau \in \Theta$.

(2) If $\mathbf{T}^s_1(\boldsymbol{\Phi})$ holds, then $\mathbf{T}^s_2(\boldsymbol{\Phi})$ is satisfied iff $\boldsymbol{\Phi}$ is of bounded variation from above. In this case, $\boldsymbol{\Phi}^s$ is the integral $\boldsymbol{\Phi}^m$ of $\boldsymbol{\Phi}$ as defined in § 4.

(3) If $\mathbf{T}^i_1(\boldsymbol{\Phi})$ is satisfied, then $\mathbf{T}^i_2(\boldsymbol{\Phi})$ is equivalent to $\mathbf{T}^s_1(\boldsymbol{\Phi})$. In this case, $\boldsymbol{\Phi}^i$ is the integral $\boldsymbol{\Phi}^m$ of $\boldsymbol{\Phi}$ as defined in § 4.

Proof. (1) This is evident.

(2) The "only if" part follows from Prop. 5.1. Conversely, if $\boldsymbol{\Phi}$ is of bounded variation, then $\boldsymbol{\Phi}^m$ as defined in § 4 is a martingale of bounded variation from above such that (Prop. 4.2) $\boldsymbol{\Phi} \leqslant \boldsymbol{\Phi}^m$, which implies $\mathbf{T}^m_2(\boldsymbol{\Phi})$. The identity of $\boldsymbol{\Phi}^s$ with the integral $\boldsymbol{\Phi}^m$ defined in § 4 follows immediately from (5.0.1).

(3) If $\mathbf{T}^i_1(\boldsymbol{\Phi})$ and $\mathbf{T}^i_2(\boldsymbol{\Phi})$ both hold, then for any $\tau \in \Theta$ there exists η such that $\tau \ll \eta$ and $\sup_{\eta \ll \rho}(-\varphi_\rho|\mathscr{B}_\tau) < +\infty$; thus $\varphi_\rho(E) > -\infty$ holds for a terminal set of values of ρ, and so $\mathbf{T}^s_1(\boldsymbol{\Phi})$ is valid.

Conversely, suppose that $\mathbf{T}^i_1(\boldsymbol{\Phi})$ and $\mathbf{T}^s_1(\boldsymbol{\Phi})$ both hold. Then there exists by (1) a value of $\rho \in \Theta$ such that $\varphi_\rho(E) > -\infty$. If τ is any member of Θ then there exists $\eta \in \Theta$ such that $\rho \ll \eta$, $\tau \ll \eta$. Since the δ-family $(\varphi_\delta|\mathscr{B}_\tau : \eta \ll \delta)$ is increasing for fixed τ and η, then

$$\varphi_\rho(E) \leqslant \inf_{\eta \ll \delta}(\varphi_\delta|\mathscr{B}_\tau)\ (E),$$

which implies $\mathbf{T}^i_2(\boldsymbol{\Phi})$. The identity of $\boldsymbol{\Phi}^i$ with the integral defined in § 4 follows from the definition of $\boldsymbol{\Phi}^i$ and analogy with (5.0.1).

5.5. Corollaries. 1. For a submartingale $\boldsymbol{\Phi}$, the conjunction "$\mathbf{T}^s_1(\boldsymbol{\Phi})$, $\mathbf{T}^s_2(\boldsymbol{\Phi})$, $\mathbf{T}^i_1(\boldsymbol{\Phi})$, $\mathbf{T}^i_2(\boldsymbol{\Phi})$" is equivalent to "$\mathbf{T}^s_1(\boldsymbol{\Phi})$ and $\boldsymbol{\Phi}$ is of bounded variation from above" and also "$\boldsymbol{\Phi}$ is of bounded variation".

2. If $\boldsymbol{\Phi}$ is a premartingale satisfying $\mathbf{T}^s_1(\boldsymbol{\Phi})$, then a sufficient condition for the validity of $\mathbf{T}^s_2(\boldsymbol{\Phi})$ is the existence of a submartingale of bounded variation from above that majorizes $\boldsymbol{\Phi}$.

We return to the case of an arbitrary premartingale $\boldsymbol{\Phi} = (\varphi_\tau)$ and assume that $\boldsymbol{\Phi}$ is of bounded variation from below and satisfies $\mathbf{T}^s_2(\boldsymbol{\Phi})$.

Since $\boldsymbol{\Phi}^s$ is a submartingale of bounded variation from below (Props. 5.2 and 5.1), its integral in the sense of § 4 exists. We denote it by $\boldsymbol{\Phi}^{sm} = (\varphi_\tau^{sm})$. Explicitly, we have

$$\varphi_\tau^{sm} = \sup_{\tau \ll \zeta}\Big(\inf_{\zeta \ll \eta}\Big(\sup_{\eta \ll \delta}(\varphi_\delta|\mathscr{B}_\zeta)\Big)\Big)|\mathscr{B}_\tau = \operatorname{env}\sup_{\tau \ll \zeta}\Big(\operatorname{env}\inf_{\zeta \ll \eta}\Big(\sup_{\eta \ll \delta}(\varphi_\delta|\mathscr{B}_\zeta)\Big)\Big)|\mathscr{B}_\tau .$$

$\boldsymbol{\Phi}^{sm}$ is a martingale of bounded variation from below (Prop. 4.2).

Similarly, one defines $\boldsymbol{\Phi}^{im}$ if $\boldsymbol{\Phi}$ is of bounded variation from above and satisfies $\mathbf{T}_2^i(\boldsymbol{\Phi})$; $\boldsymbol{\Phi}^{im}$ is then a martingale of bounded variation from above. We have the inequalities

$$\boldsymbol{\Phi}^{im} \leqslant \boldsymbol{\Phi}^i \leqslant \boldsymbol{\Phi}^s \leqslant \boldsymbol{\Phi}^{sm}.$$

These relations imply the following assertion.

5.6. Proposition. A premartingale is integrable iff $\boldsymbol{\Phi}^{im} = \boldsymbol{\Phi}^{sm}$; the value of this integral is the common value of $\boldsymbol{\Phi}^{im}$ and $\boldsymbol{\Phi}^{sm}$.

6. Martingales and additive functions. The definitions and propositions formulated in §§ 4 and 5 are based on the results of § 2 concerning σ-additive functions on a Boolean σ-algebra with unit E. In what follows we shall establish relations between martingales and additive functions on a Boolean algebra with unit E.

Let $(\mathscr{B}_\tau)$ be an increasing stochastic basis. The set $\mathscr{A} = \bigcup_{\tau \in \Theta} \mathscr{B}_\tau$ is a Boolean subalgebra of $\mathscr{B}$ with unit E. Any element of $\mathscr{A}$ belongs to $\mathscr{B}_\tau$ for a terminal set of parameters τ. Let $\boldsymbol{\Phi} = (\varphi_\tau)$ denote a martingale of basis $(\mathscr{B}_\tau)$. We write

$$\varphi^\vee(A) = \varphi_\tau(A) \quad \text{whenever} \quad A \in \mathscr{B}_\tau \quad \text{and} \quad \tau \in \Theta; \tag{6.0.1}$$

it is easily seen that the function $\varphi^\vee$ is defined on $\mathscr{A}$ in a unique manner and possesses the following properties:

(I) $\varphi^\vee$ is additive on $\mathscr{A}$.

(II) The restriction of $\varphi^\vee$ to each family $\mathscr{B}_\tau$, $\tau \in \Theta$, is σ-additive.

We note that if $\varphi^\vee$ is known, then so is Φ; in fact,

$$\varphi_\tau = \varphi^\vee|\mathscr{B}_\tau \quad \text{for} \quad \tau \in \Theta. \tag{6.0.2}$$

On the other hand, to each function φ^∇ defined on $\mathscr{A}$ and satisfying (I) and (II) there corresponds a martingale defined by $\varphi_\tau = \varphi^\nabla|\mathscr{B}_\tau, \tau \in \Theta$, such that the function $\varphi^\vee$ defined by (6.0.1) coincides with φ^∇. Consequently, the correspondence $\boldsymbol{\Phi} \leftrightarrow \varphi^\vee$ defined by (6.0.1) and (6.0.2) is a one-to-one application of the set $\mathscr{M}_0$ of the functions satisfying (I) and (II) on $\mathfrak{M}$. We represent it by

$$\varphi^\vee = Z(\Phi), \quad \Phi = Z^{-1}(\varphi^\vee). \tag{6.0.3}$$

Z is an isomorphism with respect to order, addition, and multiplication by a scalar. The proposition "$\boldsymbol{\Phi}$ is of bounded variation" is equivalent to "$\varphi^\vee$ is of bounded variation"; similar results hold if we add "from below" or "from above". If $\mathfrak{C}$ denotes any set of martingales, $Z(\mathfrak{C})$ satisfies condition $\mathbf{S}^s$ iff $\mathfrak{C}$ contains a martingale of bounded variation from below, which implies $\mathbf{S}^s_{\mathfrak{C}}$. Assuming $\mathbf{S}^s$, the supremum of $Z(\mathfrak{C})$ in $\mathcal{M}_0$ is equal to $Z(\sup_{\mathfrak{M}} \mathfrak{C})$. Additionally, the following stronger result holds:

6.1. Proposition. For every set $\mathfrak{C}$ of martingales containing a martingale of bounded variation from below,

$$Z\left(\sup_{\mathfrak{M}} \mathfrak{C}\right) = \sup_{\mathcal{M}_{\mathscr{A}}} Z(\mathfrak{C}).$$

(N. B. The proposition is not a consequence of the isomorphism of Z because the supremum on the right refers to $\mathcal{M}_{\mathscr{A}}$ and not to $\mathcal{M}_0$).

Proof. Let $\tau \in \Theta$ be fixed, $\delta \in \Theta$ and $\tau \ll \delta$, $A \in \mathscr{B}_\tau$, and let $\mathscr{R}_\delta$ denote the δ-section of $\mathfrak{C}$. Set $\xi = \sup_{\mathcal{M}_{\mathscr{A}}} Z(\mathfrak{C})$ and $(\xi^m_\tau) = \sup_{\mathfrak{M}} \mathfrak{C}$.

In accordance with Prop. 1.2,

$$\xi(A) = \text{l.u.b.} \sum_i \varphi^\vee_i(A_i)$$

where $A = \bigvee_{i=1}^{k} A_i$, $A_i \in \mathscr{A}$, $\varphi^\vee_i \in Z(\mathfrak{C})$, and the somas A_i are pairwise disjoint, $1 \leqslant i \leqslant k$. Since $\mathscr{A}$ is the union of an increasing family $(\mathscr{B}_\delta)$, we have, for any $\tau \in \Theta$,

$$\xi(A) = \lim_{\tau \ll \delta} \left(\text{l.u.b} \sum_i \varphi^\vee_i(A_i)\right),$$

where $A_i \in \mathscr{B}_\delta$, $\delta \in \Theta$. Consequently, according to Prop. 1.2,

$$\xi(A) = \lim_{\tau \ll \delta} (\sup \mathscr{R}_\delta)(A).$$

Recalling Props. 4.1, 4.2, and 4.3, it is now clear that $\xi^m_\tau(A) = \xi(A)$, whence $Z(\sup_{\mathfrak{M}} \mathfrak{C}) = \sup_{\mathcal{M}_{\mathscr{A}}} (Z(\mathfrak{C}))$.

Special cases. For a martingale $\boldsymbol{\Phi}$ we have (Prop. 4.3) $\varphi^{\vee T} = Z(\boldsymbol{\Phi}^T)$, $\varphi^{\vee +} = Z(\boldsymbol{\Phi}^+)$, $\varphi^{\vee -} = Z(\boldsymbol{\Phi}^-)$.

Remark. For a martingale $\boldsymbol{\Phi} = (\varphi_\tau)$ of bounded variation, $\|\boldsymbol{\Phi}\| = \|\varphi^\vee\| = \varphi^{\vee T}(E)$.

7. σ-additive martingales. A martingale $\boldsymbol{\Phi}$ is called *σ-additive* or *purely simply additive* in accordance with whether $\varphi^\vee = Z(\boldsymbol{\Phi})$ is σ-additive or purely simply additive (cf. § 2 for the appropriate definitions). Similarly, one may define these concepts *from above* and *from below*.

7.1. Proposition. Each martingale $\boldsymbol{\Phi}$ of bounded variation can be represented in a unique manner as the sum of a σ-additive martingale $\boldsymbol{\Phi}_c$ and a purely simply additive martingale $\boldsymbol{\Phi}_p$. If $\boldsymbol{\Omega} \leqslant \boldsymbol{\Phi}$, then $\boldsymbol{\Phi}_c$ is the greatest of all σ-additive martingales $\boldsymbol{\Psi}$ satisfying $\boldsymbol{\Omega} \leqslant \boldsymbol{\Psi} \leqslant \boldsymbol{\Phi}$. We have

$$(\boldsymbol{\Phi}^+)_c = (\boldsymbol{\Phi}_c)^+, \quad (\boldsymbol{\Phi}^+)_p = (\boldsymbol{\Phi}_p)^+ ,$$

and corresponding relations obtained by replacing "+" by "−" or "T".

Proof. Let $\varphi^\vee = \varphi_c^\nabla + \varphi_p^\nabla$, where this equation expresses the decomposition of $\varphi^\vee = Z(\boldsymbol{\Phi})$ into a σ-additive and a purely simply additive function. φ_c^∇ and φ_p^∇ satisfy (I) of § 6. Since $\boldsymbol{\Phi}_c^\nabla$ is σ-additive on $\mathscr{A}$, it is *a fortiori* σ-additive on each set $\mathscr{B}_\tau$, $\tau \in \Theta$. Also, since $\varphi_p^\nabla | \mathscr{B}_\tau$ is the difference of two σ-additive functions $\varphi_p^\vee | \mathscr{B}_\tau$ and $\varphi_c^\nabla | \mathscr{B}_\tau$ of bounded variation, it is also σ-additive. Thus φ_c^∇ and φ_p^∇ both satisfy (II) of § 6. We need only write $\boldsymbol{\Phi}_c = Z^{-1}(\varphi_c^\nabla)$ and $\boldsymbol{\Phi}_p = Z^{-1}(\varphi_p^\nabla)$ to obtain the desired decomposition of $\boldsymbol{\Phi}$. The uniqueness of this representation follows from that of φ_c^∇ and φ_p^∇.

7.2. Definition. The deficiency of σ-additivity of a martingale $\boldsymbol{\Phi} = (\varphi_\tau)$ of bounded variation is defined as $\|\boldsymbol{\Phi}_p\| = \|\varphi_p^\vee\| = \varphi_p^{\vee T}(E)$.

8. Induced martingales. We denote by Θ_ω the basic filtering set completed by an "element at infinity", ω; thus $\Theta_\omega = \Theta \cup \{\omega\}$ and $\tau \ll \omega$ for all $\tau \in \Theta$. Let $\mathscr{B}_\omega$ be the smallest Boolean σ-algebra that includes $\mathscr{A} = \bigcup_{\tau \in \Theta} \mathscr{B}_\tau$; $\mathscr{B}_\omega$ is the Borel extension of $\mathscr{A}$ (in $\mathscr{B}$). Given a martingale $\boldsymbol{\Phi} = (\varphi_\tau : \tau \in \Theta)$ of basis $(\mathscr{B}_\tau : \tau \in \Theta)$, a σ-additive function φ_ω defined on $\mathscr{B}_\omega$ is an extension of $\varphi^\vee = Z(\boldsymbol{\Phi})$ if and only if $(\varphi_\tau : \tau \in \Theta_\omega)$ is a martingale of basis $(\mathscr{B}_\tau : \tau \in \Theta_\omega)$. Also, from a σ-additive function φ_ω on $\mathscr{B}_\omega$, we obtain in a unique manner a martingale $\boldsymbol{\Phi} = (\varphi_\tau : \tau \in \Theta)$ by setting $\varphi_\tau = \varphi_\omega | \mathscr{B}_\tau$, $\tau \in \Theta$; in fact $(\varphi_\tau : \tau \in \Theta_\omega)$ is a martingale by this definition. In effect, we define $\boldsymbol{\Phi} = Z^{-1}(\varphi_\omega | \mathscr{A})$. We say that $\boldsymbol{\Phi}$ is the martingale *induced by* φ_ω. In order that a function φ^∇ defined on $\mathscr{A}$ may be extended to a σ-additive function on $\mathscr{B}_\omega$, it is necessary and sufficient that φ^∇ be of bounded variation from above or below and σ-additive, which leads to the following result:

8.1. Proposition. A martingale $\boldsymbol{\Phi} = (\varphi_\tau : \tau \in \Theta)$ is σ-additive and of bounded variation from above or below iff $\boldsymbol{\Phi}$ is induced by a σ-additive function φ_ω defined on $\mathscr{B}_\omega$. If φ_ω is finite, then $\boldsymbol{\Phi}$ is of bounded variation and conversely.

8.2. Proposition. Let $\boldsymbol{\Phi} = (\varphi_\tau : \tau \in \Theta)$ be a submartingale such that $\varphi_\rho(E) > -\infty$ for a certain $\rho \in \Theta$. In order that $\boldsymbol{\Phi}^m$ be of bounded variation from above and σ-additive from above, it is necessary and sufficient that there exists a function φ_ω defined on $\mathscr{B}_\omega$ such that $(\varphi_\tau : \tau \in \Theta_\omega)$ is a submartingale and $\varphi_\omega(E) < +\infty$.

Proof. Suppose φ_ω exists, satisfying the stated conditions. Let $\boldsymbol{\Psi}$ be the martingale induced by φ_ω. Then $\boldsymbol{\Phi}^m \leqslant \boldsymbol{\Psi}$; hence $Z(\boldsymbol{\Phi}^m) \leqslant Z(\boldsymbol{\Psi}) = \varphi_\omega | \mathscr{A}$. Consequently, in accordance with Cor. 2.2, the relation $\varphi_\omega(E) <$

$+\infty$ implies that $\boldsymbol{\Phi}^m$ is of bounded variation from above and σ-additive from above.

Conversely, if $\boldsymbol{\Phi}^m$ is of bounded variation from above and σ-additive from above, then $\varphi^m = Z(\boldsymbol{\Phi}^m)$ enjoys the same properties. Thus φ^{m+} can be extended to a positive and σ-additive function φ_ω of bounded variation defined on $\mathscr{B}_\omega$. Since $\varphi_\tau \leqslant \varphi_\tau^m = \varphi^m|\mathscr{B}_\tau \leqslant \varphi^{m+}|\mathscr{B}_\tau = \varphi_\omega|\mathscr{B}_\tau$ for each $\tau \in \Theta$, then $(\varphi_\tau\colon \tau \in \Theta_\omega)$ is a submartingale satisfying $\varphi_\omega(E) < +\infty$.

9. Premartingales and cell functions. If $\mathscr{A}$ is an arbitrary Boolean subalgebra of $\mathscr{B}$ with unit E, then every function $\varphi^\nabla \in \mathscr{M}_{\mathscr{A}}$ can be generated by a martingale in the manner described in § 6. In fact, we may take for Θ the set of all finite $\mathscr{A}$-partitions $J = \{J_1, J_2, \ldots, J_k\}$ of E, i.e., $J_i \in \mathscr{A}$, $J_i \neq O$, $J_i \wedge J_j = \bigcirc$ for $i \neq j$, $1 \leqslant i, j \leqslant k$; and $\bigvee_{i=1}^{k} J_i = E$. We fix for $\ll$ the relation of *partition fineness*, denoted by $\sqsubset$; i.e., $\rho \ll \tau$ iff τ is a refinement of ρ. In other words, each element of ρ is the supremum of a subset of τ. $\mathscr{B}_\tau$ is defined as the Boolean algebra generated by τ. Because each $\tau \in \Theta$ consists of finitely many elements, it follows that $\mathscr{B}_\tau$ is a Boolean σ-algebra. We note that $\rho \ll \tau$ is equivalent to $\mathscr{B}_\rho \subset \mathscr{B}_\tau$, so that the filtering family $(\mathscr{B}_\tau)$ is an increasing stochastic basis and $\mathscr{A} = \bigcup_{\tau \in \Theta} \mathscr{B}_\tau$. With respect to this basis, φ^∇ possesses properties (I) and (II) of § 6. We set $\varphi_\tau = \varphi^\nabla|\mathscr{B}_\tau$ for each $\tau \in \Theta$. Then $\boldsymbol{\Phi} = (\varphi_\tau\colon \tau \in \Theta)$ is a martingale and $\varphi^\nabla = Z(\boldsymbol{\Phi})$.

The martingales obtained this way are particular instances of martingales occurring in the theory of *cell functions*. The setting consists of a Boolean σ-algebra $\mathscr{B}$ with null element $\bigcirc$ and unit element E, and of a non-empty family $\Theta = \mathfrak{T}$ of enumerable $\mathscr{B}$-partitions $\tau = \mathscr{T}$ of E, not containing the null element, filtering with respect to the fineness of the partitions. If $\mathscr{T} = \{J_1, J_2, \ldots\}$, then $\mathscr{B}_{\mathscr{T}}$ consists of those elements of $\mathscr{B}$ that can be represented as suprema of subsets of $\mathscr{T}$. The *constituents* (or *components*) J_i of $\mathscr{T}$ are the *atoms* of $\mathscr{B}_{\mathscr{T}}$, which is atomic. If we choose as our order relation $\ll$ the relation $\sqsubset$, then the family $(\mathscr{B}_{\mathscr{T}}: \mathscr{T} \in \mathfrak{T})$ is a stochastic basis called a *cell basis*. By *cell* we mean any element of an arbitrary partition in $\mathfrak{T}$. The set $\mathscr{J}$ of all cells is thus $\bigcup(\mathscr{T}: \mathscr{T} \in \mathfrak{T})$. A *complex* is defined as an enumerable (possibly finite or empty) set of disjoint cells; an *active complex* is defined as a subset of a partition in $\mathfrak{T}$. Therefore, the Boolean algebra $\mathscr{A} = \bigcup_{\mathscr{T} \in \mathfrak{T}} \mathscr{B}_{\mathscr{T}}$ consists of all elements of the form $\bigvee \mathscr{J}$, where $\mathscr{J}$ denotes an active complex. To each finite complex $\mathscr{J}$ there corresponds at least one active complex $\mathscr{J}'$ such that $\bigvee \mathscr{J} = \bigvee \mathscr{J}'$.

Examples. (*IV.* 8).

(1) $\mathscr{B}$ is the set of Borel subsets of the unit interval $E = [0,1[$, closed to the left and open to the right, and $\mathfrak{T}$ consists of all finite partitions

of E into intervals $[u,v[$. The cells are these intervals. Each active complex is finite; conversely, each finite set of disjoint intervals is an active complex. The supremum (union) of an active complex is not in general a cell.

In the following examples $\mathscr{B}$ denotes an arbitrary Boolean σ-algebra of unit E.

(2) $\mathscr{A}$ is a Boolean subalgebra of $\mathscr{B}$ with unit E and $\mathfrak{T}$ consists of all finite $\mathscr{A}$-partitions of E. Each active complex is finite and conversely. The supremum of an active complex is a cell.

(3) $\mathfrak{T}$ consists of all $\mathscr{B}$-partitions of E. Each complex is active and each non-null element of $\mathscr{B}$ is a cell.

(4) De la Vallée Poussin's nets. $\mathfrak{T}$ is a sequence $\mathscr{P}_1, \mathscr{P}_2, \ldots,$ of $\mathscr{B}$-partitions of E such that $\mathscr{P}_1 \underset{\wedge}{\sqsubset} \mathscr{P}_2 \underset{\wedge}{\sqsubset} \ldots$. A finite complex is not necessarily active. The supremum of a finite complex is not in general a cell.

9.1. Definitions. In the setting $(\mathscr{B},\mathfrak{T})$ a *cell function* is, by definition, an extended real-valued function φ, defined on the set $\mathscr{I}$ of all cells, such that for each partition $\mathscr{T}$ of $\mathfrak{T}$ the sum $\varphi(\mathscr{T})$, defined as $\sum_{J\in\mathscr{T}} \varphi(J)$, exists; that is, the sum of the positive terms and the sum of the negative terms are not both infinite. Then, since each active complex $\mathscr{J}$ is included in a partition of $\mathfrak{T}$, it follows that $\varphi(\mathscr{J})$, defined as $\sum_{J\in\mathscr{J}} \varphi(J)$, exists.

Let φ be a cell function and $\mathscr{T} \in \mathfrak{T}$. Each element A from $\mathscr{B}_{\mathscr{T}}$ admits a unique representation $A = \vee \mathscr{J}$, where $\mathscr{J} \subset \mathscr{T}$. We write

$$G\colon \varphi_{\mathscr{T}}(A) = \varphi(\mathscr{J}).$$

It is clear that $\varphi_{\mathscr{T}}$ is σ-additive on $\mathscr{B}_{\mathscr{T}}$, consequently $\boldsymbol{\Phi} = (\varphi_{\mathscr{T}})$ is a premartingale. We write $\boldsymbol{\Phi} = G(\varphi)$ and read this as "$\boldsymbol{\Phi}$ is generated by φ". For each constituent J of a partition $\mathscr{T}$ belonging to $\mathfrak{T}$, we have $\varphi(J) = \varphi_{\mathscr{T}}(J)$. Thus there is a unique relationship between $\boldsymbol{\Phi}$ and φ. Because of this correspondence, each proposition concerning premartingales in general implies, as a special case, a proposition for cell functions. To each concept concerning premartingales there corresponds one related to cell functions. For instance, to the martingales $(\varphi_{\mathscr{T}}^{T})$, $(\varphi_{\mathscr{T}}^{+})$, and $(\varphi_{\mathscr{T}}^{-})$ there correspond the cell functions whose values on a cell J are $|\varphi(J)|$, $(\varphi(J))^{+}$, and $(\varphi(J))^{-}$, respectively. However, we observe in the next proposition that an arbitrary premartingale on a cell basis $(\mathscr{B}_{\mathscr{T}})$ is not necessarily generated by a cell function according to the transfer law G.

9.2. Proposition. In order that a premartingale $\boldsymbol{\Phi} = (\varphi_{\mathscr{T}})$ on the cell basis $(\mathscr{B}_{\mathscr{T}})$ can be generated from a cell function φ in accordance with G, it is necessary and sufficient that for any cell J, the number $\varphi_{\mathscr{T}}(J)$ be the same for each partition $\mathscr{T}$ containing J.

9.3. Definition. We call the criterion of Prop. 9.2 *condition* (**B**).

Condition (**B**) is satisfied for all martingales on the cell basis $(\mathscr{B}_{\mathscr{T}})$ if each cell J belongs to only one partition $\mathscr{T}$ in $\mathfrak{T}$, as is the case for certain DE LA VALLÉE POUSSIN nets.

Example. $E = [0,1[, \mathscr{P}_n$ is the partition of E into intervals $[(i-1)\cdot 2^{-n}, i\cdot 2^{-n}[$, where $i = 1,2,\dots,2^n$.

A cell function φ is said to be of *bounded variation (from above, from below)* if the premartingale $\Phi = G(\varphi)$ is of bounded variation (from above, from below). If φ is of bounded variation, then the norm $\|\varphi\|$ is defined as

$$\|G(\varphi)\| = \limsup_{\mathscr{T}\in\mathfrak{T}} \varphi^T_{\mathscr{T}}(E).$$

A cell function is said to be *subadditive*, *superadditive*, or *additive* according to whether the premartingale $\Phi = G(\varphi)$ is a submartingale, supermartingale, or martingale, respectively. Thus subadditivity is equivalent to the following property:

For any partitions $\mathscr{T}$ and $\mathscr{S}$ of $\mathfrak{T}$ such that $\mathscr{T} \sqsubset \mathscr{S}$, and any (active) complex $\mathscr{J} \subset \mathscr{S}$ such that $\vee \mathscr{J}$ is a cell in $\mathscr{T}$, we have $\varphi(\vee \mathscr{J}) \leqslant \varphi(\mathscr{J})$.

Additivity of φ is equivalent to the following property: For any active complex $\mathscr{J}$, the supremum of which is a cell, we have $\varphi(\vee \mathscr{J}) = \varphi(\mathscr{J})$.

If Φ is an arbitrary martingale on the cell basis $(\mathscr{B}_{\mathscr{T}})$, then condition (**B**) (Def. 9.3) holds; therefore Φ is generated in accordance with G by an additive cell function, namely, the restriction of $\varphi^\vee = Z(\Phi)$ to $\mathscr{I}$ (§ 6). This function $\varphi^\vee$ is the extension of φ on $\mathscr{A} = \bigcup_{\mathscr{T}\in\mathfrak{T}} \mathscr{B}_{\mathscr{T}}$, which satisfies conditions (I) and (II) of § 6. Also, if a function φ^∇, defined on $\mathscr{A}$, satisfies conditions (I) and (II), then its restriction φ to $\mathscr{I}$ is an additive cell function and $\varphi^\nabla = Z(G(\varphi))$.

Prop. 9.2 can be interpreted as follows: the (single-valued) cell functions on a cell basis $(\mathscr{B}_{\mathscr{T}})$ generate those premartingales possessing the "martingale property for atoms". Hence it is not surprising that martingale theorems can be partially transferred to non-additive cell functions.

By *total*, *positive*, and *negative variations* of the additive cell function φ we mean the cell functions φ^T, φ^+, and φ^- corresponding to the martingales Φ^T, Φ^+, and Φ^-, respectively, where $\Phi = G(\varphi)$; thus we mean the restrictions to $\mathscr{I}$ of the functions $Z(\Phi^T)$, $Z(\Phi^+)$, and $Z(\Phi^-)$, respectively. Therefore, on a cell I, the values of these functions are, respectively, the upper bounds of

$$\sum_{J\in\mathscr{J}} |\varphi(J)|, \quad \sum_{J\in\mathscr{J}} (\varphi(J))^+, \quad \text{and} \quad \sum_{J\in\mathscr{J}} (\varphi(J))^-$$

for the active complexes $\mathscr{J}$ whose supremum $\vee\mathscr{J} = I$.

The additive function φ is said to be *σ-additive* or *purely simply additive*, if the martingale $\Phi = G(\varphi)$ is σ-additive or purely simply additive, respectively. We note that σ-additivity of φ is equivalent to the property:

For each (not necessarily active) complex $\mathscr{J}$ whose supremum is a cell, $\varphi(\mathscr{J})$ exists and equals $\varphi(\vee \mathscr{J})$, where $\varphi(\mathscr{J})$ is defined as $\sum_{J \in \mathscr{J}} \varphi(J)$.

It follows that φ is purely simply additive iff any σ-additive cell function ψ whose absolute value is majorized by φ^T vanishes identically.

If φ is additive and of bounded variation, then we define the σ-additive part φ_c as $Z(\Phi_c)|\mathscr{I}$ and the purely simply additive part φ_p of φ as $Z(\Phi_p)|\mathscr{I}$ (Prop. 7.1). For $0 \leqslant \varphi, \varphi_c$ is the supremum of the σ-additive cell functions majorized by φ. If φ is of variable sign, then $\varphi_c = (\varphi^+)_c - (\varphi^-)_c$. The function φ_c admits the following individual definition that we give without proof: For $I \in \mathscr{I}$, take $\mathfrak{K}$ as the set of (not necessarily active) complexes $\mathscr{J}$ whose supremum coincides with I, filtering with respect to $\sqsubset$; then $\varphi_c(I) = \lim_{\mathscr{J} \in \mathfrak{K}} \varphi(\mathscr{J})$.

10. Integrals of cell functions. Let φ be a cell function and $\Phi = G(\varphi)$. We assume first that Φ satisfies $\mathbf{T}_1^s(\Phi)$ and $\mathbf{T}_2^s(\Phi)$ (§ 5). According to Prop. 5.1, the integral Φ^s is a submartingale. For $\mathscr{T} \in \mathfrak{T}$ we have, by definition,

$$\varphi_{\mathscr{T}}^s = \inf_{\mathscr{T} \sqsubset \mathscr{H}} \left(\sup_{\mathscr{H} \sqsubset \mathscr{X}} (\varphi_{\mathscr{X}} | \mathscr{B}_{\mathscr{T}}) \right).$$

Let I be a cell of $\mathscr{T}$. Since I is an atom of $\mathscr{B}_{\mathscr{T}}$ we obtain, with reference to Cor. 1.5 and § 5,

$$\left(\sup_{\mathscr{H} \sqsubset \mathscr{X}} (\varphi_{\mathscr{X}} | \mathscr{B}_{\mathscr{T}}) \right)(I) = \underset{\mathscr{H} \sqsubset \mathscr{X}}{\text{l.u.b.}}\, \varphi_{\mathscr{X}}(I),$$

hence

$$\varphi_{\mathscr{T}}^s(I) = \underset{\mathscr{T} \sqsubset \mathscr{H}}{\text{g.l.b.}} \left(\underset{\mathscr{H} \sqsubset \mathscr{X}}{\text{l.u.b.}}\, \varphi_{\mathscr{X}}(I) \right) = \limsup_{\mathscr{T} \sqsubset \mathscr{X}} \varphi_{\mathscr{X}}(I)$$

and, according to G,

$$\varphi_{\mathscr{T}}^s(I) = \limsup_{\mathscr{X} \in \mathfrak{D}_I} \varphi(\mathscr{I}_{\mathscr{X}}),$$

where $\mathfrak{D}_I$ denotes the set of partitions $\mathscr{X} \in \mathfrak{T}$ such that $I \in \mathscr{B}_{\mathscr{X}}, \mathscr{I}_{\mathscr{X}}$ is the complex included in $\mathscr{X}$ such that $I = \vee \mathscr{I}_{\mathscr{X}}$, and the limit superior is taken with respect to $\sqsubset$. This expression of $\varphi_{\mathscr{T}}^s(I)$ shows that Φ^s satisfies condition (**B**). Thus we can define φ^s as $G^{-1}(\Phi^s)$ and we have at hand the *individual definition*

$$\varphi^s(I) = \limsup_{\mathscr{X} \in \mathfrak{D}_I} \varphi(\mathscr{I}_{\mathscr{X}}). \tag{10.0.1}$$

Similarly, assuming $\mathbf{T}_1^i(\Phi)$ and $\mathbf{T}_2^i(\Phi)$, we may define $\varphi^i = G^{-1}(\Phi^i)$ and

$$\varphi^i(I) = \liminf_{\mathscr{X} \in \mathfrak{D}_I} \varphi(\mathscr{I}_{\mathscr{X}}). \tag{10.0.2}$$

We observe that φ^s is a subadditive cell function of bounded variation from below and φ^i is a superadditive cell function of bounded variation from above.

The martingales $\boldsymbol{\Phi}^{sm} = (\varphi^{sm}_{\mathscr{T}})$ and $\boldsymbol{\Phi}^{im} = (\varphi^{im}_{\mathscr{T}})$ can be defined individually by

$$\varphi^{sm}_{\mathscr{T}}(I) = \underset{\mathscr{T} \sqsubset \mathscr{X}}{\text{l.u.b.}}\ \varphi^s_{\mathscr{X}}(I) = \lim_{\mathscr{T} \sqsubset \mathscr{X}} \varphi^s_{\mathscr{X}}(I)\,,$$

$$\varphi^{im}_{\mathscr{T}}(I) = \underset{\mathscr{T} \sqsubset \mathscr{X}}{\text{g.l.b.}}\ \varphi^s_{\mathscr{X}}(I) = \lim_{\mathscr{T} \sqsubset \mathscr{X}} \varphi^i_{\mathscr{X}}(I)\,.$$

We define $\varphi^{sm} = G^{-1}(\boldsymbol{\Phi}^{sm})$ and $\varphi^{im} = G^{-1}(\boldsymbol{\Phi}^{im})$. Then φ^{sm} and φ^{im} are additive cell functions of bounded variation from above and from below, respectively. Using G, (10.0.1), and (10.0.2), we obtain the following individual definitions:

$$\begin{aligned} \varphi^{sm}(I) &= \underset{\mathscr{X} \in \mathfrak{D}_I}{\text{l.u.b.}}\ \varphi^s(\mathscr{I}_{\mathscr{X}}) = \lim_{\mathscr{X} \in \mathfrak{D}_I} \varphi^s(\mathscr{I}_{\mathscr{X}})\,; \\ \varphi^{im}(I) &= \underset{\mathscr{X} \in \mathfrak{D}_I}{\text{g.l.b.}}\ \varphi^i(\mathscr{I}_{\mathscr{X}}) = \lim_{\mathscr{X} \in \mathfrak{D}_I} \varphi^i(\mathscr{I}_{\mathscr{X}})\,. \end{aligned} \tag{10.0.3}$$

Discarding our hypotheses $\mathbf{T}^s_1(\boldsymbol{\Phi})$, $\mathbf{T}^s_2(\boldsymbol{\Phi})$, $\mathbf{T}^i_1(\boldsymbol{\Phi})$, and $\mathbf{T}^i_2(\boldsymbol{\Phi})$, we define φ^s, φ^i, φ^{sm}, and φ^{im} by the relations (10.0.1), (10.0.2), and (10.0.3). From the definition follows

$$\varphi^{im} \leqslant \varphi^i \leqslant \varphi^s \leqslant \varphi^{sm}\,.$$

The functions φ^s and φ^i are called *Burkill-Kolmogoroff integrals of* φ. They are called *Burkill integrals* if the partitions used are finite; more precisely, if the cell basis satisfies the following condition:

(**P**). For any cell I, each active complex $\mathscr{I}$ such that $I = \bigvee \mathscr{I}$ is finite.

We may note that if every partition of $\mathfrak{T}$ consists of only a finite number of cells, then condition "φ is of bounded variation from above (from below)" implies $\mathbf{T}^s_2(\boldsymbol{\Phi})(\mathbf{T}^i_2(\boldsymbol{\Phi}))$.

Example. Let E denote the unit interval $[0,1[$ and $\mathscr{B}$ the σ-algebra of its Borel subsets. $\mathfrak{T}$ is the sequence of the regular binary partitions; thus

$$\mathscr{P}_n = ([(k-1)\cdot 2^{-n}, k\cdot 2^{-n}[: k = 1,2,\ldots,2^n)\,.$$

Let $(p_n : n = 1,2,\ldots)$ be a sequence of numbers dense in E and $\varphi_n = \varphi_{p_n}$ the measure defined on $\mathscr{B}_{\mathscr{P}_n}$ by placing a unit mass at p_n. The premartingale (φ_n) satisfies condition (**B**) (cf. Def. 9.3 and the comments following it), so that it can be generated by a cell function; namely, by φ satisfying $\varphi(I) = 1$ or 0 according to whether $p_n \in I$ or $p_n \notin I$, respectively, where n is the index such that $I \in \mathscr{B}_{\mathscr{P}_n}$. Then, from (10.0.1), (10.0.2), and (10.0.3), we have

$$\varphi^s(I) = 1\,, \quad \varphi^i(I) = 0\,, \quad \varphi^{sm}(I) = +\infty\,, \quad \varphi^{im}(I) = 0$$

for each cell $I \in \mathscr{I}$. Notice that $\Phi = G(\varphi)$ is of bounded variation and satisfies the conditions $\mathbf{T}_1^s(\Phi)$, $\mathbf{T}_2^s(\Phi)$, $\mathbf{T}_1^i(\Phi)$ and $\mathbf{T}_2^i(\Phi)$. However, Φ^s is not of bounded variation from above.

In the following, we shall limit ourselves to finite functions φ with finite integrals φ^s and φ^i [*39*, 209]. We shall say that φ is *integrable in the sense of Burkill-Kolmogoroff* iff $\varphi^s = \varphi^i$; and we write

$$\varphi^m(I) = \varphi^s(I) = \varphi^i(I).$$

In case φ is of bounded variation and satisfies $\mathbf{T}_2^s(\Phi)$ and $\mathbf{T}_2^i(\Phi)$, Prop. 5.6 tells us that φ is integrable if and only if $\varphi^{sm} = \varphi^{im}$.

Remark. We use the notations $\varphi^s(I)$ and $\varphi^i(I)$ to represent the integrals of Burkill-Kolmogoroff for cells I only. The upper and lower Burkill-Kolmogoroff integrals on E are defined as

$$\limsup_{\mathscr{T} \in \mathfrak{T}} \varphi(\mathscr{T}) \quad \text{and} \quad \liminf_{\mathscr{T} \in \mathfrak{T}} \varphi(\mathscr{T}), \quad [36, 258]$$

respectively. This is legitimate because $\varphi(\mathscr{T})$ is meaningful for all partitions $\mathscr{T}$ of $\mathfrak{T}$. If φ is of bounded variation, then its norm $\|\varphi\|$ (defined in § 9) is equal to the upper Burkill-Kolmogoroff integral of ψ on E, where

$$\psi(I) = |\varphi(I)| \quad \text{for each} \quad I \in \mathscr{I}.$$

Properties of the integrals φ^s and φ^i. As we saw above, if φ satisfies $\mathbf{T}_1^s(\Phi)$, $\mathbf{T}_2^s(\Phi)$, $\mathbf{T}_1^i(\Phi)$, and $\mathbf{T}_2^i(\Phi)$, then φ^s and φ^i are semi-additive. These properties hold also if the cell basis satisfies the condition (**P**) above [*10*, 138; *39*, 209] but, contrary to the behavior of cell functions in the classical setting, φ^s and φ^i are not necessarily additive in the present general setting, as is shown by the example. In order to obtain additive cell functions φ^{sm} and φ^{im} we must repeat the process of integration. This situation has already been observed; for instance, in [*10*]. It is only under rather strict hypotheses such as the condition (**E**) below that φ^s and φ^i are additive themselves (Prop. 10.1 below). This property (**E**) does not possess an interesting counterpart in the theory of general martingales with general bases, although many cell bases possess it. We now state it.

Condition (**E**): For any $\mathscr{T}$ in $\mathfrak{T}$ and any finite complex $\mathscr{J}$, each constituent of which is included in one constituent of $\mathscr{T}$, there exists in $\mathfrak{T}$ a partition $\mathscr{Z}$ such that $\mathscr{J} \subset \mathscr{Z}$ and $\mathscr{T} \sqsubset \mathscr{Z}$.

From (**P**) and (**E**) we deduce the additivity of φ^s and φ^i, always assuming φ, φ^s, and φ^i to be finite [*10*, 138 and 141; *39*, 209].

10.1. Proposition. If the cell basis $(\mathscr{B}_{\mathscr{T}})$ possesses properties (**P**) and (**E**), φ is a cell function of bounded variation and $\Phi = G(\varphi)$, then Φ^s and Φ^i are martingales, hence $\Phi^{sm} = \Phi^s$ and $\Phi^{im} = \Phi^i$. For any element

A of $\mathscr{B}_{\mathscr{T}}$ that is the supremum of a finite complex included in $\mathscr{T}$, the following equations hold:

$$\varphi_{\mathscr{T}}^{sm}(A) = \limsup_{\mathscr{T} \sqsubset \mathscr{X}} \varphi_{\mathscr{X}}(A) ; \qquad \varphi_{\mathscr{T}}^{im}(A) = \liminf_{\mathscr{T} \sqsubset \mathscr{X}} \varphi_{\mathscr{X}}(A) . \quad (10.1.1)$$

In particular, if every partition of $\mathfrak{T}$ is finite and the condition (**E**) holds, then the relations (10.1.1) hold for each element A of $\mathscr{B}_{\mathscr{I}}$.

Relativization of $\mathfrak{T}$. In the eqs. (10.0.1), (10.0.2), (10.0.3) and (10.1.1), the set of parameters consists of partitions of the unit E. However, it may happen that we require only the partitions of a single cell I or perhaps, more generally, the supremum A of an active complex. To this end, we denote by $\mathfrak{T}_A$ the set, filtering with respect to $\sqsubset$, of those cell partitions $\mathscr{I}$ of A enjoying the following property: There exists a partition $\mathscr{X}$ in $\mathfrak{T}$ such that $\mathscr{I} \subset \mathscr{X}$. In other words, $\mathfrak{T}_A$ is the $\sqsubset$-filtering set of those active complexes $\mathscr{I}$ such that $\bigvee \mathscr{I} = A$. Let us assume that $(\mathscr{B}_{\mathscr{T}})$ possesses the following property:

(**F**): Whenever $\mathscr{T} \in \mathfrak{T}$, $I \in \mathscr{B}_{\mathscr{T}}$, and $\mathscr{I}$ is an active complex such that $\bigvee \mathscr{I} = I$, there exists in $\mathfrak{T}$ a partition $\mathscr{Z}$ with $\mathscr{I} \subset \mathscr{Z}$ and $\mathscr{T} \sqsubset \mathscr{Z}$.

(**F**) implies the equations

$$\limsup_{\mathscr{X} \in \mathfrak{D}_I} \varphi(\mathscr{I}_{\mathscr{X}}) = \limsup_{\mathscr{I} \in \mathfrak{T}_I} \varphi(\mathscr{I}) ; \quad \liminf_{\mathscr{X} \in \mathfrak{D}_I} \varphi(\mathscr{I}_{\mathscr{X}}) = \liminf_{\mathscr{I} \in \mathfrak{T}_I} \varphi(\mathscr{I})$$

for any cell function φ, where the limits on the right hand sides are taken with respect to $\sqsubset$ in $\mathfrak{T}_I$.

If (**P**) and (**E**) hold, then so does (**F**), and the eq. (10.1.1) can be written

$$\varphi^{sm}(A) = \limsup_{\mathscr{I} \in \mathfrak{T}_A} \varphi(\mathscr{I}) ; \qquad \varphi^{im}(A) = \liminf_{\mathscr{I} \in \mathfrak{T}_A} \varphi(\mathscr{I})$$

for the supremum A of any finite active complex.

A set $\mathfrak{T}$ of partitions of E, totally ordered by $\sqsubset$, satisfies (**F**) but not necessarily (**E**). The condition (**E**) is satisfied in the examples (1), (2), and (3) of § 9; the condition (**F**) holds in all examples of § 9.

11. Convergence theorems for martingales of bounded variation when $\mathscr{B}$ is a measure algebra. We assume that $\mathscr{B}$ is a measure algebra; i.e., $\mathscr{B}$ admits at least one σ-additive, σ-finite and strictly positive measure defined on it. $\mathscr{B}$ is then a Boolean algebra and every subset $\mathscr{K}$ of $\mathscr{B}$ includes an enumerable set $\mathscr{L}$ such that

$$\bigvee \mathscr{L} = \bigvee \mathscr{K} \quad \text{and} \quad \bigwedge \mathscr{L} = \bigwedge \mathscr{K} \quad [59] .$$

Consequently, every Boolean σ-subalgebra $\mathscr{C}$ of $\mathscr{B}$ is a complete subalgebra of $\mathscr{B}$; i.e., $\mathscr{C}$ contains the supremum and the infimum of each subset of $\mathscr{C}$, enumerable or not.

If $\mathscr{C}$ denotes a Boolean σ-subalgebra of $\mathscr{B}$ with unit E, by *$\mathscr{C}$-measurable function* or simply *$\mathscr{C}$-function* we mean an "*Ortsfunktion*" *(place function)* of Carathéodory [5, 72–98] relative to $\mathscr{C}$, with values in $[-\infty, +\infty]$. Such a function can be defined by its *spectral scale* $E_\alpha = [f \leqslant \alpha]$, $-\infty \leqslant \alpha \leqslant \infty$. The spectral scales of the $\mathscr{C}$-functions are characterized by the following properties: $E_\alpha \in \mathscr{C}$; $\alpha \leqslant \beta$ implies $E_\alpha \leqslant E_\beta$; $E_{+\infty} = E$; if (α_n) is a non-increasing sequence converging to α, then $\lim_n E_{\alpha_n} = E_\alpha$.

These functions obey the ordinary rules of measurable point functions; their properties are expounded in [5, 72–98; 50]. The space $\mathscr{F}_\mathscr{C}$ of the $\mathscr{C}$-functions with natural ordering is a complete lattice, a complete sublattice of the lattice $\mathscr{F}_\mathscr{B}$. If $B \in \mathscr{B}$, then the $\mathscr{B}$-function that takes on the value 1 on B and the value 0 on $E - B$ is called the *indicatrix* (or *characteristic function*) of B and is denoted by c_B. If α denotes an arbitrary finite number, then αc_E will be denoted more briefly by α. The complete sublattice of $\mathscr{F}_\mathscr{B}$ consisting of the indicatrix functions of the elements (somas) of $\mathscr{C}$ is isomorphic to $\mathscr{C}$ by the correspondence $B \leftrightarrow c_B$, $B \in \mathscr{C}$. Thus we represent the infima and suprema in $\mathscr{F}_\mathscr{C}$ and $\mathscr{F}_\mathscr{B}$ by the same signs $\wedge$, $\bigwedge$, $\vee$, and $\bigvee$ as in $\mathscr{C}$ and $\mathscr{B}$. In particular,

$$|f| = f \vee (-f), \quad f^+ = f \vee 0, \quad \text{and} \quad f^- = (-f) \vee 0 .$$

By *function* we shall mean a $\mathscr{B}$-function; their family $\mathscr{F}_\mathscr{B}$ will also be denoted by $\mathscr{F}$. For a function f and a soma B, we represent by $f|B$ the restriction of f to B.

We denote by $\mathscr{V}_1$ or $\mathscr{V}$ the Banach space of those real (finite) functions ψ that are defined on the Boolean subalgebra $\mathscr{A}$ of $\mathscr{B}$ with unit E, σ-additive and of bounded variation (cf. § 1), equipped with the norm $\|\psi\|_1$ or $\|\psi\|$ equal to the total variation of ψ on E. The family of these functions is thus $\mathscr{M}_\mathscr{A}^{T;\sigma}$ introduced at the beginning of § 2. Because each of these functions admits a unique σ-additive extension on $\mathscr{B}_\omega$ (§ 8) with the same total variation we can, in the definition of $\mathscr{V}$, replace "defined on $\mathscr{A}$" by "defined on $\mathscr{B}_\omega$". The dual space $\mathscr{V}_1'$ or $\mathscr{V}'$ is the vector space of the bounded $\mathscr{B}_\omega$-functions f with $\|f\|' = \sup|f|$.

11.1. Proposition. For a martingale $\boldsymbol{\Phi} = (\varphi_\tau)$ of bounded variation the following conditions are equivalent:

(1) $\boldsymbol{\Phi}$ is σ-additive (defined in § 7);

(2_e) for each $\tau \in \Theta$ the function φ_τ can be extended to a function ψ_τ of $\mathscr{V}$ such that the sequence (ψ_τ) converges strongly in $\mathscr{V}$;

(3_e) for each $\tau \in \Theta$ the function φ_τ can be extended to a function ψ_τ of $\mathscr{V}$ such that the sequence (ψ_τ) converges weakly in $\mathscr{V}$.

Proof. That (1) implies (2_e) and (2_e) implies (3_e) is clear. It remains to be shown that (3_e) implies (1). Accordingly, let us assume the existence of the extensions ψ_τ converging weakly to ψ_ω in $\mathscr{V}$. Letting B represent an arbitrary soma of $\mathscr{B}_\omega$, the functional λ defined on $\mathscr{V}$ by $\lambda(\psi) = \psi(B)$ is linear and continuous on $\mathscr{V}$, hence the weak convergence of the

ψ_τ to ψ_ω implies $\psi_\omega(B) = \lim_\tau \psi_\tau(B)$; in particular, $\psi_\omega(A) = \lim_\tau \psi_\tau(A)$. Thus, it follows that

$$\psi_\omega(A) = \lim_{\tau \in \Delta_A} \varphi_\tau(A) = \varphi^\vee(A).$$

(Δ_A is defined in § 3 and $\varphi^\vee$ in § 6.)

Since $\varphi^\vee | A$ admits a σ-additive extension, ψ_ω is σ-additive; therefore condition (1) is satisfied.

11.2. Definition. [*16*]. The functions ψ of a subset $\mathscr{W}$ of $\mathscr{V}$ are called *uniformly σ-additive* iff for any decreasing null sequence $B_1, B_2, \ldots, B_n, \ldots$, of elements of $\mathscr{B}_\omega$ (cf. § 2), the sequence $\psi^T(B_1), \psi^T(B_2), \ldots, \psi^T(B_n), \ldots$, converges to zero uniformly on $\mathscr{W}$.

11.3. Theorem. *Let $\Phi = (\varphi_\tau)$ be a σ-additive martingale and for each $\tau \in \Theta$, let ψ_τ denote a σ-additive extension of φ_τ on $\mathscr{B}_\omega$. Assume that the functions ψ_τ are uniformly σ-additive and that the set of their norms $\|\psi_\tau\|$ is bounded.*

Then the sequence (ψ_τ) converges weakly on $\mathscr{V}$ to the function φ_ω that induces Φ in $(\mathscr{B}_\tau)$ (§ 8).

Proof. Since $\mathscr{B}_\omega$ is the Borelian extension of $\mathscr{A}$, the norm of φ_ω equals that of its restriction to $\mathscr{A}$. Let us denote by ε an arbitrary positive number. Then there exists a decomposition of E into disjoint elements $A_1, A_2, \ldots, A_k$ of $\mathscr{A}$ such that

$$\sum_{i=1}^{k} |\varphi_\omega(A_i)| \geqslant \|\varphi_\omega\| - \varepsilon.$$

For each $A \in \mathscr{A}$, we have

$$\lim_\tau \psi_\tau(A) = \varphi_\omega(A).$$

Consequently, whenever $\rho \in \Theta$, there exists $\tau \in \Theta$ with $\tau \gg \rho$ and

$$\sum_{i=1}^{k} |\psi_\tau(A_i)| \geqslant \sum_{i=1}^{k} |\varphi_\omega(A_i)| - \varepsilon.$$

Since $\|\psi_\tau\| \geqslant \sum_{i=1}^{k} \psi_\tau(A_i)$, the combination of the preceding inequalities yields $\|\psi_\tau\| \geqslant \|\varphi_\omega\| - 2\varepsilon$. Hence, because of the arbitrary nature of ρ and ε, we have

$$\|\varphi_\omega\| \leqslant \limsup_\tau \|\psi_\tau\|.$$

Writing $\kappa = \sup\{\|\psi_\tau\|\}$, we have $\kappa \geqslant \|\varphi_\omega\|$. Let ε again be an arbitrary positive number, f any function in $\mathscr{V}'$. The Boolean algebra $\mathscr{A}$ is dense in $\mathscr{B}_\omega$, therefore there exists a sequence $s_1, \ldots, s_j, \ldots$ of $\mathscr{A}$-measurable functions each taking only a finite number of finite values such that if we define γ by

$$\gamma = \sup\{\|f\|', \|s_j\|', j = 1,2,\ldots\},$$

then γ is finite and the functions s_j, $j = 1,2,\ldots$, converge to f in accordance with the order relation; i.e.,

$$f = \bigvee_k \bigwedge_{k \ll j} s_j = \bigwedge_k \bigvee_{k \ll j} s_j .$$

By virtue of Egoroff's theorem, there exists an increasing sequence $E_1, E_2, \ldots, E_i, \ldots$ of $\mathscr{B}_\omega$-somas that converge to E and, on each soma E_i, the sequence (s_j) converges uniformly to f. Thanks to the uniform σ-additivity of the functions ψ_τ and the σ-additivity of φ_ω, and keeping in mind that these functions are of bounded variation, there is a positive integer i_0 such that, putting $E_{i_0} = E_0$, we have

$$\psi_\tau^T(E - E_0) < \varepsilon/8\gamma$$

for all $\tau \in \Theta$ and

$$\varphi_\omega^T(E - E_0) < \varepsilon/8\gamma .$$

The uniform convergence of (s_j) to f on E_0 ensures the existence of a positive integer n_0 such that, setting $s_{n_0} = s_0$, we have

$$|f - s_0| < \varepsilon/4\kappa \quad \text{on} \quad E_0 .$$

We consider the inequality

$$|\langle\psi_\tau, f\rangle - \langle\varphi_\omega, f\rangle| \leqslant |\langle\psi_\tau, f\rangle - \langle\psi_\tau, s_0\rangle| + |\langle\psi_\tau, s_0\rangle - \langle\varphi_\omega, s_0\rangle| + |\langle\varphi_\omega, s_0\rangle - \langle\varphi_\omega, f\rangle| .$$

On any soma on which s_0 is constant, there exists $\rho \in \Theta$ such that ψ_τ and φ_ω coincide if $\tau \gg \rho$. Therefore, for $\tau \gg \rho$, $\langle\psi_\tau, s_0\rangle - \langle\varphi_\omega, s_0\rangle$ vanishes. Next, let us observe that

$$\begin{aligned}|\langle\psi_\tau, f\rangle - \langle\psi_\tau, s_0\rangle| &\leqslant \Big|\int_{E_0} (f - s_0)\, d\psi_\tau\Big| + \Big|\int_{E - E_0} f\, d\psi_\tau\Big| + \Big|\int_{E - E_0} s_0\, d\psi_\tau\Big| \\ &< \varepsilon\kappa/4\kappa + \varepsilon\gamma/8\gamma + \varepsilon\gamma/8\gamma = \varepsilon/2 .\end{aligned}$$

Similarly $|\langle\varphi_\omega, s_0\rangle - \langle\varphi_\omega, f\rangle| < \varepsilon/2$. Combining these results yields $|\langle\psi_\tau, f\rangle - \langle\varphi_\omega, f\rangle| < \varepsilon$. Thus $\langle\psi_\tau, f\rangle$ converges to $\langle\varphi_\omega, f\rangle$ for any f in $\mathscr{V}'$. By definition, ψ_τ converges weakly to φ_ω in $\mathscr{V}$.

Complements. 1. In the formulation of the preceding theorem, one can replace the hypothesis "$\boldsymbol{\Phi} = (\varphi_\tau)$ is a σ-additive martingale" by "$\boldsymbol{\Phi}$ is a premartingale such that for each $A \in \mathscr{A}$, $\lim_\tau \psi_\tau(A)$ exists". (cf. [*54*]). This limit is, thanks to the uniform σ-additivity of the functions ψ_τ and the finiteness of the l.u.b. of the norms $\|\psi_\tau\|$, an element of $\mathscr{V}$ that we can denote by φ_ω. The preceding proof can be applied if we substitute for the assertion of the existence of ρ such that "$\tau \gg \rho$ implies $\langle\psi_\tau, s_0\rangle - \langle\varphi_\omega, s_0\rangle = 0$" the following one: "to each $\delta > 0$ there corresponds ρ such that $\tau \gg \rho$ implies $|\langle\psi_\tau, s_0\rangle - \langle\varphi_\omega, s_0\rangle| < \delta$. We conclude that the functions ψ_τ converge weakly to φ_ω in $\mathscr{V}$.

2. The uniform σ-additivity of the functions ψ_τ and the finiteness of the l.u.b. of the norms $\|\psi_\tau\|$ follow in case the functions ψ_τ^T are uniformly majorized by a function belonging to $\mathscr{V}$.

3. We conjecture that the supplementary hypothesis "$\|\varphi_\tau\| = \|\psi_\tau\|$ for all τ" in the formulation of the theorem ensures the strong convergence of the sequence (ψ_τ) in $\mathscr{V}$ and that for any martingale Φ of bounded variation and σ-additive, this hypothesis implies the uniform σ-additivity of the functions ψ_τ.

Chapter II

Theory in a Measure Space without Vitali Conditions

Henceforth we assume that $\mathscr{B}$ is the domain of a measure μ that is σ-additive, σ-finite and strictly positive; hence $\mu(B) = 0$ implies $B = \bigcirc$.

1. Preliminaries. Let φ be a function defined on a Boolean subalgebra $\mathscr{A}$ of $\mathscr{B}$ with unit E, additive and vanishing at $\bigcirc$; i.e., $\varphi \in \mathscr{M}_{\mathscr{A}}$ (I, § 1). φ is said to be *absolutely continuous with respect to* μ or *μ-continuous* if, corresponding to any $\varepsilon > 0$, there exists a soma H of finite μ-measure and a positive number δ such that

$$A \in \mathscr{A} \quad \text{and} \quad \mu(H \cdot A) < \delta \quad \text{imply} \quad \varphi^T(A) < \varepsilon .$$

φ is called *μ-singular* if, whenever $\varepsilon > 0$ and H is a soma of finite μ-measure, there exists an element C of $\mathscr{A}$ such that

$$\mu(C \cdot H) < \varepsilon \quad \text{and} \quad \varphi^T(E - C) < \varepsilon .$$

Replacing φ^T by φ^+ and φ^-, we obtain the definitions of a μ-absolutely continuous or a μ-singular function from above or from below, respectively. In the definitions of μ-absolute continuity, we can replace $\varphi^T(A) < \varepsilon$, $\varphi^+(A) < \varepsilon$, or $\varphi^-(A) < \varepsilon$ by $|\varphi(A)| < \varepsilon$, $\varphi(A) < \varepsilon$, or $\varphi(A) > -\varepsilon$, respectively. If $H_1, H_2, \ldots, H_n, \ldots$ is an increasing sequence of somas of finite μ-measure with E as their supremum (union), it suffices to formulate the preceding definitions for only the somas H_n, $n = 1, 2, \ldots$, instead of all somas of finite μ-measure. In particular, if $\mu(E)$ is finite, it suffices to consider the case $H = E$.

1.1. Proposition. Each function in $\mathscr{M}_{\mathscr{A}}$ of bounded variation can be represented in a unique manner as the sum of a μ-absolutely continuous function and a μ-singular function.

1.2. Proposition. A function in $\mathscr{M}_{\mathscr{A}}$ of bounded variation is μ-absolutely continuous iff it is σ-additive and μ-singular iff it is purely simply additive.

Proof. It follows readily from the definition and from Prop. I. 2.1, that every μ-absolutely continuous function of bounded variation is σ-additive. Conversely, suppose that φ is of bounded variation and σ-additive. According to Prop. I. 2.6, φ possesses a σ-additive extension φ^0 on the Boolean σ-algebra $\mathscr{C}$ generated by $\mathscr{A}$. Since μ is strictly positive, then for any set $B \in \mathscr{C}$ with $\mu(B) = 0$, we have $B = \bigcirc$, and so $\varphi^0(B) = 0$. By a classical argument it follows that φ^0 is μ-absolutely continuous; hence so is φ.

Next, we assume that φ is of bounded variation and purely simply additive. We may assume that $\varphi \geqslant 0$ without loss of generality. We may write $\varphi = \varphi_1 + \varphi_2$, where φ_1 is μ-absolutely continuous, and φ_2 is μ-singular. Since φ_1 is σ-additive and also $0 \leqslant \varphi_1 \leqslant \varphi$, then it follows from the purely simply additive nature of φ that $\varphi_1 = 0$; hence $\varphi = \varphi_2$ is μ-singular. Finally, suppose that φ is of bounded variation and μ-singular. According to Prop. I. 2.7, we can write $\varphi = \varphi_c + \varphi_p$ where φ_c is σ-additive and φ_p is purely simply additive. By what has already been shown, φ_c is μ-absolutely continuous and φ_p is μ-singular. Since φ is itself μ-singular, the uniqueness of its decomposition as the sum of a μ-absolutely continuous function and a μ-singular function implies $\varphi_c = 0$; thus $\varphi = \varphi_p$ is purely simply additive.

Remark. Prop. 1.1, and the implications "μ-absolute continuity implies σ-additivity" and "purely simple additivity implies μ-singularity" for functions in $\mathscr{M}_{\mathscr{A}}$ of bounded variation, hold even when μ is not strictly positive.

1.3. Definitions. A $\mathscr{B}$-measurable function f (I. 11) is said to be *semi-integrable* if $\int_E f\,d\mu$ exists; f is *integrable from above (from below)* if $\int_E f\,d\mu < +\infty \left(\int_E f\,d\mu > -\infty\right)$.

Let $\mathscr{C}$ be a Boolean σ-subalgebra of $\mathscr{B}$ with unit E and let f be a $\mathscr{B}$-measurable function. If f is integrable from above, then the function defined on $\mathscr{C}$ by

$$\varphi(C) = \int_C f\,d\mu, \quad C \in \mathscr{C}, \tag{1.3.1}$$

is σ-additive, absolutely continuous from above, and of bounded variation from above. On the other hand, if φ is a σ-additive function defined on $\mathscr{C}$, then there exists one and only one function g that is $\mathscr{C}$-measurable, semi-integrable, and with $\varphi(C) = \int_C g\,d\mu$ for each $C \in \mathscr{C}$.

If the function f is generated by (1.3.1), then g is called the *conditional expectation of the random variable f for a given $\mathscr{C}$*, and is denoted by

$\mathscr{E}(f|\mathscr{C})$. Similar remarks apply to the case of a subalgebra $\mathscr{A}$ of $\mathscr{B}$, instead of $\mathscr{C}$, taking into account Prop. I. 2.6; g must then be taken measurable with respect to the σ-algebra generated by $\mathscr{A}$.

2. Absolutely continuous and singular premartingales. We return to the stochastic bases $(\mathscr{B}_\tau)$ of Ch. I. and assume henceforth that the restriction of μ to each family $\mathscr{B}_\tau$ is σ-finite. We set $\mathscr{A} = \bigcup_{\tau \in \Theta} \mathscr{B}_\tau$.

A premartingale $\boldsymbol{\Phi} = (\varphi_\tau)$ is called *μ-absolutely continuous* if, for each $\varepsilon > 0$, there exists a terminal subset Δ of Θ, a soma H of finite μ-measure and a positive number δ such that if $A \in \mathscr{B}$ and $\mu(H \cdot A) < \delta$, then $\varphi_\tau^T(A) < \varepsilon$.

Replacing the last inequality by '$\varphi_\tau^+(A) < \varepsilon$' or by '$\varphi^-(A) < \varepsilon$' yields the definitions for a premartingale μ-absolutely continuous from above or from below, respectively. The inequalities $|\varphi_\tau(A)| < \varepsilon$, $\varphi_\tau(A) < \varepsilon$, or $\varphi_\tau(A) > -\varepsilon$, respectively, may also be used. In the case of a martingale, Δ may be taken equal to Θ; the same is true for a semi-martingale in the definition of μ-absolute continuity from above. In fact, under these hypotheses, the sequences $(\varphi_\tau^T(A))$ and $(\varphi_\tau^+(A))$ are increasing for any $A \in \mathscr{A}$.

A premartingale $\boldsymbol{\Phi} = (\varphi_\tau)$ is called *μ-singular* if, for any $\varepsilon > 0$ and any soma H of finite μ-measure, there exists a parameter $\rho \in \Theta$ and an element C of $\mathscr{B}$ with $\mu(H \cdot C) < \varepsilon$, such that whenever $\tau \in \Theta$ and $\tau \gg \rho$, we have $\varphi_\tau^T(E - C) < \varepsilon$. The definitions of μ-singularity from above or from below are obtained by replacing 'φ_τ^T' by 'φ_τ^+' or 'φ_τ^-', respectively.

If $H_1, H_2, \ldots$ denotes an arbitrary increasing sequence of somas of finite μ-measure whose supremum is E, then it suffices to formulate the foregoing definitions for the somas H_n only, instead of all somas H of finite μ-measure. In particular, in the definition of μ-absolute continuity, once the parameter ρ is prescribed, H can always be chosen in $\mathscr{B}_\rho$.

Unlike the concepts of σ-additivity and purely simple additivity, the concepts of μ-absolute continuity and μ-singularity are defined for arbitrary premartingales and not only for martingales. This observation will enable us to give (Prop. 2.7) a simpler (in a sense) version of Prop. I. 8.2. First we deduce immediately the following proposition.

2.1. Proposition. Let $\boldsymbol{\Phi} = (\varphi_\tau)$ be a submartingale such that $\varphi_\rho(E) > -\infty$ for some $\rho \in \Theta$. Then $\boldsymbol{\Phi}^m$ is μ-absolutely continuous iff $\boldsymbol{\Phi}$ is μ-absolutely continuous.

2.2. Proposition. A martingale $\boldsymbol{\Phi} = (\varphi_\tau)$ is μ-absolutely continuous or μ-singular iff the additive function $\varphi^\vee = Z(\boldsymbol{\Phi})$ (I.6), defined on $\mathscr{A}$, possesses the corresponding property. Consequently, for martingales of bounded variation, the concepts of μ-absolute continuity and σ-additivity are equivalent, and so are the concepts of μ-singularity and purely simple additivity. Corresponding statements with respect to the same concepts "from above" and "from below" are valid.

Proof. It is sufficient to observe that $\varphi_\tau^T(A) \leqslant \varphi^{\vee T}(A)$ and $\lim_\tau \varphi_\tau^T(A) = \varphi^{\vee T}(A)$ holds for each $\tau \in \Theta$ and each $A \in \mathscr{A}$.

From Prop. I. 8.1 comes the following result.

2.3. Proposition. A martingale of bounded variation is μ-absolutely continuous iff it is induced by a σ-additive function defined on $\mathscr{B}_\omega$.

From Props. 2.1, 2.2, and I. 8.2, we deduce the following proposition.

2.4. Proposition. Let $\boldsymbol{\Phi} = (\varphi_\tau)$ be a submartingale such that there exists $\rho \in \Theta$ satisfying $\varphi_\rho(E) > -\infty$. Then a necessary and sufficient condition for the existence of a function φ_ω defined on $\mathscr{B}_\omega$, such that $(\varphi_\tau : \tau \in \Theta_\omega)$ is a submartingale and $\varphi_\omega(E) < +\infty$, is that $\boldsymbol{\Phi}$ be of bounded variation from above and μ-absolutely continuous from above.

3. Stochastic processes. Let $\boldsymbol{\Phi} = (\varphi_\tau)$ be a premartingale of basis $(\mathscr{B}_\tau)$. To each $\tau \in \Theta$ there corresponds a unique $\mathscr{B}_\tau$-measurable, semi-integrable function f_τ such that

$$\varphi_\tau(A) = \int_A f_\tau \, d\mu \tag{3.0.1}$$

for each $A \in \mathscr{B}_\tau$. To the premartingale $\boldsymbol{\Phi}$ there corresponds the sequence (f_τ) of the integrands, or *stochastic process.* (f_τ) is called the *integrand μ-representation*, or simply the *integrand representation, of* $\boldsymbol{\Phi}$. We write $\boldsymbol{\Phi} : (f_\tau)$. Conversely, to each stochastic process (f_τ), where f_τ denotes a $\mathscr{B}_\tau$-measurable and semi-integrable function, there corresponds the premartingale $\boldsymbol{\Phi} = (\varphi_\tau)$ defined by (3.0.1), which is called the *integral representation* of the stochastic process; and we shall sometimes write $\boldsymbol{\Phi} = (f_\tau)$.

Many definitions and propositions of Ch. I. are expressed in simple form by means of the integrand representation. For instance, the condition for $\boldsymbol{\Phi}$ to be a submartingale or a martingale becomes: $\rho \ll \tau$ implies $f_\rho \leqslant \mathscr{E}(f_\tau | \mathscr{B}_\rho)$ or $f_\rho = \mathscr{E}(f_\tau | \mathscr{B}_\rho)$, respectively.

Let $\mathfrak{C}$ denote the set of integrand μ-representations (f_τ) of a family of premartingales, where f_τ is used as a generic notation, and suppose that for each τ_0 at least one of the functions f_{τ_0} is integrable from below. Then the process (g_τ), where $g_{\tau_0} = \vee \{f_{\tau_0} : (f_\tau) \in \mathfrak{C}\}$, is the integrand μ-representation of the smallest premartingale majorizing the premartingales in $\mathfrak{C}$. We may observe that the function g_τ is defined independently of the integrand version of $\mathbf{S}_{\mathfrak{C}}^s$ (I. 4), which ensures only the integrability of g_τ from below. The notion "smallest premartingale majorizing a set of premartingales" is clearly intrinsic, meaning that it depends on the measure algebra $\mathscr{B}$, not on the measure μ itself.

For a single premartingale $\boldsymbol{\Phi} = (\varphi_\tau)$ with integrand representation (f_τ), to the premartingales (f_τ^T), (f_τ^+), and (f_τ^-) there correspond the sequences of integrands $(|f_\tau|)$, (f_τ^+), and (f_τ^-), respectively. These premartingales are submartingales if $\boldsymbol{\Phi}$ is a martingale.

A martingale $\boldsymbol{\Phi} = (\varphi_\tau)$ with (f_τ) as its integrand representation is induced by a σ-additive function φ_ω defined on the Borel extension $\mathscr{B}_\omega$ of $\mathscr{A}$ iff there exists a $\mathscr{B}_\omega$-measurable and semi-integrable function f_ω for which $f_\tau = \mathscr{E}(f_\omega|\mathscr{B}_\tau)$ whenever $\tau \in \Theta$. More generally, if f is a semi-integrable function and $f_\tau = \mathscr{E}(f|\mathscr{B}_\tau)$ for each $\tau \in \Theta$, then the sequence (f_τ) represents a martingale that is said to be *induced by* f; (f_τ) is then also induced by $f_\omega = \mathscr{E}(f|\mathscr{B}_\omega)$. The martingale $\boldsymbol{\Phi}$ whose integrand representation is $(f_\tau\colon \tau \in \Theta)$ is induced by a $\mathscr{B}_\omega$-measurable function f_ω iff $(f_\tau : \tau \in \Theta_\omega)$ represents a martingale. Props. I. 8.1, and II. 2.6 give us conditions ensuring the existence of such a function f_ω.

An additional example of the translation of a property of $\boldsymbol{\Phi}$ using (f_τ) is now given.

3.1. Proposition. A necessary and sufficient condition that the premartingale of integrand representation (f_τ) be of bounded variation (from above, from below) is that there exists a terminal subset Δ of Θ such that the set of numbers $\varphi_\tau^\wedge(B) = \int_B f_\tau \, d\mu$, where $\tau \in \Theta$ and $B \in \mathscr{B}$, be bounded (from above, from below, respectively).

In this proposition, we have considered the indefinite integral of f_τ on the Boolean σ-algebra $\mathscr{B}$ and not only on $\mathscr{B}_\tau$. We call this the *natural extension* of φ_τ relative to μ [37, 323]. In order to obtain an analogous criterion for μ-absolutely continuous premartingales of bounded variation, we give the following definitions, in which $(\varphi_\tau^\wedge)$ denotes the sequence of the indefinite integrals of a sequence (f_τ) of functions considered on the whole σ-algebra $\mathscr{B}$:

(φ_τ) is *terminally uniformly bounded* [37, 316] iff there exists a terminal set Δ in Θ such that the set of numbers $\varphi_\tau^\wedge(B)$, for all $B \in \mathscr{B}$ and $\tau \in \Delta$, is bounded;

(φ_τ) is *terminally uniformly μ-absolutely continuous* [37, 316] iff, for any positive number ε, there exists a positive number δ, a terminal set Δ in Θ, and an element H in $\mathscr{B}$ of finite μ-measure such that $|\varphi_\tau^\wedge(B)| < \varepsilon$ whenever $\tau \in \Delta$, $B \in \mathscr{B}$, and $\mu(H \cdot B) < \delta$.

(f_τ) is *terminally uniformly integrable* [37, 316) iff, for any $\varepsilon > 0$, there exists a terminal subset Δ of Θ, an element H of $\mathscr{B}$ of finite μ-measure and a positive number δ such that $\int_{G_\tau} |f_\tau| \, d\mu < \varepsilon$ and $\int_G |f_\tau| \, d\mu < \varepsilon$ whenever $\tau \in \Delta$, where $G_\tau = \{|f_\tau| \geqslant \delta\}$ and $G = E - H$.

The above definitions can be formulated *from above* and *from below* analogously.

We see that (φ_τ) is terminally uniformly bounded and terminally uniformly absolutely continuous iff the sequence (f_τ) is terminally uniformly integrable.

3.2. Proposition. A necessary and sufficient condition for a premartingale to be of bounded variation and μ-absolutely continuous is that its integrand representation be terminally uniformly integrable. Analogous results hold with respect to the same concepts "from above" and "from below".

In I. 3, we defined the trace Φ^A of a premartingale Φ on a soma A of $\mathscr{A}$ as the sequence $(\varphi_\tau | A \wedge \mathscr{B}_\tau)$ restricted to those values of $\tau \in \Theta$ for which $A \in \mathscr{B}_\tau$. If Φ is a premartingale whose integrand representation is (f_τ) and $A \in \mathscr{B}$, then we define the *μ-trace* of Φ as that premartingale with basis $(A \wedge \mathscr{B}_\tau)$ and whose integrand representation with respect to the measure $\mu | A \wedge \mathscr{B}_\tau$ is (f_τ^A), where $f_\tau^A = f_\tau | A$ for each $\tau \in \Theta$.

4. Stochastic convergence. A sequence (A_τ) of somas is said to *converge stochastically* to $\bigcirc$ if $\lim_\tau \mu(H \cdot A_\tau) = 0$ for each soma H of finite μ-measure. To a sequence (f_τ) of functions we associate the set $\mathscr{H}$ of those functions h such that the sequence $([h < f_\tau])$ converges stochastically to $\bigcirc$, and the set $\mathscr{G}$ of those functions g such that the sequence $([f_\tau < g])$ converges stochastically to $\bigcirc$. The *stochastic upper limit* $s \limsup_\tau f_\tau$ is defined as $\wedge \mathscr{H}$, and the *stochastic lower limit* $s \liminf_\tau f_\tau$ as $\vee \mathscr{G}$. These definitions and that given in [*38*, § 2] are equivalent. They are also intrinsic, i.e., they depend on the measure algebra $\mathscr{B}$ (I. 11) but not on the strictly positive, finite or σ-finite, measure μ on $\mathscr{B}$. Thus the algebra $\mathscr{B}$ can be replaced by a σ-subalgebra $\mathscr{C}$ without changing the stochastic limits, provided the functions f_τ, $\tau \in \Theta$, are $\mathscr{C}$-measurable. In case the two limit functions are equal, their common value is called the *stochastic limit* $s \lim_\tau f_\tau$ of the sequence (f_τ), which is then said to *converge stochastically*. This form of convergence is also known under the names of convergence in probability, convergence in measure, and asymptotic convergence. However, their definitions are usually given for a finite measure and for a FRÉCHET sequence of finite functions converging stochastically to a finite function. The stochastic convergence admits a localization principle: A sequence (f_τ) converges stochastically iff, for every soma A of finite measure, $(f_\tau | A)$ converges stochastically.

4.1. Proposition [*38*, 2.9]. A sequence (f_τ) of finite functions converges stochastically to a finite function f iff for each positive number η, the sequence $([|f_\tau - f| > \eta])$ converges stochastically to $\bigcirc$.

4.2. Proposition [*42*, Th. 3]. Let $\bar{\mathbf{R}}$ denote the extended real line, p a positive integer, and a a function taking values in $\bar{\mathbf{R}}$, defined and continuous on a subset of the topological product $\times_{i=1}^{p} \bar{\mathbf{R}}$. Suppose also that $(f_{i\tau})$ denotes sequences for $i = 1, 2, \ldots, p$, $\tau \in \Theta$, such that

$$s \lim_\tau f_{i\tau} = f_i, \quad i = 1, 2, \ldots, p,$$

and the values taken by $(f_1, f_2, \dots, f_p)$ and $(f_{1\tau}, f_{2\tau}, \dots, f_{p\tau})$, $\tau \in \Theta$, are in S. Then $s \lim_\tau a(f_{1\tau}, f_{2\tau}, \dots, f_{p\tau}) = a(f_1, f_2, \dots, f_p)$.

4.3. Fatou's lemmas for stochastic limits. We first formulate two conditions:

$(\mathrm{S})^+$ $((\mathrm{S})^-)$ For each sequence (A_τ) in $\mathscr{B}$ converging to $\bigcirc$ stochastically we have the relation

$$\limsup_\tau \int_{A_\tau} f_\tau \, d\mu \leqslant 0 \quad \left(\liminf_\tau \int_{A_\tau} f_\tau \, d\mu \geqslant 0\right).$$

The conjunction of $(\mathrm{S})^+$ and $(\mathrm{S})^-$ will be denoted by (S).

4.3.1. Proposition [*38*, 480]. If (f_τ) is a sequence of semi-integrable functions satisfying $(\mathrm{S})^+$ and if $s \limsup f$ is semi-integrable, then for each element A in $\mathscr{B}$ we have

$$\limsup_\tau \int_A f_\tau \, d\mu \leqslant \int_A (s \limsup_\tau f_\tau) \, d\mu . \tag{4.3.1.1}$$

Analogously, under the condition $(\mathrm{S})^-$, if $s \liminf_\tau f_\tau$ is semi-integrable, then

$$\int_A \left(s \liminf_\tau f_\tau\right) d\mu \leqslant \liminf_\tau \int_A f_\tau \, d\mu . \tag{4.3.1.2}$$

Proof. It is sufficient to consider the case $A = E$. We abbreviate by setting $h^\star = s \limsup_\tau f_\tau$. We assume first that

$$-\infty < \int_E h^\star \, d\mu . \tag{4.3.1.3}$$

Then $-\infty < h^\star$ and so, given $\varepsilon > 0$, there exists an integrable majorant h' of $h^\star$ such that

$$\int_B h' \, d\mu \leqslant \int_B h^\star \, d\mu + \varepsilon \tag{4.3.1.4}$$

holds for each B in $\mathscr{B}$. We define

$$A_\tau = [h' < f_\tau] \quad \text{and} \quad B_\tau = E - A_\tau = [f_\tau \leqslant h'] .$$

Since (A_τ) converges stochastically to $\bigcirc$, we can invoke $(\mathrm{S})^+$ to find $\tau_1 \in \Theta$ so that $\int_{A_\tau} f_\tau \, d\mu < \varepsilon$ holds for each $\tau \in \Theta$, $\tau_1 \ll \tau$. Also, by virtue of (4.3.1.3), there exists $\tau_2 \in \Theta$ such that

$$-\varepsilon < \int_{A_\tau} h^\star \, d\mu$$

whenever $\tau_2 \ll \tau$. Combining these inequalities, we see that if $\tau_1 \ll \tau$ and $\tau_2 \ll \tau$, then

$$\int_E f_\tau \, d\mu = \int_{B_\tau} f_\tau \, d\mu + \int_{A_\tau} f_\tau \, d\mu \leqslant \int_{B_\tau} h' \, d\mu \leqslant \int_{B_\tau} h^\star \, d\mu + 2\varepsilon \leqslant \int_E h^\star \, d\mu + 3\varepsilon .$$

The assumption (4.3.1.3) can now be dropped. To this end, we consider the functions $f_{\tau,n} = f_\tau \vee g_n$, where the Fréchet sequence (g_n) is monotone decreasing, converges to $-\infty$, and each function g_n is negative and integrable. One may now use the relation $s\lim_\tau \sup f_\tau = \left(s\lim_\tau \sup f_\tau\right) \vee g_n$ which follows from Prop. 4.2, with $p = 2$ and $a(x,y) = x \vee y$.

The inequality (4.3.1.2) is obtained from (4.3.1.1) by taking $-f$ in place of f.

Remark. An example of AMERIO [*18*, 14] shows that the hypothesis of the semi-integrability of $s\lim_\tau \sup f_\tau$ cannot be dropped, even if (f_τ) is a Fréchet sequence and we have $(\mathbf{S})^+$ and $(\mathbf{S})^-$ both holding.

4.4. Proposition. If the premartingale $\boldsymbol{\Phi}$ with integrand representation (f_τ) is of bounded variation from above, μ-absolutely continuous from above, and $s\lim_\tau \sup f_\tau$ is semi-integrable, then for each soma B we have

$$\lim_\tau \sup \int_B f_\tau d\mu \leqslant \int_B \left(s\lim_\tau \sup f_\tau\right) d\mu .$$

Analogous results hold when "from below" replaces "from above" in the hypotheses, and $s\lim_\tau \inf f_\tau$ is assumed to be semi-integrable.

If $\boldsymbol{\Phi}$ is of bounded variation and μ-absolutely continuous, and if $f_\omega = s\lim_\tau f_\tau$ exists, then f_ω is integrable, $\lim_\tau \int_B f_\tau d\mu = \int_B f_\omega d\mu$, and $\lim_\tau \int_B |f_\tau - f_\omega| d\mu = 0$ for any soma B.

Proof. If (f_τ) is of bounded variation from above and μ-absolutely continuous from above, then it follows from Proposition 3.2 that (f_τ) is terminally uniformly integrable from above and $(\mathbf{S})^+$ holds. Application of Prop. 4.3.1 completes the proof.

4.5. Lemma. [*38*, 483]. *If $(-f_i\colon i = 1,2,\dots,r)$ is a negative submartingale with $(\mathscr{B}_i)$ as basis, β is a fixed real number, and $S = [\beta < \bigvee_{i=1}^{r} f_i]$, then we have*

$$\beta\mu(S) \leqslant \int_E f_1 d\mu . \tag{4.5.1}$$

Proof. The functions f_i, $i = 1,2,\dots,r$ are non-negative. We define

$$K_1 = \{\beta < f_1\};\quad K_i = \{\beta \geqslant f_j, 1 \leqslant j < i; \beta < f_i\},\quad i = 1,2,\dots,r. \tag{4.5.2}$$

Then $S = \bigvee_{i=1}^{r} K_i$. Taking account of the submartingale nature of the sequence $(-f_i)$ we have

$$\begin{aligned} \int_A (-f_i) d\mu \leqslant \int_A (-f_{i+1}) d\mu, \quad &\text{and} \\ \int_A f_i d\mu \geqslant \int_A f_{i+1} d\mu, \quad &\text{whenever } A \in \mathscr{B}_i,\ 1 \leqslant i \leqslant r. \end{aligned} \tag{4.5.3}$$

We define the functions $\bar{f}_i$ as follows:

$$\bar{f}_i|K_j = f_j|K_j, \quad j = 1,2,\ldots,i;$$

$$\bar{f}_i|(E-S) = f_i|(E-S_i), \tag{4.5.4}$$

where $S_i = \bigvee_{j=1}^{i} K_j$, $i = 1,2,\ldots,r$.

If $\int_E f_1\, d\mu = +\infty$ then (4.5.1) holds trivially, so we may suppose that $\int_E f_1\, d\mu < +\infty$. We see from (4.5.3) that the functions f_i, $1 \leqslant i \leqslant r$, are integrable. Now

$$\int_{E-S_i} (\bar{f}_i - \bar{f}_{i+1})\, d\mu = \int_{E-S_i} (f_i - f_{i+1})\, d\mu, \tag{4.5.5}$$

since the integrands are identical on $E-S_i$. Thus these integrals are non-negative because of (4.5.3).

Also

$$\int_{S_i} (\bar{f}_i - \bar{f}_{i+1})\, d\mu = 0; \tag{4.5.6}$$

thus, combining (4.5.5) and (4.5.6) we obtain

$$\int_E (\bar{f}_i - \bar{f}_{i+1})\, d\mu \geqslant 0, \quad \text{or} \quad \int_E \bar{f}_i\, d\mu \geqslant \int_E \bar{f}_{i+1}\, d\mu. \tag{4.5.7}$$

Therefore,

$$\int_E f_1\, d\mu = \int_E \bar{f}_1\, d\mu \geqslant \int_E \bar{f}_r\, d\mu \geqslant \int_S \bar{f}_r\, d\mu \geqslant \beta\mu(S),$$

and the proof is complete.

4.6. Lemma. [*38*, 483]. *Suppose that $A \in \mathscr{B}$, $\mu(A) < +\infty$, and $\zeta \in \Theta$; also (K_τ) is a sequence in $\mathscr{B}$ such that for each B satisfying $\bigcirc < B \leqslant A$ we have*

$$\limsup_\tau \mu(B \cdot K_\tau) > 0. \tag{4.6.1}$$

Then there exists a Fréchet sequence (ξ_i) in Θ satisfying $\zeta \ll \xi_1 \ll \xi_2 \ll \ldots$ and $A \leqslant \bigvee_{i=1}^{\infty} K_{\xi_i}$.

Proof. (By an exhaustion process as in [*23*, 207]). For each $B \in \mathscr{B}$ with $B \leqslant A$, define $v(B) = \limsup_\tau \mu(B \cdot K_\tau)$. If $B \leqslant C \leqslant A$, $B \in \mathscr{B}$, and $C \in \mathscr{B}$ then $v(B) \leqslant v(C)$; and $v(B) = 0$ is equivalent to $B = \bigcirc$.

We put $B_1 = A$. We may assume $B_1 > \bigcirc$ or there would be nothing to prove. We choose ξ_1 such that $\zeta \ll \xi_1$ and $\mu(B_1 \cdot K_{\xi_1}) > 2^{-1} v(B_1)$. We construct the sequence (ξ_i) inductively. We assume that $\xi_1, \xi_2, \ldots, \xi_r$ have already been obtained, with

$$\zeta \ll \xi_1 \ll \ldots \ll \xi_r, \quad \mu(B_i \cdot K_{\xi_i}) > 2^{-1} v(B_i), \quad i = 1,2,\ldots,r, \tag{4.6.2}$$

where $B_i = A - A\cdot(K_{\xi_1} \vee K_{\xi_2} \dots \vee K_{\xi_{i-1}})$. We put

$$B_{r+1} = A - A\cdot(K_{\xi 1} \vee K_{\xi 2} \dots \vee K_{\xi r}).$$

If $B_{r+1} = \bigcirc$, then we have attained our goal. If $B_{r+1} > \bigcirc$, then there is an element ξ_{r+1} with $\xi_r \ll \xi_{r+1}$ and $\mu(B_{r+1}\cdot K_{\xi_{r+1}}) > 2^{-1} v(B_{r+1})$. Thus we need consider only the case where $B_s \neq \bigcirc$ for $s = 1, 2, \dots$. Since the elements $B_i \cdot K_{\xi_i}$ are pairwise disjoint and are parts of A, for $i = 1, 2, \dots$, it follows that $\lim_i \mu(B_i \cdot K_{\xi_i}) = 0$; hence by (4.6.2), we also have $\lim_i v(B_i) = 0$. Therefore, $v\left(\bigwedge_{i=1}^{\infty} B_i\right) = 0$ and so $\bigwedge_{i=1}^{\infty} B_i = \bigcirc$. However, $\bigwedge_{i=1}^{\infty} B_i = A - A \cdot \bigvee_{i=1}^{\infty} K_{\xi_i}$, and this proves the lemma.

4.7. Theorem [42, 490]. *If Φ is a submartingale of bounded variation and Φ^m its integral as defined in* I. 4, *then the corresponding integrand representations (f_τ) and (f_τ^m) converge stochastically to a function f_ω that is integrable. If Φ denotes a martingale of bounded variation from above or from below, then (f_τ) converges stochastically to a function f_ω that is integrable from above or from below, respectively, and we have*

$$\int_B f_\omega^+ \, d\mu \leqslant \lim_\tau \int_B f_\tau^+ \, d\mu, \qquad \int_B f_\omega^- \, d\mu \leqslant \lim_\tau \int_B f_\tau^- \, d\mu$$

and

$$\int_B |f_\omega| \, d\mu \leqslant \lim_\tau \int_B |f_\tau| \, d\mu, \tag{4.7.1}$$

for any soma B in $\mathscr{A}$; the first inequality remains true in the case of a submartingale of bounded variation.

Proof. We prove first the existence and integrability of the stochastic limits of the submartingale (f_τ) of bounded variation. Each submartingale (f_τ) of bounded variation can be written for terminally many indices τ in the form $f_\tau^m + (f_\tau - f_\tau^m)$, where (f_τ^m) is a martingale of bounded variation and $(f_\tau - f_\tau^m)$ is a negative submartingale. According to Prop. I. 4.5, each martingale of bounded variation is the difference of two positive, and hence also of two negative, martingales. Thus it is sufficient to establish the stochastic convergence for a sequence (f_τ) with $0 \leqslant f_\tau$, $\tau \in \Theta$, where $(-f_\tau)$ represents a submartingale.

We suppose that (f_τ) does not converge stochastically. We denote by $\mathscr{B}_\omega$ the smallest Boolean σ-algebra of $\mathscr{B}$ that includes every set $\mathscr{B}_\tau$, $\tau \in \Theta$. Then there exists an element $A \in \mathscr{B}_\omega$ of finite μ-measure and two positive finite numbers α and β such that

$$s \lim_\tau \inf f_\tau < \alpha < \beta < s \lim_\tau \sup f_\tau \ [A], \tag{4.7.2}$$

where $[A]$ denotes the restriction of these expressions to A.

For each element B, $\bigcirc < B \leqslant A$, we have

$$\limsup_{\tau} \mu(B \cdot [f_\tau < \alpha]) > 0, \limsup_{\tau} \mu(B \cdot [\beta < f_\tau]) > 0. \quad (4.7.3)$$

In particular, we have

$$\eta = 2^{-1} \limsup_{\tau} \mu(A \cdot [f_\tau < \alpha]) > 0. \quad (4.7.4)$$

Let ε satisfy $0 < \varepsilon < 1$. Since $A \in \mathscr{B}_\omega$, there exists an index ρ and an element C in $\mathscr{B}_\rho$ such that

$$\mu(A - C) < \varepsilon\eta, \quad (4.7.5)$$

where $-$ denotes symmetric difference. Then, using (4.7.4), there is an index ζ with $\rho \ll \zeta$ and

$$\mu(A \cdot [f_\zeta < \alpha]) > \eta. \quad (4.7.6)$$

We define $L = C \cdot [f_\zeta < \alpha]$, so that

$$L \in \mathscr{B}_\zeta, \; f_\zeta < \alpha \, [L]. \quad (4.7.7)$$

Moreover, we shall show that

$$\mu(L \cdot A) > (1-\varepsilon)\mu(L). \quad (4.7.8)$$

In fact, if we put $\delta = \mu(A \cdot [f_\zeta < \alpha])$, $\delta_1 = \mu((A - A \cdot C) \cdot [f_\zeta < \alpha])$, and $\delta_2 = \mu((C - C \cdot A) \cdot [f_\zeta < \alpha])$, then

$$\mu(L) = \delta - \delta_1 + \delta_2 \quad \text{and} \quad \mu(L \cdot A) = \delta - \delta_1.$$

By virtue of (4.7.6) and (4.7.5), and $\delta_1 + \delta_2 < \varepsilon\eta$ and $\delta > \eta$; thus $\delta_1 < \varepsilon\delta$ and $\delta_2 < \varepsilon\delta$. Therefore,

$$\frac{\mu(L)}{\mu(L \cdot A)} = \frac{\delta - \delta_1 + \delta_2}{\delta - \delta_2} = 1 + \frac{\delta_2}{\delta - \delta_2} < 1 + \frac{\varepsilon}{1-\varepsilon} = \frac{1}{1-\varepsilon},$$

and so (4.7.8) is established.

From the second inequality in (4.7.3) and Lemma 4.6, we infer the existence of a Fréchet sequence (ξ_i) in Θ satisfying

$$\zeta \ll \xi_1 \ll \xi_2 \ll \ldots \quad \text{and} \quad A \leqslant \bigvee_{i=1}^{\infty} [\beta < f_{\xi_i}].$$

Hence there exists a positive integer r such that

$$K = L \cdot \bigvee_{i=1}^{r} [\beta < f_{\xi_i}] = L \cdot \left[\beta < \bigvee_{i=1}^{r} f_{\xi_i}\right] \quad (4.7.9)$$

satisfies the inequality $\mu(A \cdot L - A \cdot K) < \varepsilon\mu(A \cdot L)$. Using (4.7.8) we obtain

$$\mu(K) \geqslant \mu(A \cdot K) > \mu(A \cdot L)(1-\varepsilon) > (1-\varepsilon)^2 \mu(L). \quad (4.7.10)$$

Because of (4.7.7) and since $\zeta \ll \xi_1 \ll \xi_1 \ldots \ll \xi_r$, the restrictions to L of the functions $-f_{\xi_1}, -f_{\xi_2}, \ldots, -f_{\xi_r}$ constitute a negative submartingale. From Lemma 4.5 and (4.7.9) it follows that

$$\beta\mu(K) \leqslant \int_L f_{\xi_1} d\mu .$$

Because $\zeta \ll \xi_1$ and we are dealing with a submartingale we may infer

$$\int_L f_{\xi_1} d\mu \leqslant \int_L f_\zeta d\mu \leqslant \alpha\mu(L),$$

whence $\beta\mu(K) \leqslant \alpha\mu(L)$ and thus, using (4.7.10), we obtain $\beta(1-\varepsilon)^2 < \alpha$. By taking ε sufficiently small, this leads to a contradiction of the fact that $\alpha < \beta$. This establishes the existence and the integrability of the stochastic limit of (f_τ).

We wish now to show that $s \lim_\tau f_\tau = s \lim_\tau f_\tau^m$. We consider the submartingale $\Phi - \Phi^m$, with $\eta_\tau = f_\tau - f_\tau^m$, $\tau \in \Theta$, as its integrand representation. Thanks to Prop. 4.2,

$$h_\omega = s \lim_\tau h_\tau = s \lim_\tau f_\tau - s \lim_\tau f_\tau^m .$$

These limits are integrable and so are finite because Φ and Φ^m are both of bounded variation. Clearly $h_\tau \leqslant 0$, $\tau \in \Theta$; therefore, $h_\omega \leqslant 0$. Since (h_τ) is μ-absolutely continuous from above, we have, by Prop. 4.4,

$$\lim_\tau \int_E h_\tau d\mu \leqslant \int_E h_\omega d\mu \leqslant 0 .$$

The definition of Φ^m implies $\lim_\tau \int_E h_\tau d\mu = 0$; thus $\int_E h_\omega d\mu = 0$ and so $h_\omega = 0$.

The existence of the right-hand limits in (4.7.1) for any $B \in \mathscr{A}$ follows because under our assumptions (f_τ^+), (f_τ^-), and (f_τ^T) are submartingales. To obtain the inequalities themselves it is sufficient to apply Prop. 4.4.

4.8. Corollary. If Φ is a martingale of bounded variation then the sequences $(f_\tau^+)^m$, $(f_\tau^-)^m$, and $(f_\tau^T)^m$ converge stochastically to f_ω^+, f_ω^-, and f_ω^T, respectively.

Proof. Applying Th. 4.7 to the submartingale (f_τ^+), which is of bounded variation, and Prop. 4.2, for $n = 1$, $a(s) = x^+$, we obtain

$$s \lim_\tau f_\tau^{+m} = s \lim_\tau f_\tau^+ = \left(s \lim_\tau f_\tau\right)^+ .$$

Similar results hold for (f_τ^{-m}) and (f_τ^{Tm}).

4.9. Theorem [*42*, 491]. *If $\Phi = (\varphi_\tau)$ is a martingale of bounded variation with integrand μ-representation (f_τ), then Φ admits the decomposition*

$$\varphi_\tau(A) = \varphi_{p,\tau}(A) + \int_A f_\omega d\mu, \qquad \tau \in \Theta, \quad A \in \mathscr{B}_\tau ,$$

where $f_\omega = s\lim_\tau f_\tau$ *and* $\Phi_p = (\varphi_{p,\tau})$ *is the purely simply additive part of* Φ*; the* σ*-additive part* Φ_c *of* Φ *is then the martingale induced by* f_ω.

Proof. We bear in mind Props. 2.2 and 2.3, as we proceed.

Since the martingale Φ^0 induced by f_ω is σ-additive, it suffices to prove that $\Phi^{00} = \Phi - \Phi^0$ is purely simply additive. By virtue of Cor. 4.8, we need to treat only the case $\Phi \geqslant \Omega$. According to the inequalities (4.7.1) of Th. 4.7, we have $\Phi^{00} \geqslant \Omega$. We denote the stochastic limit of the integrand sequence of Φ^0 by f^0_ω, and the integrand representation of Φ^{00} by (f^{00}_τ). Props. 2.3 and 4.4 yield

$$\int_B f^0_\omega d\mu = \lim_\tau \int_B f^0_\tau d\mu = \int_B f_\omega d\mu$$

for each soma B of $\mathscr{A}$; consequently, $f^0_\omega = f_\omega$ and $s\lim_\tau f^{00}_\tau = 0$. Let Ψ be a σ-additive martingale with (g_τ) as its integrand representation, such that

$$\Omega \leqslant \Psi \leqslant \Phi .$$

Since

$$0 \leqslant s\lim_\tau g_\tau \leqslant s\lim_\tau f^{00}_\tau ,$$

we have $s\lim_\tau g_\tau = 0$.

From Props. 2.2 and 4.4, we obtain

$$\int_B g_\rho d\mu = \lim_\tau \int_B g_\tau d\mu = \int_B \left(s\lim_\tau g_\tau\right) d\mu = 0$$

for each $B \in \mathscr{B}_\rho$; thus $\Psi = \Omega$.

4.10. Corollaries. In the terminology of Th. 4.9, we have: (1) A necessary and sufficient condition for Φ to be σ-additive is that Φ be induced by f_ω; i.e., that the sequence $(f_\tau : \tau \in \Theta_\omega)$ represent a martingale.

(2) Φ is purely simply additive iff $f_\omega = 0$.

Proof. These are immediate consequences of Th. 4.9.

Complement to Th. 4.9, *and* Cor. 4.10. According to Cor. 4.8, Φ_c^+, Φ_c^-, and Φ_c^T are induced by f^+_ω, f^-_ω, and f^T_ω, respectively.

4.11. Theorem. *Suppose that* Φ *denotes a submartingale of bounded variation,* Φ^m *its integral,* (f_τ) *and* (f^m_τ) *their respective integrand representations, and* $f_\omega = s\lim_\tau f_\tau$. *Then the martingale induced by* f_ω *is the* μ*-absolutely continuous part* Φ^m_c *of* Φ^m. Φ *is* μ*-absolutely continuous from above iff* $(f_\tau : \tau \in \Theta_\omega)$ *represents a submartingale.*

Proof. The first assertion follows from Ths. 4.7 and 4.9. The sufficiency of the condition in the second assertion follows from Th. 2.4.

Finally, we assume that Φ is absolutely continuous from above. For each $\rho \in \Theta$ and each $A \in \mathscr{B}_\rho$ we have, by virtue of Prop. 4.4,

$$\int_A f_\rho d\mu \leqslant \lim_\tau \int_A f_\tau d\mu \leqslant \int_A f_\omega d\mu .$$

These ensure that $(f_\tau : \tau \in \Theta_\omega)$ represents a submartingale.

Remark. Ths. 4.9 and 4.11 are convergence theorems in the full sense (cf. "full differentiation theorems" according to [*25*, 221]); i.e., we not only prove the convergence of the sequence (f_τ), but in addition we interpret the integral of the limit f_ω.

5. Mean convergence of order 1. The norm of order 1 of a function f is defined as $\|f\|_1 = \int_E |f| \, d\mu$. We denote by $L_1 = L_1(\mathscr{B})$ the Banach space of those functions satisfying $\|f\|_1 < +\infty$, of norm $\|f\|_1$. If φ denotes the indefinite integral of f on $\mathscr{B}$, then this norm is the total variation of φ, and so it is the norm defined in (I. 11). The transformation $f \to \varphi$ maps $L_1(\mathscr{B}_\omega)$ isomorphically and isometrically onto $\mathscr{V}_1$. For a premartingale $\boldsymbol{\Phi}$ with the integrand representation (f_τ), the norm $\|\boldsymbol{\Phi}\|_1$, defined at the end of (I. 3), is $\limsup_\tau \|\varphi_\tau\|_1$. Thus $\boldsymbol{\Phi}$ is of bounded variation iff this limit superior is finite. In the case of a martingale

$$\|\boldsymbol{\Phi}\|_1 = \lim_\tau \|f_\tau\|_1 = \sup_\tau \|f_\tau\|_1 .$$

(cf. definitions following Cor. I. 4.4.)

Let $\boldsymbol{\Phi}$ be a martingale of bounded variation, $\boldsymbol{\Phi}_c$ the σ-additive part, $\boldsymbol{\Phi}_p$ the purely simply additive part, and $f_\omega = s \lim_\tau f_\tau$. We put

$$\varphi^{\nu T} = Z(\boldsymbol{\Phi}^T), \quad \varphi_c^{\nu T} = Z(\boldsymbol{\Phi}_c^T), \quad \text{and} \quad \varphi_p^{\nu T} = Z(\boldsymbol{\Phi}_p^T). \tag{I. 6, I. 7}$$

We recall that the operators Z and ^{T} are commutable, as are Z and $_c$; also Z and $_p$. In accordance with the definition of $\boldsymbol{\Phi}^T$ (I. 4.4), we have $\|\boldsymbol{\Phi}\|_1 = \varphi^{\nu T}(E)$. Moreover, we infer from Cor. 4.10 that $\|f_\omega\|_1 = \varphi_c^T(E)$. Since $\boldsymbol{\Phi}_p^T = \boldsymbol{\Phi}^T - \boldsymbol{\Phi}_c^T$, we obtain

$$\varphi_p^{\nu T}(E) = \|\boldsymbol{\Phi}\|_1 - \|f_\omega\|_1 . \tag{DS}$$

This is an expression for the deficiency of σ-additivity of $\boldsymbol{\Phi}$ (Def. I. 7.2).

5.1. Proposition [*43*, 494]. Let $\boldsymbol{\Phi}$ be a martingale of bounded variation, (f_τ) its integrand representation, and $f_\omega = s \lim_\tau f_\tau$. Then the following conditions are equivalent:

(1) $\boldsymbol{\Phi}$ is σ-additive;
(2) (f_τ) converges strongly to f_ω in L_1;
(3) (f_τ) converges weakly to f_ω in L_1;
(4) $\|\boldsymbol{\Phi}\|_1 = \|f_\omega\|_1$.

Proof. (1) implies (2) on account of Props. 2.2 and 4.4. Next, we prove that (3) implies (1). We consider a martingale such that the sequence (f_τ) converges weakly in L_1 to a function f. For each soma $A \in \mathscr{B}_\tau$, $\tau \in \Theta$, we have

$$\int_A f_\tau \, d\mu = \lim_{\tau \ll \delta} \int_A f_\delta \, d\mu = \int_A f \, d\mu ,$$

since the integral over A of a function in L_1 is a linear functional in L_1. Thus $\boldsymbol{\Phi}$ is induced by f. In accordance with Cor. 4.10, $\boldsymbol{\Phi}$ is of bounded variation and σ-additive; consequently the sequence (f_τ) converges weakly in L_1 to $s \lim_\tau f_\tau$. Finally, (4) is equivalent to (1) by virtue of (DS).

The equivalence of the four conditions: "$\boldsymbol{\Phi}$ is uniformly integrable", "$\boldsymbol{\Phi}$ is a martingale induced by a function in L_1", "$\boldsymbol{\Phi}$ converges strongly or weakly in L_1" was proved in [*26*]. Whereas, in the proof of Prop. 5.1, the limit function (f_τ) in L_1 appears as a stochastic limit that exists by virtue of Th. 4.7, HELMS introduces it by referring to the weak compactness of the uniformly integrable subsets of L_1. A third method of obtaining the limit function proceeds from Prop. I. 8.1; in fact, for induced martingales, strong convergence in L_1 is established quite easily. The equivalence of (1) and (3) can also be deduced from Prop. I. 11.1 and Th. I. 11.3 applied to ψ_τ, the indefinite integral of f_τ on $\mathscr{B}_\omega$.

We observe that $\|\varphi_\tau\| = \|\psi_\tau\|$ on $\mathscr{B}_\omega$, whenever $\tau \in \Theta$. Thus, if the conjecture in Complement 3^0 at the end of (I. 11) proves to be true, then the corresponding proposition, Prop. I. 11.1, and Th. I. 11.3, could be envisaged as an intrinsic version (without reference to μ) of Prop. 5.1, at least for (1), (2), and (3). The intrinsic formulation of (4) is $\|\boldsymbol{\Phi}\|_1 = \|\boldsymbol{\Phi}_c\|_1$ (Def. I. 7.2).

Let $\mathscr{C}$ denote a Boolean σ-subalgebra of $\mathscr{B}$, (cf. opening of I. 11), ξ a real function defined on $\mathscr{C}$, σ-additive and of bounded variation. Then to each measure $\mu|\mathscr{B}$ finite and strictly positive, there corresponds a σ-additive extension of ξ on $\mathscr{B}$, having the same norm as ξ, namely the *natural extension* $\xi^\wedge|\mathscr{B}$ (cf. remarks following (Prop. 3.1)). We conjecture that every σ-additive norm-preserving extension of ξ on $\mathscr{B}$ can be interpreted as the natural extension of ξ with respect to a suitable measure μ. In this connection, CAIROLI [*4*, Th. 4] has proved that the conjecture is true if $\mathscr{B}$ and $\mathscr{C}$ are Boolean σ-algebras of sets (tribes). His proofs hold if $\mathscr{B}$ is a measure algebra, in which case the measure obtained is not necessarily strictly positive.

A theorem by DUBROVSKIJ [*15*] concerning subsets $\mathscr{W}$ of $\mathscr{V}$ asserts the equivalence of the following properties: (i) The $\mathscr{W}$-functions are uniformly σ-additive; (ii) There exists a strictly positive, finite and σ-additive measure μ defined on $\mathscr{B}$ such that the functions of $\mathscr{W}$ are uniformly μ-continuous. We consider now a σ-additive martingale $\boldsymbol{\Phi} = (\varphi_\tau)$ of bounded variation and for each τ a σ-additive, norm-preserving extension $\psi_\tau|\mathscr{B}_\tau$. One may conjecture the possibility that there exists a finite positive measure μ such that for each τ, ψ_τ is the natural extension of φ_τ on $\mathscr{B}_\omega$ relative to μ. If this should be so, then according to Prop. 5.1, the sequence would converge strongly to φ_ω.

The preceding considerations regarding mean convergence can be applied to the theorems of stochastic convergence of martingales in § 4 and later to the theorems on order convergence in (III. 3); these convergence notions are of intrinsic nature.

6. Convergence in Orlicz spaces [*61*, 78–85]. We denote by x and y two functions defined in $[0, +\infty[$, taking non-negative values, including possibly $+\infty$, non-decreasing, not vanishing identically, and finite at least on $[0, \delta]$ with $0 < \delta < +\infty$. We assume that x and y are inverses of each other. Thus the "saltuses" of x correspond to the intervals of constancy of y, and vice versa. If $x(u) \leqslant \lambda$ for some finite λ and all $u \in [0, +\infty[$, then $y(v) = +\infty$ whenever $\lambda < y$. Since x determines y, except for at most an enumerable set, and conversely, then the functions

$$X(u) = \int_0^u x(\bar{u})\,d\bar{u} \quad \text{and} \quad Y(v) = \int_0^v y(\bar{v})\,d\bar{v}$$

are determined in a unique manner by either x or y. We have $X(0) = Y(0) = 0$, X and Y are non-negative, non-decreasing, not identically zero, finite on the interval $[0, \delta]$, and convex on every interval where they are finite. For any μ-measurable function f on E we put

$$\|f\|_X = \text{l.u.b.}\{\int_E |fg|\,d\mu : \int_E Y(|g|)\,d\mu \geqslant 1\}, \tag{6.0.1}$$

$$\|f\|_Y = \text{l.u.b.}\{\int_E |fg|\,d\mu : \int_E X(|g|)\,d\mu \geqslant 1\}.$$

The set L_X of those functions f with $\|f\|_X < +\infty$ is a Banach space with respect to the norm $\|\ \|_X$. L_X and L_Y are called *complementary* (real) Orlicz spaces. We have

$$\int_E |fg|\,d\mu \leqslant \|f\|_X \cdot \|g\|_Y. \tag{6.0.2}$$

6.1. Special cases. (i) $x(u) = 1$ for all $u \in [0, +\infty[$. Then $X(u) = u$, $y(v) = Y(v) = 0$ if $0 \leqslant v \leqslant 1$; $y(v) = Y(v) = +\infty$ if $1 < v$. Also $\|f\|_X = \|f\|_1$, $\|f\|_Y = \|f\|_{+\infty}$, $L_X = L_1$, and $L_Y = L_{+\infty}$.

If A is a subelement of E of finite μ-measure, then there exist two finite positive constants α and β such that

$$\alpha \int_A |f|\,d\mu \leqslant \|f c_A\|_X \leqslant \beta \ \underset{A}{\text{l.u.b.}}\, |f| \qquad \text{for each } f.$$

Therefore, we have $L_{+\infty} \subset L_X \subset L_1$, and strong convergence in $L_{+\infty}$ or L_X implies convergence in L_X or L_1, respectively.

(ii) p and q are two finite, positive constants such that $p^{-1} + q^{-1} = 1$, $x(u) = u^{p-1}$, then $X(u) = p^{-1}u^p$; $y(v) = v^{q-1}$, $Y(v) = q^{-1}v^q$, $\|f\|_X = q^q \|f\|_p$, and $\|f\|_Y = p^{p-1}\|f\|_q$, $L_X = L_p$, and $L_Y = L_q$.

6.2. General case. Concepts "at finite measure". In case $\mu(E) = +\infty$, a function f in L_X need not be semi-integrable (on E); however, it is "integrable at finite measure", meaning that f is integrable on every set of finite measure. This term is used in a manner analogous to the common expression "at finite distance". Accordingly, we define a *premartingale at finite measure* (with the basis $(\mathscr{B}_\tau)$) as a sequence $\boldsymbol{\Phi} = (\varphi_\tau | \mathscr{B}_\tau^f)$, where $\mathscr{B}_\tau^f$ denotes the Boolean subalgebra of the somas in $\mathscr{B}$ of finite measure, and φ_τ is a σ-additive function on $\mathscr{B}_\tau^f$ taking finite values. $\boldsymbol{\Phi}$ is called a *submartingale (martingale) at finite measure* iff, whenever $\rho \leqslant \tau$ and $A \in \mathscr{B}_\rho^f$, we have

$$\varphi_\rho(A) \leqslant \varphi_\tau(A) \quad (\varphi_\rho(A) = \varphi_\tau(A)).$$

To each function $\varphi_\tau | \mathscr{B}_\tau^f$ there corresponds a $\mathscr{B}_\tau$-measurable function f_τ whose integral on any soma $A \in \mathscr{B}_\tau^f$ is just $\varphi_\tau(A)$. The sequence (f_τ) is said to be the *integrand representation* of $\boldsymbol{\Phi}$. A martingale at finite measure $\boldsymbol{\Phi} = (\varphi_\tau)$ is called σ-additive if for each $\rho \in \Theta$ and each $A \in \mathscr{B}_\rho^f$, the martingale $(\varphi_\tau | A \wedge \mathscr{B}_\tau)$, defined for those $\tau \in \Theta$ such that $A \in \mathscr{B}_\tau$ (cf. the trace of $\boldsymbol{\Phi}$ on A as defined in I. 3), is σ-additive.

These notions "at finite measure" are not required for stochastic or order (III. 1) convergence because such convergence on each soma of finite measure implies the same convergence. We need these notions only in the present section.

By *martingale in* L_X we shall mean a martingale at finite measure, whose integrand representation (f_τ) satisfies $f_\tau \in L_X$ for each $\tau \in \Theta$.

6.3. Theorem [*43*, 499]. *Let* $\boldsymbol{\Phi}$ *denote a martingale in* L_X *of integrand representation* (f_τ). *If the sequence* (f_τ) *is bounded in* L_X, *then it converges stochastically and the limit* f_ω *is in* L_X.

Moreover, the following conditions are equivalent:

(1) $\boldsymbol{\Phi}$ is σ-additive at finite measure and bounded in L_X;

(2) The sequence (f_τ) converges stochastically, its limit f_ω belongs to L_X and $\boldsymbol{\Phi}$ is induced by f_ω;

(3) $\boldsymbol{\Phi}$ is induced by a function f in L_X.

If these conditions are satisfied, then the following relations hold:

$$\|\boldsymbol{\Phi}\|_X = \|f_\omega\|_X \leqslant \|f\|_X,$$

where, for a premartingale $\boldsymbol{\Phi}$, the norm $\|\boldsymbol{\Phi}\|_X$ is defined as $\limsup_\tau \|f_\tau\|_X$.

6.4. Theorem [*43*, 500]. *Let* $\boldsymbol{\Phi}$ *denote a martingale in* L_X, (f_τ) *its integrand representation. Then the following conditions are equivalent:*

(1) (f_τ) *converges strongly in* L_X;

(2) (f_τ) *converges weakly in* L_X;

(3) *The conditions* (1), (2), *and* (3) *of Th.* 6.3 *hold and, corresponding to any* $\varepsilon > 0$ *there exists a parameter* ρ *and a* $\mathscr{B}_\rho$*-measurable function* h *in* L_X *such that* $\|f_\omega - h\|_X < \varepsilon$.

7. Cell functions. We take over the setting developed in (I. 9); then $\mathscr{B}_{\mathscr{T}}$ is the atomic Boolean σ-algebra generated by the partition $\mathscr{T}$ of E,

$\preccurlyeq$ is the order relation $\sqsubset$ (partition fineness). The hypothesis of σ-finiteness of μ on each family $\mathscr{B}_{\mathscr{T}}$ is expressed by saying that the constituents have finite μ-measure. We recall that the constituents are $\neq \bigcirc$.

A cell function φ is called *μ-absolutely continuous* or *μ-singular* if the premartingale $\Phi = G(\varphi)$ generated by φ possesses the same property; these properties are defined *from above* and *from below* analogously. As an example, we state explicitly the definition of μ-absolute continuity: given any $\varepsilon > 0$, there exist a terminal subset $\mathfrak{D}$ of $\mathfrak{T}$, a soma H of finite measure, and a positive number δ such that $|\varphi(\mathscr{J})| < \varepsilon$ whenever $\mathscr{J}$ is a subset of a partition $\mathscr{T}$ of $\mathfrak{D}$ satisfying $\mu(H \wedge \vee \mathscr{J}) < \delta$. For a prescribed partition $\mathscr{Y}$ of E, it is always possible to choose $H \in \mathscr{B}_{\mathscr{Y}}$ and, if $\mu(E)$ is finite, to take $H = E$. According to Prop. 2.2, an additive cell function is μ-absolutely continuous or μ-singular if the additive extension $\varphi^{\vee}$ of φ to the algebra $\mathscr{A}$ enjoys the property, respectively. An additive cell function φ of bounded variation is μ-absolutely continuous or μ-singular in accordance with whether φ, or equivalently $\varphi^{\vee}$, is σ-additive or purely simply additive, respectively. In an analogous manner, Props. 2.2, 2.3, and 2.4 can be readily transferred to cell functions.

For a cell function φ, the integrand representation $(f_{\mathscr{T}})$ (relative to μ) of the premartingale $\Phi = G(\varphi)$ is defined as follows: $f_{\mathscr{T}}$ is the (measurable) function taking on each constituent J of $\mathscr{T}$ the value $\varphi(J)/\mu(J)$. In other words, $f_{\mathscr{T}}$ is the $\mathscr{T}$-derivate of φ which we denote by $D_{\mathscr{T}}\varphi$. The *upper* and *lower stochastic derivates*, denoted by $sD^s\varphi$ and $sD^i\varphi$, respectively, are defined as the stochastic superior and inferior limits of $(D_{\mathscr{T}}\varphi : \mathscr{T} \in \mathfrak{T})$, respectively. φ is said to be *stochastically derivable* or *derivable* L_X in accordance with whether $(D_{\mathscr{T}}\varphi)$ converges stochastically (meaning that $sD^s\varphi = sD^i\varphi$) or strongly in L_X. The limit is then called the *stochastic derivative* or *L_X-derivative* of φ and represented by $sD\varphi$ and L_X-$D\varphi$ If there is no ambiguity from the context, we denote both cases by $D\varphi$ for brevity. Props. 4.3.1, 5.1, and Ths. 4.7, 4.9, 4.11, 6.3, and 6.4 are readily translated into our present setting. For instance, taking (I. 10) into account, particularly the definitions (10.0.1) and (10.0.2), and referring to Ths. 4.7 and 4.9, we obtain the following result.

7.1$_a$. Theorem. *(The subscript a stands for "atomic".) Any subadditive cell function of bounded variation possesses a stochastic integrable derivative that is also the stochastic derivative of the integrable cell function φ^m. For any cell I we have*

$$\int_I D\varphi\, d\mu = (\varphi^m)_c(I),$$

where $(\varphi^m)_c$ denotes the σ-additive part of φ^m.

To transfer Th. 6.3, the L_X-norm of an additive cell function is defined as the L_X-norm of the martingale generated by it.

Finally, we enunciate some typically cellular theorems.

7.2. Theorem [39, 210]. *Let $\mathfrak{T}$ fulfill the conditions* (**E**) *and* (**P**) *(cf.* I. 10, *opening discussion), and suppose that φ is of bounded variation. Then*

$$sD^s\varphi = sD^s\varphi^s \quad \text{and} \quad sD^i\varphi = sD\varphi^i.$$

Consequently, φ is derivable stochastically iff $D\varphi^s = D\varphi^i$ which is the case if, in particular, φ is integrable in the sense of Burkill-Kolmogoroff.

7.3. Theorem [39, 208]. *Let $\mathfrak{T}$ satisfy condition* (**E**), *and suppose that φ is of bounded variation and μ-absolutely continuous. Then*

$$\varphi^s(E) = \int_E sD^s\varphi\, d\mu \quad \text{and} \tag{7.3.1 s}$$

$$\varphi^i(E) = \int_E sD^i\varphi\, d\mu. \tag{7.3.1 i}$$

7.4. Consequences [38, 482]. For each $B \in \mathscr{B}$ we have

$$\limsup_{\mathscr{T}} \int_B D_{\mathscr{T}}\varphi\, d\mu = \int_B sD^s\varphi\, d\mu \tag{7.4.1 s}$$

$$\liminf_{\mathscr{T}} \int_B D_{\mathscr{T}}\varphi\, d\mu = \int_B sD^i\varphi\, d\mu. \tag{7.4.1 i}$$

Hence, for $B = I \in \mathscr{I}$, taking account of the individual definitions of $\varphi^s(I)$ and $\varphi^i(I)$ in (I. 10.0.1) and (I. 10.0.2), and the fact that

$$\varphi(\mathscr{I}_{\mathfrak{X}}) = \int_I D_{\mathfrak{X}}\varphi\, d\mu,$$

we have

$$\varphi^s(I) = \int_I sD^s\varphi\, d\mu \quad \text{and} \tag{7.4.2 s}$$

$$\varphi^i(I) = \int_I sD^i\varphi\, d\mu. \tag{7.4.2 i}$$

We thus recognize the σ-additivity and hence

$$\varphi^s = \varphi^{sm}, \quad \varphi^i = \varphi^{im}.$$

7.5. Extreme derivates in the mean. Keeping the setting of Prop. 4.3.1 and applying this proposition with $f_\tau = D_{\mathscr{T}}\varphi$, we see that

$$\limsup_{\mathscr{T}} \int_B D_{\mathscr{T}}\varphi\, d\mu \leqslant \int_B sD^s\varphi\, d\mu. \tag{7.5.1 s}$$

An analysis of the proof of Prop. 4.3.1 enables us to assert the following implication:

For each $\varepsilon > 0$, there exists $\mathscr{R} \in \mathfrak{T}$ such that for each $B \in \mathscr{B}$ and each $\mathscr{T} \in \mathfrak{T}$ satisfying $\mathscr{R} \sqsubset \mathscr{T}$, we have

$$\int_B (f^s - D_{\mathscr{T}}\varphi)\, d\mu > -\varepsilon, \tag{7.5.2 s}$$

where $f^s = sD^s\varphi$.

This implication is stronger than (7.5.1 s) since the choice of $\mathscr{R}$ depends only on ε and not on B.

Similarly, an analysis of the proof of the relation

$$\lim_{\mathscr{T}} \sup \int_B D_{\mathscr{T}} \varphi \, d\mu \geqslant \int_B D^s \varphi \, d\mu \tag{7.5.3 s}$$

[*39*, 206] leads to:

For each $\varepsilon > 0$ and each $\mathscr{R} \in \mathfrak{T}$, there exists $\mathscr{T} \in \mathfrak{T}$ such that $\mathscr{R} \sqsubset \mathscr{T}$ and, for any $B \in \mathscr{B}$, the relation

$$\int_B (f^s - D_{\mathscr{T}} \varphi) \, d\mu < \varepsilon \tag{7.5.4 s}$$

holds.

For any cell function φ we define an *upper derivate in the mean of* φ as a function f^s satisfying (7.5.2 s) and (7.5.4 s). If such a function exists, then it is unique

Thus we have proved that, under the assumptions of Th. 7.3, the upper stochastic derivate can be regarded as an upper derivate in the mean. Similar considerations hold for the lower stochastic derivate and the lower derivate in the mean.

Chapter III

Theory in a Measure Space with Vitali Conditions

1. Preliminaries and definitions. We assume the setting described in Chapter II. in the following theory.

1.1. Order limits. For a stochastic process $(f_\tau \colon \tau \in \Theta)$, we define the extreme order limits by

$$\limsup_{\tau} f_\tau = \bigwedge_{\rho \in \Theta} \bigvee_{\rho \ll \tau} f_\tau \quad \text{and} \quad \liminf_{\tau} f_\tau = \bigvee_{\rho \in \Theta} \bigwedge_{\rho \ll \tau} f_\tau$$

[*36*, 225; *37*, 316]. In case both limit functions are equal, their common value is the *order limit of* (f_τ), which is then said to be *order convergent*. In analogous fashion one defines the order limits for a Moore-Smith sequence of somas by means of the indicatrix function.

If Θ admits a cofinal countable subsequence (τ_n) then

$$\limsup_{\tau} f_\tau = \bigwedge_{n} \bigvee_{\tau_n \ll \tau} f_\tau \quad \text{and} \quad \liminf_{\tau} f_\tau = \bigvee_{n} \bigwedge_{\tau_n \ll \tau} f_\tau .$$

The proof of the following inequalities is simple [*38*, 478]:

$$\liminf_{\tau} f_\tau \leqslant s \liminf_{\tau} f_\tau \leqslant s \limsup_{\tau} f_\tau \leqslant \limsup_{\tau} f_\tau .$$

Consequently, if $\lim_{\tau} f_\tau$ exists, so does $s \lim_{\tau} f_\tau$, and both limits are equal.

1.2. Overlap of order Y. The *excess function* $e_{\mathscr{L}}$ of an ordinary finite or infinite sequence $\mathscr{L} = (L_i)$, $i = 1,2,\ldots$ (or of the corresponding M-family) of somas is defined as

$$e_{\mathscr{L}} = \sum_i c_{L_i} - c_L, \quad \text{where} \quad L = \bigvee_i L_i .$$

(I. II. 1.2). If E is an Orlicz space with norm $\| \quad \|_Y$, then the number (possibly infinite)

$$\omega_Y(\mathscr{L}) = \|e_{\mathscr{L}}\|_Y$$

is called the *overlap of order* Y *of* $\mathscr{L}$. The inequality (II. 6.0.2), applied to the function f and $e_{\mathscr{L}}$, yields

$$|\sum_i \int_{L_i} f d\mu - \int_L f d\mu| \leqslant \|f\|_X \omega_Y(\mathscr{L}) \tag{1.2.1}$$

provided $\|f\|_X$ and $\omega_Y(\mathscr{L})$ are finite.

We have as a special case $\omega_q(\mathscr{L}) = \|e_{\mathscr{L}}\|_q$ (II. 6.1).

1.3. Fine coverings. By a *fine covering of a soma* M with respect to the stochastic basis $(\mathscr{B}_\tau)$ we mean a sequence $(K_\tau : \tau \in \Theta)$ such that the following condition holds:

$C(M,(K_\tau))$: $K_\tau \in \mathscr{B}_\tau$ for all τ and $M \leqslant \limsup_\tau K_\tau$.

Now $M \leqslant \limsup_\tau K_\tau$ iff $M \leqslant \bigvee \{K_\tau : \tau \in \Delta\}$ whenever Δ is a terminal subset of Θ. Therefore, if (K_τ) is a fine covering of M and Δ is a terminal subset of Θ, then the sequence $(K_\tau : \tau \in \Delta)$ is a fine covering of M with respect to the basis $(\mathscr{B}_\tau : \tau \in \Delta)$. If (K_τ) is a fine covering of M and $A \in \mathscr{A} = \bigcup_{\tau \in \Theta} \mathscr{B}_\tau$, then for any terminal subset Δ of Θ such that $A \in \mathscr{B}_\tau$ for $\tau \in \Delta$, (this is Δ_A as defined in (I. 3)), the sequence $(K_\tau \cdot A : \tau \in \Delta)$ is a fine covering of $M \cdot A$ with respect to the basis $(\mathscr{B}_\tau : \tau \in \Delta)$.

If (f_τ) is a stochastic process with basis $(\mathscr{B}_\tau)$ and if $\alpha < \limsup_\tau f_\tau$ on the soma M, then the sequence (K_τ) defined by $K_\tau = [\alpha < f_\tau]$ is a fine covering of M.

2. Vitali conditions.

2.1. A stochastic basis $(\mathscr{B}_\tau)$ is said to possess the Vitali property of order Y if the following condition holds:

(V_Y): For each soma M of finite measure, each fine covering (K_τ) of M, and each $\varepsilon > 0$, there exists a finite sequence $(\xi_1, \xi_2, \ldots, \xi_r)$ of Θ-members and a finite sequence $\mathscr{L} = (L_1, L_2, \ldots, L_r)$ of somas such that

$$L_i \in \mathscr{B}_{\xi_i} \quad \text{and} \quad L_i \leqslant K_{\xi_i} \quad \text{for} \quad i = 1,2,\ldots,r ;$$

$$\omega_Y(\mathscr{L}) < \varepsilon \quad \text{(overlap limitation)} ;$$

$$\mu(M - M \cdot L) < \varepsilon, \text{ where } L = \bigvee \mathscr{L} \quad \text{(deficiency of covering limitation)}.$$

As particular cases (II. 6.1), using $\omega_q(\mathscr{L})$ instead of $\omega_Y(\mathscr{L})$, we define conditions $(\mathbf{V}_q)$ when $1 \leqslant q \leqslant \infty$. $(\mathbf{V}_{+\infty})$ [$(\mathbf{V}_0)$ in *37*, 325] is obtained from $(\mathbf{V}_Y)$ by replacing the overlap condition by "$\mathscr{L}$ is disjoint". If $Y_1(u) = O(Y_2(u))$ as u tends to infinity, where O is Landau's symbol, then $(\mathbf{V}_{Y_2})$ implies $(\mathbf{V}_{Y_1})$; thus if $q_1 < q_2$, then $(\mathbf{V}_{q_2})$ implies $(\mathbf{V}_{q_1})$. By analogy with the preceding notions in the case of pointwise derivation, $(\mathbf{V}_{+\infty})$ is sometimes called the *strong Vitali property* [*25*, 233; *41*, 282; *12*] and $(\mathbf{V}_1)$ the *weak Vitali property* [*25*, 256; *41*, 282].

If Δ denotes a terminal subset of Θ and if $(\mathscr{B}_\tau : \tau \in \Theta)$ possesses property $(\mathbf{V}_Y)$, then so does the stochastic basis $(\mathscr{B}_\tau : \tau \in \Delta)$.

We obtain an equivalent condition if, in the formulation of $(\mathbf{V}_Y)$, we limit ourselves to somas from $\mathscr{B}_\omega$.

2.2. Proposition [*43*, *507*]. The overflow condition (I. II. 1.3) $\mu(L - L \cdot M) < \varepsilon$ can be added to the requirements on $\mathscr{L}$ in the definition of $(\mathbf{V}_Y)$ without effectively strengthening these requirements, so long as in their formulation we restrict ourselves to somas $M \in \mathscr{B}_\omega$.

2.3. Definition. A stochastic basis is said to possess the property $(\mathbf{W}_Y)$ if for each soma M, each fine covering (K_τ) of M, and each positive number ε, there exist an ordinary sequence $(\xi_1, \xi_2, \ldots)$ of Θ-members and a sequence $\mathscr{L} = (L_1, L_2, \ldots,)$ of somas such that

$$L_i \in \mathscr{B}_{\xi_i} \quad \text{and} \quad L_i \leqslant K_{\xi_i} \quad \text{for} \quad i = 1,2,\ldots,$$
$$\omega_Y(\mathscr{L}) < \varepsilon, \quad \text{and} \quad M \leqslant \bigvee_i L_i.$$

2.4. Proposition [*43*, 508]. Conditions $(\mathbf{V}_Y)$ and $(\mathbf{W}_Y)$ are equivalent.

2.5. Remark. We envisage $(\mathbf{V}_Y)$ and $(\mathbf{W}_Y)$ as two versions of the same property, which we call the ε-version and the 0-version, where ε and 0 refers to the deficiency of the covering. As in the enunciation of $(\mathbf{V}_Y)$, one can add the overflow condition, namely $\mu(L - M \cdot L) < \varepsilon$, in the formulation of $(\mathbf{W}_Y)$ for $M \in \mathscr{B}_\omega$. Thus one obtains the formulation of the Vitali condition of order Y in complete form and in the 0-version. Condition $(\mathbf{W}_{+\infty})$ does not depend on the measure μ; consequently neither does $(\mathbf{V}_{+\infty})$ [*37*, 325].

3. Order convergence of martingales.

3.1. Theorem. *Let* $\| \ \|_X$ *and* $\| \ \|_Y$ *denote norms of two complementary Orlicz spaces, and suppose that the stochastic basis* $(\mathscr{B}_\tau : \tau \in \Theta)$ *satisfies the condition* $(\mathbf{V}_Y)$. *Let* Φ *be a martingale of basis* $(\mathscr{B}_\tau)$ *of bounded variation.*

Then the integrand representation (f_τ) *of* Φ *is order convergent. Moreover, if* $(\mathscr{B}_\tau)$ *satisfies* $(\mathbf{V}_{+\infty})$ *then the same conclusion applies to each martingale* Φ *of semi-bounded variation.*

3.2. Theorem, *(Order density theorem). Let* $(\mathscr{B}_\tau)$ *possess the weak Vitali property* (V_1). *Let* B *be a soma from* $\mathscr{B}$ *and* $(c_{B,\tau})$ *the integrand* *representation of the martingale induced by the indicatrix function of B. Then* $(c_{B,\tau})$ *is order convergent to* $\mathscr{E}(c_B|\mathscr{B}_\omega)$. *If* $B \in \mathscr{B}_\omega$, *then* $\lim_\tau c_{B,\tau} = c_B$.

4. Necessity of the Vitali conditions.

4.1. Theorem [*41*, § 2]. *If, for any soma B of finite measure from* $\mathscr{B}_\omega$, *the sequence* $(c_{B,\tau})$ *converges with respect to the order to* c_B, *then the stochastic basis satisfies* $(\mathbf{V}_1)$.

4.2. Theorem [*41*, § 3]. *Let* $(\mathscr{B}_\tau)$ *be a stochastic basis such that* Θ *admits a countable cofinal subset. Suppose that for each* $\mathscr{B}_\omega$*-measurable function f in* L_p, *the sequence* (f_τ) *induced by f converges to f with respect to the order. Then the basis* $(\mathscr{B}_\tau)$ *satisfies* $(\mathbf{V}_{q'})$ *for* $1 \leqslant q' < q$.

Remarks. We have no property of convergence for martingales implying $(\mathbf{V}_{+\infty})$; however, this was not to be expected [*25*, 261].

We conjecture the validity of the following theorem, more general than Theorem 4.2: If Θ admits a countable cofinal subset and if $Y_0(v) = o(v)$ (o is Landau's symbol), when v tends to $+\infty$, then convergence with respect to the order of the martingales induced by the $\mathscr{B}_\omega$-measurable functions f of L_X to f implies the validity of $(\mathbf{V}_{Y_0})$.

5. Order convergence of submartingales.

5.1. Theorem. *If the stochastic basis* $(\mathscr{B}_\tau)$ *possesses the property* $(\mathbf{V}_Y)$, *then for any submartingale of bounded variation of order X whose integrand representation is* (f_τ), *we have*

$$\limsup_\tau f_\tau = s \lim_\tau f_\tau .$$

5.2. Definition. We define the *Vitali property* (introduced in [*38*, 486] and rectified in [*41*, 288]):

$(\mathbf{V}')$: For each soma M of finite measure, each fine covering (K_τ) of M, and each $\varepsilon > 0$, there exists a finite sequence $(\xi_1, \xi_2, \ldots, \xi_r)$ of members of Θ such that

$$\xi_1 \ll \cdots \ll \xi_i \ll \cdots \ll \xi_r \quad \text{and} \quad \mu\left(M - M \cdot \bigvee_{i=1}^{r} K_{\xi_i}\right) < \varepsilon .$$

As is the case for $(\mathbf{V}_Y)$, there corresponds to the ε-version of $(\mathbf{V}')$ a 0-version, and we observe that $(\mathbf{V}')$ does not depend on the measure μ.

5.3. Proposition. Condition $(\mathbf{V}')$ implies $(\mathbf{V}_{+\infty})$. $(\mathbf{V}')$ always holds if Θ is totally ordered by $\ll$.

5.4. Theorem. [*38*, 486]. *If the basis* $(\mathscr{B}_\tau)$ *satisfies* $(\mathbf{V}')$, *then the integrand representation of any submartingale of bounded variation converges with respect to the order.*

6. Order convergence of cell functions. We adopt here the setting of (II. 7).

6.1. Definition. The *extreme order derivates* $D^s\varphi$ and $D^i\varphi$ are defined as the upper and lower limits, taken with respect to order, of $(D_{\mathcal{T}}\varphi: \mathcal{T} \in \mathfrak{T})$. We have

$$D^i\varphi \leqslant sD^i\varphi \leqslant sD^s\varphi \leqslant D^s\varphi,$$

and we say that φ is *order derivable* if $D^i\varphi = D^s\varphi$. The common value $D\varphi$ is called the *order derivative.*

6.2. To any set $\mathcal{L}$ of cells there corresponds a sequence $(K_{\mathcal{T}})$, where $K_{\mathcal{T}} \in \mathcal{B}_{\mathcal{T}}$, $\mathcal{T} \in \mathfrak{T}$ and

$$K_{\mathcal{T}} = \vee (\mathcal{L} \cdot \mathcal{T}). \tag{6.2.1}$$

On the other hand to each sequence $(K_{\mathcal{T}})$, where $K_{\mathcal{T}} \in \mathcal{B}_{\mathcal{T}}$, $\mathcal{T} \in \mathfrak{T}$, there corresponds a set $\mathcal{K}$ of cells defined as follows:

For each $\mathcal{T} \in \mathfrak{T}$, denote by $\mathcal{J}_{\mathcal{T}}$ that subset of $\mathcal{T}$ whose supremum is $K_{\mathcal{T}}$ and let $\mathcal{K} = \bigcup_{\mathcal{T} \in \mathfrak{T}} \mathcal{J}_{\mathcal{T}}$. (6.2.2)

From this definition we obtain

$$K_{\mathcal{T}} \leqslant \vee (\mathcal{K} \cdot \mathcal{T}). \tag{6.2.3}$$

(6.2.4) **Proposition.** If $(K_{\mathcal{T}})$ is a sequence of somas such that $K_{\mathcal{T}} \in \mathcal{B}_{\mathcal{T}}$, and if $\mathcal{K}$ is defined by (6.2.2), then the following conditions are equivalent:

(1) Equality holds in (6.2.3).

(2) There exists a set $\mathcal{L}$ of cells such that (6.2.1) is satisfied for any $\mathcal{T}$.

(3) The premartingale $(c_{K_{\mathcal{T}}})$ is generated by a cell function φ by means of the transfer law G (I. 9).

If these conditions are satisfied, then $\mathcal{K} = \mathcal{L}$, $\varphi(I) = 0$ for $I \notin \mathcal{L}$ and $\varphi(I) = \mu(I)$ for $I \in \mathcal{L}$. Consequently, if we start with a set $\mathcal{L}$ of cells and define $(K_{\mathcal{T}})$ by (6.2.1), we obtain $\mathcal{L}$ as the set given by (6.2.2).

6.3 Definition. By a *fine cell covering* [*36*, 243] *of a soma*, we mean a set $\mathcal{L}$ of cells such that the sequence $(K_{\mathcal{T}})$ defined by (6.2.1) is a fine covering of M; i.e., such that the condition

$$\mathbf{C}^0(M,\mathcal{L}): M \leqslant \limsup_{\mathcal{T}} \vee (\mathcal{L} \cdot \mathcal{T})$$

holds.

If $(K_{\mathcal{T}})$ is an arbirary fine covering of M, then the set $\mathcal{K}$ defined by (6.2.2) is a fine cell covering $\mathcal{L}$ of M by virtue of (6.2.3).

6.4. To each of the Vitali conditions ($\mathbf{V}_Y$) and $\mathbf{V}'$) there corresponds the weaker conditions ($\mathbf{aV}_Y$) and ($\mathbf{aV}'$) if we admit in their formulations only those fine coverings $(K_{\mathcal{T}})$ of M that satisfy the conditions of Prop. 6.2.4; i.e., that correspond by virtue of (6.2.1) to the fine cell coverings $\mathcal{L}$ of M.

6.4.1. Proposition. The condition ($\mathbf{a}\mathbf{V}_Y$) is equivalent to ($\mathbf{V}_Y$) and can be formulated as follows:

For each soma M of finite measure, each fine cellular covering $\mathscr{Z}$ of M, and each $\varepsilon > 0$, there exists a finite subset $\mathscr{P}$ of $\mathscr{Z}$ such that

$$\omega_Y(\mathscr{P}) < \varepsilon \quad \text{and} \quad \mu(M - M \cdot \vee \mathscr{P}) < \varepsilon .$$

Condition ($\mathbf{a}\mathbf{V}'$) is strictly weaker than ($\mathbf{V}'$), but it implies ($\mathbf{a}\mathbf{V}_{+\infty}$), and therefore also ($\mathbf{V}_{+\infty}$).

By transferring Th. 3.1 to the present setting we obtain

6.4.2. Theorem. *Let $\| \quad \|_X$ and $\| \quad \|_Y$ be complementary Orlicz norms, and suppose that the stochastic cell basis ($\mathscr{B}_{\mathscr{T}}$) has the property* ($\mathbf{a}\mathbf{V}_Y$). *Then every additive cell function φ of bounded variation of order X, i.e., with $\sup_{\mathscr{T}} \| D_{\mathscr{T}} \varphi \|_X < +\infty$, is order derivable. If ($\mathscr{B}_{\mathscr{T}}$) satisfies* ($\mathbf{a}\mathbf{V}_{+\infty}$), *then each additive cell function of semi-bounded variation has an order derivative.*

In analogous fashion, Ths. 3.2, 4.1, 4.2, 5.1, and 5.4 can be transferred to the atomic setting.

6.5. Specifically cellular theorems. We begin by introducing the following Vitali conditions [*41*, 288 and 290).

($\mathbf{a}\mathbf{V}''$): For any fine cellular covering $\mathscr{Z}$ of a soma $M \neq \bigcirc$, $\lim_{\mathscr{T}} \sup \mu(M \cdot \vee (\mathscr{Z} \cdot \mathscr{T}))$ is positive.

($\mathbf{a}\mathbf{V}'''$): For each soma M of finite measure, each fine cellular covering $\mathscr{Z}$ of M and each $\varepsilon > 0$, there exists a partition $\mathscr{H}$ in $\mathscr{T}$ such that $\mu(M - M \cdot \vee (\mathscr{Z} \cdot \mathscr{H})) < \varepsilon$.

Clearly ($\mathbf{a}\mathbf{V}'''$) implies ($\mathbf{a}\mathbf{V}''$) and ($\mathbf{a}\mathbf{V}''$) implies ($\mathbf{a}\mathbf{V}'$). The converse statements are not true [*41*, examples 4 and 5]. However, (**E**) (I. 10) and ($\mathbf{a}\mathbf{V}_{+\infty}$) imply ($\mathbf{a}\mathbf{V}'''$).

6.5.1 Theorem. *The condition* ($\mathbf{a}\mathbf{V}''$) *is necessary and sufficient for $D^s \varphi = s D^s \varphi$ and $D^i \varphi = s D^i \varphi$ for any cell function φ.*

6.5.2. Theorem. *Condition* ($\mathbf{a}\mathbf{V}'''$) *is necessary and sufficient for*

$$\varphi^s(I) = \int_I D^s \varphi \, d\mu \quad \text{and} \quad \varphi^i(I) = \int_I D^i \varphi \, d\mu .$$

Chapter IV

Applications

1. Pointwise setting.

1.1. Transfer of the somatic theory. L. H. Loomis proved [*48*, 757] that every Boolean σ-algebra can be represented isomorphically as the quotient of a Boolean σ-algebra of sets by a σ-ideal of the latter algebra.

If the abstract σ-algebra bears a strictly positive measure, then it can be transferred to the "concrete" σ-algebra; the dividing σ-ideal consists of the nullsets.

In fact, in most applications the primary datum is a Boolean σ-algebra of sets provided with a measure. A strictly positive measure is then obtained by dividing by the σ-ideal of the nullsets. In order to take this situation into account, we shall modify the notations used thus far and introduce new ones as follows:

$\mathscr{B}$ denotes a Boolean σ-algebra of subsets of a set E, the unit of $\mathscr{B}$, $\emptyset$ the empty set, μ a σ-finite measure defined on $\mathscr{B}$, and $\mathscr{N}$ the σ-ideal in $\mathscr{B}$ of the μ-nullsets.

The results of the preceding chapters in Part II are readily applicable to the Boolean σ-algebra $\mathscr{B}/\mathscr{N}$ and to the measure μ^e obtained by transferring μ onto $\mathscr{B}/\mathscr{N}$. It is often suitable to formulate them in the setting $(\mathscr{B}, \mathscr{N}, \mu)$ rather than $(\mathscr{B}/\mathscr{N}, \mu^e)$. The somas of $\mathscr{B}/\mathscr{N}$ are the classes of $\mathscr{B}$-sets that are equivalent mod $\mathscr{N}$. The functions that are $\mathscr{B}/\mathscr{N}$-measurable (spectral scales) are represented by classes of $\mathscr{B}$-measurable functions defined on E.

The notions of infimum and supremum of a set of functions (I. 11), of stochastic limits (II. 4), of order limits (III. 1), of L_X-limits (II. 6) of a sequence (filtering family), transferred to a set $\mathscr{G}$ of $\mathscr{B}$-measurable functions, and to a family $(f_\tau\colon \tau \in \Theta)$ of μ-measurable functions, yield the notions, respectively, of *essential infimum* $e \wedge \mathscr{G}$ and *supremum* $e \vee \mathscr{G}$ of *stochastic limits* $s \liminf_\tau f_\tau$ and $s \limsup_\tau f_\tau$, of *essential limits* $e \liminf_\tau f_\tau$ and $e \limsup_\tau f_\tau$, and of L_X-limit, regarded as functions defined mod $\mathscr{N}$; in other words, defined except on an *indeterminate μ-nullset*. For instance, $e \wedge \mathscr{G}$ denotes any $\mathscr{B}$-measurable function h with the following properties: h is a minorant mod $\mathscr{N}$ of $\mathscr{G}$, i.e., $g \in \mathscr{G}$ implies $h \leqslant g$ mod $\mathscr{N}$; and every minorant h' of $\mathscr{G}$ satisfies $h' \leqslant h$ mod $\mathscr{N}$ Modification of every function g of $\mathscr{G}$ on a μ-nullset depending possibly on g, or of every function f_τ on a μ-nullset depending possibly on τ, does not alter these extrema or these limits.

1.2. By *stochastic basis* we mean, in accordance with the general definition given in (I. 3), a sequence $(\mathscr{B}_\tau)$ of σ-subalgebras of $\mathscr{B}$ with unit E. In the present chapter we assume that $\rho \ll \tau$ implies $\mathscr{B}_\rho \subset \mathscr{B}_\tau + \mathscr{N}$, where $\mathscr{B}_\tau + \mathscr{N}$ denotes the σ-algebra of sets $B + N$ for $B \in \mathscr{B}_\tau$ and $N \in \mathscr{N}$,

1.3. A *premartingale* of basis $(\mathscr{B}_\tau)$ is a sequence (φ_τ) such that φ_τ is a real function defined and σ-additive on $\mathscr{B}_\tau$, $\tau \in \Theta$. Premartingales obtained from premartingales $\mathscr{B}/\mathscr{N}$ by inverse images and by transfer are characterized by the conjunction of the properties

(NS) $\mathscr{B}_\tau = \mathscr{B}_\tau + \mathscr{N}$,

(AC) $\varphi_\tau(N) = 0$ for each $\tau \in \Theta$ and each $N \in \mathscr{B}_\tau \cdot \mathscr{N}$.

A *stochastic process* with basis $(\mathscr{B}_\tau)$ is defined in (II. 3) as a sequence (f_τ) of real functions on E such that f_τ is $\mathscr{B}_\tau$-measurable for each $\tau \in \Theta$. We may observe that for each stochastic process with basis $(\mathscr{B}_\tau + \mathscr{N})/\mathscr{N}$, if f_τ denotes a representative of the equivalence class corresponding to the "function" of parameter τ, then f_τ is $(\mathscr{B}_\tau + \mathscr{N})$-measurable but not necessarily $\mathscr{B}_\tau$-measurable.

Condition (AC) is necessary and sufficient for the premartingale (φ_τ) to admit an integrand representation for the measure μ; i.e., for a stochastic process (f_τ) of basis $(\mathscr{B}_\tau)$ such that for any $\tau \in \Theta$, f_τ is semi-integrable and $\varphi_\tau(B) = \int_B f_\tau \, d\mu$ for any $B \in \mathscr{B}_\tau$. In the following, we shall consider premartingales satisfying (AC) but we do not demand (NS).

The definitions of submartingales and martingales of basis $(\mathscr{B}_\tau)$ are immediate; they will appear only through their integrand representations, thus as stochastic processes.

1.4. By an *essential fine covering* of a measurable set M with respect to $(\mathscr{B}_\tau)$ we mean a sequence $(K_\tau : \tau \in \Theta)$ such that $K_\tau \in \mathscr{B}_\tau$ for any $\tau \in \Theta$ and $M \subset e \limsup_\tau K_\tau$. The relation $K_\tau \in \mathscr{B}_\tau$ for each $\tau \in \Theta$ is equivalent to: (c_{K_τ}) is a stochastic process of basis $(\mathscr{B}_\tau)$. We note that this condition is more restrictive than $K_\tau \in \mathscr{B}_\tau + \mathscr{N}$ for all $\tau \in \Theta$.

1.5. The stochastic basis $(\mathscr{B}_\tau)$ is said to possess the *Vitali property* (V_Y) if it is true for $(\mathscr{B}_\tau + \mathscr{N})/\mathscr{N}$. It is easy to see that it is sufficient in its formulation to consider fine coverings in the sense just defined (1.4), and that we can replace the relations mod $\mathscr{N}$ by strict relations, hence we can give the following direct formulation:

For any measurable set M of finite measure, any essential fine covering (K_τ) of M, and any positive number ε, there exists a finite sequence $(\xi_1, \xi_2, \ldots, \xi_r)$ of Θ-members and an associated sequence $\mathscr{L} = (L_1, L_2, \ldots, L_r)$ of sets such that

$$\omega_Y(\mathscr{L}) < \varepsilon, \quad \mu(M - M \cdot L) < \varepsilon, \quad \text{where} \quad L = \bigcup \mathscr{L}.$$

We notice that for $(V_{+\infty})$, we cannot, in general, demand the strict disjunction of the $\mathscr{L}$-sets. However, a strengthening is possible if Θ is totally ordered and if the basis $(\mathscr{B}_\tau)$ is increasing, i.e., if $\rho \ll \tau$ implies $\mathscr{B}_\rho \subset \mathscr{B}_\tau$. The formulations of the other Vitali conditions are analogous to that of (V_Y).

1.6. Thus far, the question has been to pass from a "somatic" notion to the interpretation on the "concrete" model. The transfer of the results of the Chs. I, II, and III of Part II is immediate. In this manner, theorems are obtained mod $\mathscr{N}$; i.e., insensible in their formulation to a modification of each of the occurring functions on an $\mathscr{N}$-set depending on the function in question. At present, interest in the pointwise theory proper centers around those theorems where there is only one exceptional $\mathscr{N}$-set.

2. Specifically pointwise concepts and results. Convergence almost everywhere. We denote by $\mathscr{S}$ the Boolean σ-algebra of all subsets of E, by $\mathscr{N}^\star$ the set of those sets N in $\mathscr{S}$ such that the completion $\mu^\star$ of μ vanishes on N; $\mathscr{N}^\star$ is a σ-ideal in $\mathscr{S}$. Then we denote the *inferior envelope* (pointwise infimum) of a set $\mathscr{G}$ of real functions by $\wedge^\star \mathscr{G}$; it is the function defined at each point x of E by $(\wedge^\star \mathscr{G})(x) = \inf_{g \in \mathscr{G}} g(x)$. Analogously, the *superior envelope* $\vee^\star \mathscr{G}$ is defined. For a set $\mathscr{K} \subset \mathscr{S}$, we have

$$\wedge^\star \mathscr{K} = \cap \mathscr{K} \quad \text{and} \quad \vee^\star \mathscr{K} = \cup \mathscr{K}.$$

The *pointwise limits* $\liminf_\tau{}^\star f_\tau$ and $\limsup_\tau{}^\star f_\tau$ of a sequence (f_τ) of functions on E are defined as $\bigvee_{\rho \in \Theta} \bigwedge_{\rho \ll \tau} f_\tau$ and $\bigwedge_{\rho \in \Theta} \bigvee_{\rho \ll \tau} f_\tau$, respectively. Thus the *convergence almost everywhere* (abbreviated a.e.) of (f_τ); i.e., the existence of a well-determined set N of $\mathscr{N}^\star$, such that the numerical sequence $(f_\tau(x))$ converges at each point $x \in (E-N)$, can be expressed by the equation $\liminf_\tau{}^\star f_\tau = \limsup_\tau{}^\star f_\tau$ almost everywhere, meaning except for a *determinate* set $N \in \mathscr{N}^\star$.

2.1. Proposition [*37*, 333]. For any set $\mathscr{G}$ of $\mathscr{B}$-measurable functions on E we have

$$\wedge^\star \mathscr{G} \leqslant e \wedge \mathscr{G} \, (\operatorname{mod} \mathscr{N}^\star) \quad \text{and} \quad e \vee \mathscr{G} \leqslant \vee^\star \mathscr{G} \, (\operatorname{mod} \mathscr{N}^\star).$$

2.2. Proposition [*37*, 334]. If Θ admits a countable cofinal subset, then for each sequence (φ_τ) of $\mathscr{B}$-measurable functions defined on E, we have

$$\liminf_\tau{}^\star f_\tau \leqslant e \liminf_\tau f_\tau \, (\operatorname{mod} \mathscr{N}^\star) \quad \text{and}$$
$$e \limsup_\tau f_\tau \leqslant \limsup_\tau{}^\star f_\tau \, (\operatorname{mod} \mathscr{N}^\star).$$

2.3. Proposition [*13*, 62 and *37*, 335]. If Θ admits a countable cofinal set, then for each sequence (f_τ) of $\mathscr{B}$-measurable functions defined on E, there exists a sequence (g_τ) of $\mathscr{B}$-measurable functions defined on E, such that $g_\tau = f_\tau$ almost everywhere for each $\tau \in \Theta$ and

$$\liminf_\tau{}^\star g_\tau = e \liminf_\tau g_\tau, \qquad \limsup_\tau{}^\star g_\tau = e \limsup_\tau g_\tau.$$

Remark. If f_τ and g_τ are $\mathscr{B}$-measurable functions, then the set of points where they differ is a well-determined set in $\mathscr{N}$.

2.4. By a *fine covering a.e.* of a set M in $\mathscr{S}$ with respect to $(\mathscr{B}_\tau)$, we shall mean a sequence $(K_\tau : \tau \in \Theta)$ such that $K_\tau \in \mathscr{B}_\tau$ for all $\tau \in \Theta$ and $M \subset \limsup_\tau{}^\star K_\tau$ a.e. Thus the set of points of M not belonging to $\limsup_\tau{}^\star K_\tau$ is a well-determined $\mathscr{N}^\star$-set.

We now introduce conditions (**L**) and (**M**) for a stochastic basis in the pointwise setting:

(**L**): Each fine covering a.e. of a set is a fine essential covering of any measure cover of the set.

(**M**): Each fine essential covering of a $\mathscr{B}$-measurable set is a fine covering a.e. of the set.

Condition (**L**) can be expressed by saying that to each set M and each fine covering a.e. (K_τ) of M, there exists a countable sequence (τ_n) of Θ-members such that $M \subset \bigcup_n K_{\tau_n}$ a.e. In particular, (**L**) and (**M**) hold when Θ is countable.

2.5. Theorem [*41*, 303]. *If* $(\mathscr{B}_\tau)$ *is a stochastic basis in the pointwise setting, then* (**L**) *is equivalent to the inequalities*

$$e \lim_\tau \inf f_\tau \leqslant \lim_\tau \inf^\star f_\tau \,(\mathrm{mod}\,\mathscr{N}^\star),$$

$$\lim_\tau \sup^\star f_\tau \leqslant e \lim_\tau \sup f_\tau \,(\mathrm{mod}\,\mathscr{N}^\star),$$

and (**M**) *is equivalent to the opposite inequalities, for each stochastic process of basis* $(\mathscr{B}_\tau)$.

Remarks. The monotone character, mod $\mathscr{N}$, of the basis is not used in the proof.

According to Prop. 2.2, and Th. 2.5, a sufficient (but not necessary) condition for M to hold is the existence of a countable cofinal subset of Θ.

Thanks to Th. 2.5, the a.e. convergence of a stochastic process (f_τ) of basis $(\mathscr{B}_\tau)$ can be deduced from the a.e. convergence assuming (**L**) holds; and the essential convergence of (f_τ) can be inferred from the a.e. convergence assuming (**M**) holds. Consequently, if in each of the Ths. III. 3.1 and III. 5.4, carried over to the pointwise setting, we postulate (**L**) together with the Vitali condition, then we obtain the a.e. convergence theorems for martingales and semi-martingales of basis $(\mathscr{B}_\tau)$. Similarly if, in the formulation of the pointwise version of Th. III. 5.1, we postulate the conjunction of (**L**) and (**M**), then we can replace the upper essential limits by the pointwise upper limit and the equality becomes

$$\lim_\tau \sup^\star f_\tau = s \lim_\tau f_\tau \,(\mathrm{mod}\,\mathscr{N}^\star).$$

2.6. We assume that the basis $(\mathscr{B}_\tau)$ is increasing, i.e., $\mathscr{B}_\rho \subset \mathscr{B}_\tau$ whenever $\rho \ll \tau$. An analysis of the proofs of Ths. III. 3.1, and III. 5.4 shows that in each of them, instead of the conjunction of the Vitali condition and the condition (**L**) to obtain a.e. convergence theorems, we can turn to the corresponding Vitali condition for fine a.e. coverings. We denote by $(\mathbf{V}_\mathfrak{f}^\star)$, $(\mathbf{V}'^\star)$, etc., these Vitali conditions. For example,

($\mathbf{V}_Y^\star$): For each set M of finite outer measure, each a. e. fine covering (K_τ) of M and each $\varepsilon > 0$, there exists a finite sequence $(\xi_1, \xi_2, \ldots, \xi_r)$ of Θ-members and an associated sequence $\mathscr{L} = (L_1, L_2, \ldots, L_r)$ of sets such that

$$L_i \in \mathscr{B}_{\xi_i} \quad \text{and} \quad L_i \subset K_{\xi_i}, \qquad i = 1, 2, \ldots, r,$$
$$\omega_Y(\mathscr{L}) < \varepsilon, \qquad \mu(M - M \cdot L) < \varepsilon, \qquad \text{where} \quad L = \cup \mathscr{L}.$$

If we substitute ($\mathbf{V}_Y^\star$) for ($\mathbf{V}_Y$), ($\mathbf{V}_\infty^\star$) for ($\mathbf{V}_\infty$) in the pointwise version of Th. III. 3.1, we can infer the a. e. convergence of the martingales involved. We encounter the same situation in Th. III. 5.4, where the replacement of condition (**V**′) by (**V**′★) allows us to infer the a. e. convergence of the submartingales of bounded variation. If, in the pointwise version of Th. III. 5.1, we replace ($\mathbf{V}_Y$) by the conjunction of ($\mathbf{V}_Y^\star$) and (**M**), then we can substitute the pointwise upper limit for the essential upper limit in the conclusion of the theorem.

We mention the following implications: ($\mathbf{V}_Y^\star$) implies (**L**); (**L**) and ($\mathbf{V}_Y$) together imply ($\mathbf{V}_Y^\star$); and thus we see the equivalence of "(**M**) and (**L**) and ($\mathbf{V}_Y$)" with "(**M**) and ($\mathbf{V}_Y^\star$)". Similar relations hold for (**V**′) and (**V**′★).

Condition ($\mathbf{V}_\infty^\star$) has been introduced under the name ($\mathbf{V}_0$) by Y. S. Chow [*9*, 266], who established the corresponding case of Th. III. 3.1.

3. Martingales in the classical sense. The classical theory of martingales is related to sets of parameters that are subsets of the real line, therefore, totally ordered. For applications of that theory, we refer the reader to [*8*], [*13*], and [*14*].

4. Product spaces. Under suitable hypotheses, it can be shown that convergence theorems are valid in product spaces. The interested reader may find [*11*], [*17*], and [*28*] instructive.

5. The Radon-Nikodym integrand defined as a derivate. Let μ be a finite, strictly positive measure defined on the Boolean σ-algebra $\mathscr{B}$ with unit E, and φ an additive function of bounded variation defined on a Boolean subalgebra $\mathscr{A}$ of $\mathscr{B}$. We denote by $\mathfrak{T}$ the set of all the finite $\mathscr{A}$-partitions of E, filtering with respect to the relation $\sqsubset$ (cf. Example 2 of I. 9). Each soma different from $\bigcirc$ is then a cell and the restriction of φ to $\mathscr{A} - \{\bigcirc\}$ is an additive cell function of bounded variation.

Without appealing to the Radon-Nikodym theorem we can, for each $\mathscr{T} \in \mathfrak{T}$, define the $\mathscr{T}$-derivate $D_{\mathscr{T}}\varphi$ as the function taking the value $\varphi(J)/\mu(J)$ on each constituent $J \in \mathscr{T}$ (II. 7). Th. II. 7.1*a* is then applicable to the martingale generated by φ; thus its integrand representation is $(D_{\mathscr{T}}\varphi;\ \mathscr{T} \in \mathfrak{T})$. In fact, this theorem is based on Ths. II. 4.7 and II. 4.9, in which the Radon-Nikodym theorem intervenes only in defining the integrand representation of the martingale or submartingale concerned (cf. the first few lines of (II.3)). Consequently, $(D_{\mathscr{T}}\varphi)$ converges stochastically to a function $sD\varphi$, and for each soma A of $\mathscr{A}$, the equation

$$\int_A sD\varphi\, d\mu = \varphi_c(A)$$

holds, where φ_c denotes the σ-additive part of φ.

Thus the Radon-Nikodym theorem is established for additive functions, since we obtain a function $sD\varphi$ possessing the characteristic property of the Radon-Nikodym integrand. It is noteworthy that the integrand thus obtained appears as a "true" derivate, meaning a stochastic derivate [*25*, 227, Remarks]. For a σ-additive function φ, we have $\int_A D\varphi\, d\mu = \varphi(A)$ for any A in $\mathscr{A}$.

If $\mathscr{A} = \mathscr{B}$, then the proof of Th. 5.1 in [*36*, 275], making use of Zorn's lemma, is immediately applicable to our present base $(\mathscr{B}_{\mathscr{T}})$, showing that the condition $(\mathbf{a}\,\mathrm{V}_\infty)$ is satisfied. Thus, according to Th. III. 6.4.2, $(D_{\mathscr{T}}\varphi)$ converges to $sD\varphi = D\varphi$ with respect to the order relation. Consequently, the Radon-Nikodym integrand appears as a derivate with respect to the order relation (essential derivate in the pointwise version).

In the case where $\mathscr{A} = \mathscr{B}$ and φ is σ-additive, we can take for $\mathfrak{T}$ the set of all countable $\mathscr{B}$-partitions of E, and the conclusions of the finite case, in particular the validity of $(\mathbf{a}\,\mathrm{V}_\infty)$, remain valid.

We call attention to the fact that Kolmogoroff [*33*] had previously proved the Radon-Nikodym theorem for Lipschitzian functions φ while showing that the family $(D_{\mathscr{T}}\varphi)$ converges strongly in L^2.

6. Representation of the spaces L_X as spaces of cell functions. For details of this subject, we refer the reader to [*43*].

7. Pointwise derivation of cell functions. The classical theory of derivation is concerned with pointwise derivates.

7.1. Cellular pointwise theory. Let $(E, \mathscr{B}, \mu)$ be a measure space as in § 1, and $\mathscr{N}$ the σ-ideal of those $\mathscr{B}$-sets that are of μ-measure zero. We fix a non-void family $\mathfrak{T}$ of partitions of E mod $\mathscr{N}$, countable and filtering for the relation $\sqsubset$ mod $\mathscr{N}$. In other words, each element $\mathscr{T} \in \mathfrak{T}$ is a set of measurable sets called cells, pairwise disjoint mod $\mathscr{N}$ and of positive measure, such that $\bigcup \mathscr{T} = E\,(\mathrm{mod}\,\mathscr{N})$. The fineness relation $\mathscr{T} \sqsubset \mathscr{S}$ mod $\mathscr{N}$ is defined as follows: each component of $\mathscr{S}$ is included mod $\mathscr{N}$ in a component of $\mathscr{T}$. It is equivalent to the inclusion $\mathscr{B}_{\mathscr{T}} \subset \mathscr{B}_{\mathscr{S}} + \mathscr{N}$, where $\mathscr{B}_{\mathscr{T}}$ and $\mathscr{B}_{\mathscr{S}}$ denote the Boolean σ-algebras generated by $\mathscr{T}$ and $\mathscr{S}$ respectively. We see that $(\mathscr{B}_{\mathscr{T}})$ is a stochastic basis increasing with respect to the relation $\sqsubset$ mod $\mathscr{N}$ in $\mathfrak{T}$.

An active complex is a subset of a partition of $\mathfrak{T}$. A cell function is a real function φ defined on $\mathscr{I} = \bigcup \mathfrak{T}$, such that $I_1 = I_2 (\operatorname{mod} \mathscr{N})$ implies $\varphi(I_1) = \varphi(I_2)$, and that $\varphi(\mathscr{J}) = \sum_{J \in \mathscr{J}} \varphi(J)$ exists for each active complex $\mathscr{J}$. The concepts of subadditivity and of additivity for a cell function defined in (I. 9) can be readily extended to the present setting by modification mod $\mathscr{N}$.

7.2. The concepts related to the order $\ll$. We now assume that $\mathfrak{T}$ is filtering with respect to a relation $\ll$ in addition to $\sqsubset$. We use the notations $(\mathfrak{T}, \ll)$ and $(\mathfrak{T}, \sqsubset)$ for the respective filtering partition families, and $(\mathscr{B}_{\mathscr{T}}; \mathscr{T} \in \mathfrak{T}, \ll)$ and $(\mathscr{B}_{\mathscr{T}}; \mathscr{T} \in \mathfrak{T}, \sqsubset)$ for the cellular stochastic bases defined by these families.

Along with those concepts defined by passages to the limit with respect to $(\mathfrak{T}, \sqsubset)$ we now define the corresponding ones with respect to $(\mathfrak{T}, \ll)$:

the upper integral with respect to $\ll$ (I. 10):

$$\ll\text{-}\varphi^s(E) = \limsup_{\mathscr{T} \in \mathfrak{T}, \ll} \varphi(\mathscr{T})\,;$$

the total variation with respect to $\ll$:

$$\|\varphi\|_{1,\ll} = \ll\text{-}|\varphi|^s(E);$$

the function φ of bounded variation (cf. (I.9)) and the Remark in (I.10) or absolutely continuous (II.7) with respect to $\ll$:

the upper stochastic derivate with respect to $\ll$ (II. 7):

$$\ll\text{-}s D^s \varphi = s \limsup_{\mathscr{T} \in \mathfrak{T}, \ll} D_{\mathscr{T}} \varphi\,;$$

the $\mathfrak{L}_X$-derivative with respect to $\ll$, if it exists (II. 7);

the essential upper derivate with respect to:

$$\ll\text{-}e D^s \varphi = e \limsup_{\mathscr{T} \in \mathfrak{T}, \ll} D_{\mathscr{T}} \varphi \qquad ((\text{III. }6) \text{ and } (\text{IV. }1))\,;$$

the essential cellular $\ll$-fine covering of a measurable set;

the Vitali conditions $\ll$-$(\mathbf{a}\,\mathbf{V}_Y)$, $\ll$-$(\mathbf{a}\,\mathbf{V}')$, the condition **(F)** (I. 10), etc.

We observe that these derivates àre defined mod $\mathscr{N}$. We do not envisage the properties of subadditivity and additivity of a cell function, nor the notion of Burkill-Kolmogoroff integral on a cell different from E, nor property **(E)** (I. 10) as being relativizable with respect to $\ll$.

7.3. Conditions linking $\ll$- and $\sqsubset$-concepts.

(Z): $\ll$ is finer than $\sqsubset$; i.e., every subset of $\mathfrak{T}$ that is terminal relative to $\ll$ is terminal relative to $\sqsubset$.

We note that **(Z)** is always satisfied if $\mathscr{T} \ll \mathscr{X}$ whenever $\mathscr{T} \sqsubset \mathscr{X}$.

If $\mathscr{T}$ and $\mathscr{S}$ are two partitions of $\mathfrak{T}$, then we define $R_{\mathscr{S}}^{\mathscr{T}}$ as the union of the components I of $\mathscr{S}$ such that no $\mathscr{T}$-cell includes $I\,(\mathrm{mod}\,\mathscr{N})$. We now express two new conditions.

$$\text{(R): For each } \mathscr{T}\in\mathfrak{T},\ s\lim_{\mathscr{S}\in\mathfrak{T},\sqsubset} R^{\mathscr{T}}_{\mathscr{S}} = \emptyset\,.$$

$$\text{(U): For each } \mathscr{T}\in\mathfrak{T},\ e\lim_{\mathscr{S}\in\mathfrak{T},\leqslant} R^{\mathscr{T}}_{\mathscr{S}} = \emptyset\,.$$

We note that (**U**) implies (**R**).

The translation of various concepts and theorems into the language of $\leqslant$ may be found in [*43*, 530–532]. We mention this translation only for Th. II. 7.1_a.

$7.3.1_{a\leqslant}$ Theorem. *If conditions* (**Z**), (**E**), *and* (**R**) *are satisfied, then every subadditive function φ of bounded variation with respect to $\leqslant$ possesses an integrable stochastic $\leqslant$-derivative $\leqslant$-$sD\varphi$, which is also the stochastic $\leqslant$-derivative of the integral cell function φ^m. For each cell I,*

$$\int_I (\leqslant\text{-}s\,D\,\varphi)\,d\mu = \varphi^m_c(I),$$

where φ^m_c denotes the σ-additive part of φ^m.

7.4. Specifically pointwise $\leqslant$-concepts. The definition of the pointwise inferior limit with respect to $\leqslant$ of a family of functions or sets, denoted by $\liminf^{\star}_{\mathscr{T}\in\mathfrak{T},\leqslant}$ is immediate, and so is that of the superior limit.

By a $\leqslant$-fine cellular covering a.e. of a set M, we mean a set $\mathscr{X}$ of cells such that the family $(K_{\mathscr{T}}\colon\ \mathscr{T}\in\mathfrak{T})$, where $K_{\mathscr{T}} = \bigcup(\mathscr{X}\cdot\mathscr{T})$, $\mathscr{T}\in\mathfrak{T}$ satisfies

$$M \subset \limsup_{\mathscr{T}\in\mathfrak{T},\leqslant}{}^{\star} K_{\mathscr{T}} \quad \text{a.e.}$$

Here are the cellular versions of conditions (**M**) and (**L**) in (2.4) for the relation $\leqslant$:

$\leqslant$-(**a L**): Each cellular $\leqslant$-fine a.e. covering of a set M is an essential covering of the measure covers of M;

$\leqslant$-(**a M**): Each essential $\leqslant$-fine covering of a measurable set M is a $\leqslant$-fine a.e. covering of M.

We consider an arbitrary partition $\mathscr{T}$ of $\mathfrak{T}$, a cell function φ and a point x of the cell I and define

$$D^i_{\mathscr{T}}\varphi(x) = \inf\left[\frac{\varphi(I)}{\mu(I)} : x\in I, I\in\mathscr{T}\right].$$

In accordance with an established convention, $(D^i_{\mathscr{T}}\varphi)(x) = +\infty$ if there exists no $\mathscr{T}$-cell containing x; i.e., if $x\in(E-\bigcup\mathscr{T})$. $(D^s_{\mathscr{T}}\varphi)(x)$ is defined analogously. If $D^i_{\mathscr{T}}\varphi$ and $D^s_{\mathscr{T}}\varphi$ are two $\mathscr{B}_{\mathscr{T}}$-measurable representatives of $D_{\mathscr{T}}\varphi$ that are defined everywhere, then we define

$$\ll\text{-}D^{\star i}\varphi = \liminf_{\mathcal{T}\in\mathfrak{T},\ll}{}^{\star} D^i_{\mathcal{T}}\varphi\,,$$
$$\ll\text{-}D^{\star s}\varphi = \limsup_{\mathcal{T}\in\mathfrak{T},\ll} D^s_{\mathcal{T}}\varphi\,.$$

In a manner similar to that used in the proof of Th. 2.5, one can prove.

7.4.1$_a$. Theorem. *Condition $\ll$-(**a**L) is equivalent to the inequalities.*

$$\ll\text{-}eD^i\varphi \leq \ll\text{-}D^{\star i}\varphi(\operatorname{mod}\mathcal{N}^\star) \qquad \textit{and}$$
$$\ll\text{-}eD^{\star s}\varphi \leq \ll\text{-}eD^s\varphi(\operatorname{mod}\mathcal{N}^\star),$$

*and the condition $\ll$-(**a**M) is equivalent to the opposite inequalities, for any cell function φ.*

φ is said to be derivable a.e. with respect to $\ll$ iff

$$\ll\text{-}D^{\star i}\varphi = \ll\text{-}D^{\star s}\varphi \qquad \textit{a. e.}$$

We can translate into the language of $\ll$ the cellular versions of the Vitali conditions, for instance,

$\ll$-(**a**$V^\star_Y$): *For each set M of finite outer measure, each cellular $\ll$-fine a.e. covering $\mathcal{X}$ of M, and each $\varepsilon > 0$, there exists a finite subset $\mathcal{P}$ of $\mathcal{X}$ such that*

$$\omega_Y(\mathcal{P}) < \varepsilon \qquad \textit{and} \qquad \bar{\mu}(M - (M\cdot(\bigcup\mathcal{P})) < \varepsilon\,.$$

7.5. Comparison with derivation in Part I, Ch. I. In the theory of cell functions, a basis of derivation is obtained from the cellular setting as follows:

By a *deriving family at the point* x we mean a family $(I_{\mathcal{T}}\colon \mathcal{T}\in\mathfrak{R})$ where $\mathfrak{R}$ represents a cofinal subset of $\mathfrak{T}$ such that $x\in I_{\mathcal{T}}$, $I_{\mathcal{T}}\in\mathcal{T}$ for each $\mathcal{T}\in\mathfrak{R}$. The domain D of the basis of derivation is the set of those points x for which the set of $\mathcal{T}$ satisfying $x\in\mathcal{T}$ is cofinal in $\mathfrak{T}$. The domain can be empty, as the following example shows: $E = \,]0,1[$, μ is Borel measure on E, $\mathfrak{T}$ consists of the family of all finite sets of disjoint open subintervals in E covering E except for a finite set, and $\ll\, = \,\sqsubset$.

In the applications, the domain of the cellular derivation basis differs from E by only an $\mathcal{N}^\star$-set, which amounts to saying that the set $\mathcal{I}$ of all cells is a $\ll$-fine covering of E a.e. If the set $\mathcal{I}$ is assumed to be an essential a.e. covering of E, then condition $\ll$-(**a**M) implies that $(E-D)\in\mathcal{N}^\star$. If $\bigcup\mathcal{T} = E$ for each $\mathcal{T}$ in $\mathfrak{T}$, then $D = E$. Such a condition usually holds in the "concrete" bases.

At each point x of D the extreme derivates $D^i f(x)$ and $D^s f(x)$ are defined as

$$\inf_{(I_{\mathcal{T}})}\left[\liminf_{\mathcal{T},\ll}\frac{\varphi(I_{\mathcal{T}})}{\mu(I_{\mathcal{T}})}\right] \quad \text{and} \quad \sup_{(I_{\mathcal{T}})}\left[\limsup_{\mathcal{T},\ll}\frac{\varphi(I_{\mathcal{T}})}{\mu(I_{\mathcal{T}})}\right]$$

respectively, where $(I_{\mathscr{T}})$ ranges over the derivation basis at the point x, in accordance with the general definition in Part I., Ch. I. At any point $x \in (E-D)$, we put $D^i\varphi(x) = -\infty$ and $D^s\varphi(x) = +\infty$. The extreme derivates $D^i\varphi$ and $D^s\varphi$ are thus defined in a unique manner at each point $x \in E$. It is then easy to prove that $\preccurlyeq\text{-}D^{\star i}\varphi(x) = D^i\varphi(x)$ and $\preccurlyeq\text{-}D^{\star s}\varphi(x) = D^s\varphi(x)$ for each $x \in E$.

The concepts of cellular $\preccurlyeq$-fine a.e. covering of a set M and of a fine covering for a cellular basis of derivation according to [*51*, 74] and [*25*, 224] are logically equivalent.

In most applications, the relation $\preccurlyeq$ is introduced by means of a positive function ν defined on $\mathfrak{T}$ by the convention

$$\mathscr{T} \preccurlyeq \mathscr{S} \quad \text{iff} \quad \nu(\mathscr{T}) \geqslant \nu(\mathscr{S}).$$

Finally, replacing $\nu(\mathscr{T})$ by $\nu(\mathscr{T}) - \inf(\nu(\mathscr{S}) : \mathscr{S} \in \mathfrak{T})$, we can always assume that $\mathfrak{T}$ contains partitions $\mathscr{T}$ such that $\nu(\mathscr{T})$ is arbitrarily small. There exists in $\mathfrak{T}$ a countable cofinal set, $\preccurlyeq$-(**aM**) is thus satisfied, and consequently $(E-D) \in \mathscr{N}^{\star}$. Generally, the function ν is monotone, meaning that $\mathscr{T} \sqsubset \mathscr{S}$ implies $\nu(\mathscr{T}) \geqslant \nu(\mathscr{S})$; hence (**Z**) is satisfied.

For any cell I we put $\delta(I) = \inf(\nu(\mathscr{I}) : I \in \mathscr{T}, \mathscr{T} \in \mathfrak{T})$. Then it can be proved that

$$(\preccurlyeq\text{-}D^{\star i}\varphi)(x) = \sup_{\varepsilon > 0}\left[\inf\left(\frac{\varphi(I)}{\mu(I)} : x \in I, \delta(I) < \varepsilon\right)\right];$$

an analogous relation holds for the upper derivate.

For examples of concrete cellular bases, we refer to [*43*, 536–541].

8. Examples of concrete cell bases. Six examples are given in [*43*, 4.8].

9. Stochastic bases on a group [*27*]. We note briefly that the two authors of the paper cited have established convergence theorems for martingales in case μ is a Haar measure on a group E.

Bibliography

Part II

1. Bochner, S.: Additive set functions on groups. Annals of Math. **40**, 769–799 (1939).
2. –, and R. S. Phillips: Additive set functions and vector lattices. Annals of Math. **42**, 316–324 (1941).
3. – Partial ordering in the theory of martingales, Annals of Math. **62**, 162–169 (1955).
4. Cairoli, R.: Sur le prolongement naturel de functions sigma-additives. Comment. Math. Helv. **39**, 90–96 (1964).
5. Carathéodory, C.: Mass und Integral und ihre Algebraisierung. Basel, Stuttgart: Birkhäuser Verlag 1956 (Lehrbücher und Monographien. Mathematische Reihe **10**).
6. Cesari, L.: Surface area. Princeton: Princeton University Press 1956 (Annals of Mathematics Studies **35**).

7. – Area and measure. Brunswick: Summer Institute 1958.
8. CHIANG, TSE-PEI: Une remarque sur la définition de la quantité d'information [en russe]. Teorija Veroj. Prim. **3**, 99–103 (1958).
9. CHOW, Y. S.: Martingales in a σ-finite measure space indexed by directed sets. Trans. Amer. Math. Soc. **97**, 254–285 (1960).
10. COTLAR, M. y Y. FRENKEL: Una teoria general de integral basada en una extensión del concepto de limite. Univ. nac. Tucuman, Rivista, Série A, **6**, 113–159 (1947).
11. DIEUDONNÉ, J.: Sur un théorème de Jessen. Fund. Math. **37**, 242–248 (1950).
12. DOLÉANS, C.: Seminaire Choquet (Initiation à l'Analyse) 4e année **4** (1964–1965).
13. DOOB, J. L.: Stochastic processes. New York: J. Wiley; London: Chapman and Hall 1953 (Wiley Publications in Statistics).
14. – Semimartingales and subharmonic functions. Trans Amer. Math. Soc. **77**, 86–121 (1954).
15. – Notes on martingale theory, Proceedings of the Fourth Berkeley symposium on mathematical statistics and probability [**4**, 1960, Berkeley]; **2**, 95–102. Berkeley, Los Angeles: University of California Press 1961.
16. DUBROVSKIJ, V. M.: Sur la base d'une famille de fonctions d'ensemble complètement additives et sur les propriétés d'additivité uniforme et d'équicontinuite [en russe]. Doklady Akad. Nauk S. S. S. R., N. S. **58**, 737–740 (1947).
17. DUNFORD, N., and J. D. TAMARKIN: A principle of Jessen and general Fubini theorems. Duke math. J. **8**, 743–749 (1941).
18. FICHERA, G.: Intorno al passagio al limite sotto il segno d'integrale. Port. Math. **4**, 1–20 (1943).
19. FUGLEDE, B.: On a theorem of F. Riesz. Math. Scand. **3**, 283–302 (1955).
20. GOFFMAN, C., and D. WATERMAN: On upper and lower limits in measure. Notices Amer. Math. Soc. **5**, 812 (1958).
21. HAHN, H., and A. ROSENTHAL: Set functions. Albuquerque, N. M.: University of New Mexico Press 1948.
22. HALMOS, P. R.: Measure theory. Toronto, New York, London: Van Nostrand Company 1950 (The University Series in higher Mathematics).
23. HAUPT, O., G. AUMANN u. C. PAUC: Differential- und Integralrechnung, III, 2te Auflage. Berlin: W. de Gruyter 1955 (Göschens Lehrbücherei, Reine und angewandte Mathematik **26**).
24. – et C. PAUC: Propriétés de mesurabilité de bases de dérivation. Portugaliae Math. **13**, 37–54 (1954).
25. HAYES, C. A., JR., and C. PAUC: Full individual and class differentiation theorems in their relations to halo and Vitali properties. Canad. J. Math. **7**, 221–274 (1955).
26. HELMS, L. L.: Mean convergence of martingales. Trans. Amer. Math. Soc. **87**, 439–446 (1958).
27. JERISON, M., and G. RABSON: Convergence theorems obtained from induced homomorphisms of a group algebra. Annals of Math. **63**, 176–190 (1956).
28. JESSEN, B.: The theory of integration in a space of an infinite number of dimensions. Acta Math. **63**, 249–323 (1934).
29. – On strong differentiation in a space of infinitely many dimensions. Mat. Tidsskr. **B**, 54–57 (1950).
30. – On strong differentiation. Mat. Tidsskr. **B**, 90–91 (1952).
31. KAPPOS, D. A.: Strukturtheorie der Wahrscheinlichkeitsfelder und -räume. Berlin-Göttingen-Heidelberg: Springer 1960 (Ergebnisse der Mathematik 24).
32. KHINČIN, A. J.: Sur les théorèmes fondamentaux de la théorie de l'information [en russe]. Uspekhi mat. Nauk **11**, 17–75 (1956). Traduction allemande dans Arbeiten zur Informationstheorie **I**, 26–85. Berlin: VEB Deutscher Verlag der Wissenschaften 1957.

33. Kolmogoroff, A.: Untersuchungen über den Integralbegriff. Math. Annalen **103**, 654–696 (1930).
34. Krasnoselskij, A. et J. B. Rutiskij: Sur la théorie des espaces d'Orlicz [en russe]. Doklady Akad. Nauk S.S.S.R., N.S. **81**, 497–500 (1951).
35. Krickeberg, K.: La nécessité de certaines hypothèses de Vitali fortes dans la théorie de la dérivation extrême de fonctions d'intervalles. C.R. Acad. Sc. **238**, 764–766 (1954).
36. – Extreme Derivierte von Zellenfunktionen in Booleschen σ-Algebren und ihre Integration. Sitzungb. math.-naturw. Kl. Bayer. Akad. Wiss. München 217–279 (1955).
37. – Convergence of martingales with a directed index set. Trans. Amer. Math. Soc. **83**, 313–337 (1956).
38. – Stochastische Konvergenz von Semimartingalen. Math. Z. **66**, 470–486 (1957).
39. – Stochastische Derivierte. Math. Nachr. **18**, 203–217 (1958).
40. – Seminar on martingales. Aarhus: Matematisk Institut, Aarhus Universitet 58 p. (1959).
41. – Notwendige Konvergenzbedingungen bei Martingalen und verwandten Prozessen. Transactions of the Second Prague conference on information theory, statistical decision functions, random processes [1959, Prague]; 279–305. Prague: Publishing House of the Czechoslovak Academy of Sciences 1960.
42. – Bemerkungen zur stochastischen Konvergenz. Bull. Soc. Math. Grèce. Nouvelle Série, Tome 5, Fasc. 1, p. 81–92 (1964).
43. –, and C. Pauc: Martingales et dérivation. Bull. Soc. Math. France **91**, 455–544.
44. La Vallée Poussin, C. de: Sur l'intégrale de Lebesgue, Trans. Amer. Math. Soc. **16**, 435–501 (1915).
45. – Intégrales de Lebesgue, fonctions d'ensemble, classes de Baire. 2e édition. Paris: Gauthier-Villars 1934.
46. Leader, S.: The theory of L_p-space for finitely additive set functions. Annals of Math. **58**, 528–543 (1953).
47. Loéve, M.: Probability theory. Second edition. Princeton, Toronto, New York: Van Nostrand Company 1960 (The University Series in higher Mathematics).
48. Loomis, L. H.: On the representation of σ-complete Boolean algebras. Bull. Amer. Math. Soc. **53**, 757–760 (1947).
49. Moy, Shu-Teh Chen: Measure extensions and the martingale convergence theorem. Proc. Amer. Math. Soc. **4**, 902–907 (1953).
50. Olmstedt, J. M. H.: Lebesgue theory on a Boolean algebra. Trans. Amer. Math. Soc. **51**, 164–193 (1942).
51. Pauc, C.: Ableitungsbasen, Prätopologie und starker Vitalischer Satz. J. für die reine und angew. Math. **191**, 69–91 (1953).
52. – Contributions à une théorie de la différentiation de fonctions d'intervalle sans hypothèse de Vitali. C. R. Acad. Sci. **236**, 1937–1939 (1953).
53. – Dérivés et intégrants, Fonctions de cellule. Cours d'été de Varenne. Varenna: 1954 (Roma: Publ. Mat. Inst. 76).
54. Raoult, J. P.: Généralisation de la notion de sous-martingale: asympto-sous-martingale. Dérivation et théorèmes de convergence en moyenne. C. R. Acad. Sci. Paris **263**, 738–741 (1966).
55. Rutovitz, D., and C. Pauc: Theory of Ward for cell functions, I and II. Annali di Mat., Série 4, **47**, 1–57 (1959).
56. Saks, S.: Theory of the integral. Second edition. New York 1952.
57. Snell, J. L.: Applications of martingale system theorems. Trans. Amer. Math. Soc. **73**, 293–312 (1952).

58. Sparre-Andersen, E., and B. Jessen: Some limit theorems on integrals in an abstract set. Danske Vid. Selsk. Math.-Fys. Medd. **22**, n° 14, 29 p. (1946).
59. Wecken, F.: Abstrakte Integrale und fastperiodische Funktionen. Math. Z. **45**, 377–404 (1939).
60. Yosida, K., and E. Hewitt: Finitely additive measures. Trans. Amer. Math. Soc. **72**, 46–66 (1952).
61. Zaanen, A. C.: Linear analysis; measure and integral, Banach and Hilbert space, linear integral equations. Amsterdam: North-Holland publishing Company; Groningen: P. Noordhoff 1953 (Biblioteca mathematica **2**).
62. Zygmund, A.: On the differentiability of multiple integrals. Fund. Math. **23**, 143–149 (1934).

Complements

1°. Derivation of vector-valued integrals

In this complement we shall derive set functions taking their values in a Banach space $\mathfrak{X}$. The measure space is $(R, \mathcal{M}, \mu)$, μ is complete, and $\mu(R)$ is finite. $\mathfrak{B}$ denotes the derivation basis with domain $E = R$.

1. Derivation of the Bochner integral. (Fund. Math. 20, 262–276 (1933)). A function f from R to $\mathfrak{X}$ is called a *μ-simple* function if f is constant on each of a finite number of disjoint $\mathcal{M}$-sets whose union is R. f is termed *measurable* if it is μ-a. e. the strong limit of a sequence (f_n) of μ-simple functions; i.e., $\lim_n \|f_n(x)-f(x)\| = 0$ for μ-almost all $x \in R$. f is said to be *Bochner integrable over* R iff the norm $\|f(x)\|$, regarded as a real-valued function of $x \in R$, is μ-integrable over R. The Bochner integral $\int_R f d\mu$ or $\int_R f(x) d\mu(x)$ is then defined as $\lim_n \int_R f_n d\mu$, which does not depend on the choice of the approximating sequence (f_n) provided $\|f_n(x)\| \leqslant \|f(x)\|$ for each n and each $x \in R$.

Bochner integrability on R implies Bochner integrability over any set $M \in \mathcal{M}$. As a function of the set M, this integral is σ-additive and μ-continuous. The total variation of the Bochner integral is finite.

1.1. Theorem. (N. Dunford – J. Schwartz, Linear Operators, Part I, 217 (1958)). *If $\mathfrak{B}$ is a strong derivation basis, then $\mathfrak{B}$ derives the Bochner integral $\varphi = \int f d\mu$ to the value of its integrand f μ-a. e. The same holds for Lipschitzian integrals if $\mathfrak{B}$ is a weak derivation basis.*

Proof. (1) We shall prove first that f is essentially separably valued, i.e., that there exists a μ-nullset N_0 such that $f(R-N_0)$ is separable. By definition of integrability there exists a sequence of simple functions f_n converging strongly to f except on a μ-nullset N_0. The set $f_n(R-N_0)$ is finite and hence the closure in $\mathfrak{X}$ of the union of the sets $f_n(R-N_0)$ is a separable set including $f(R-N_0)$, which proves that $f(R-N_0)$ is separable.

(2) We assume the strong Vitali property for $\mathfrak{B}$. Let $\{Z_n\}$ be a countable dense subset of $f(R-N_0)$. Since $\|f\|$ is integrable, so is $\|f-z_n\|$, and it follows from I. II. 6 that for each $n = 1, 2, \ldots$ there exists a μ-nullset $N_n \in \mathcal{N}$ such that, for all deriving sequences (M_ι) in $x \in (R-N_n)$,

$$\lim_\iota \frac{1}{\mu(M_\iota)} \int_{M_\iota} \|f(y)-z_n\| \, d\mu(y) = \|f(x)-z_n\| . \tag{1.1.1}$$

The set $N = \bigcup_n N_n$, $n = 0, 1, 2, \ldots$ is a μ-nullset. Let $x \in (R-N)$ and suppose $\varepsilon > 0$. We select z_k so that $\|f(x)-z_k\| < \varepsilon$. Then

$$\sup\left[\limsup_{\iota} \frac{1}{\mu(M_\iota)} \int_{M_\iota} \|f(y)-f(x)\|\, d\mu(y)\right.$$

$$\leqslant \sup\left[\limsup_{\iota} \frac{1}{\mu(M_\iota)} \int_{M_\iota} (\|f(y)-z\| + \|z_k-f(x)\|)\, d\mu(y)\right.$$

$$= \|z_k-f(x)\| + \sup\left[\limsup_{\iota} \frac{1}{\mu(M_\iota)} \int_{M_\iota} \|f(y)-z_k\|\, d\mu(y)\right]$$

$$= 2\,\|z_k-f(x)\| < 2\varepsilon\,.$$

We have, for any set M_ι,

$$\|f(x)-\frac{1}{\mu(M_\iota)} \int_{M_\iota} f(y)\, d\mu(y)\| \leqslant \frac{1}{\mu(M_\iota)} \int_{M_\iota} \|f(y)-f(x)\|\, d\mu(y)\,;$$

thus, since ε is arbitrary in (1.1.1), it follows that this last expression tends to zero when (M_ι) is any deriving sequence converging to $x \in (R-N)$. Accordingly, $D\varphi(x)$ exists and equals $f(x)$ for each $x \in (R-N)$.

(3) If we assume the weak Vitali property for $\mathfrak{B}$, then the derivation theorem holds for integrals with bounded integrands (Def. I. II. 2.7). If f is bounded, then so is $f-z_n$ and (1.1.1) holds. Thus $D\varphi(x)$ exists and agrees with $f(x)$ for each $x \in (R-N)$.

2. Derivation of Pettis' integral. (On integration in vector spaces Trans. Amer. Math. Soc. 44, 277–304 (1938)). A function f from R into $\mathscr{X}$ is said to be *scalarly μ-measurable* iff for each element z' of the (topological) dual $\mathscr{X}'$ of $\mathscr{X}$, $\langle z', f\rangle$ is μ-measurable. The function f is said to be *Pettis integrable* iff f is scalarly μ-measurable, the scalar function $\langle z', f\rangle$ is μ-integrable, and there exists an element $\psi(R)$ of $\mathscr{X}$ such that

2.1. $\langle z', \psi(R)\rangle = \int_R \langle z', f(x)\rangle\, d\mu(x)$ for any $z' \in \mathscr{X}'$.

By definition, $\psi(R)$, which is unique, is the *integral of Pettis over* R.

If f is Pettis integrable over R, then the same is true over any set $M \in \mathscr{M}$. We denote by $\psi(M)$ the value of the integral over M. The norm $\|\psi\| = \sup_{M \in \mathscr{M}} \psi(M)$ is finite.

A function λ from the constituents of the spread of $\mathfrak{B}$ to $\mathscr{X}$ is said to be *weakly derivable* or to have a *weak derivative* $D\lambda(x)$ iff there exists a function f from R to $\mathscr{X}$ such that for every z' in $\mathscr{X}'$, the scalar function $\langle z', \lambda\rangle$ is derivable μ-a.e. in $\mathscr{X}$ to the value $\langle z', f(x)\rangle$, and we define $D\lambda(x) = f(x)$.

2.2. Proposition. (PETTIS, loc. cit., 300). If $\mathfrak{B}$ possesses the strong Vitali property, then a necessary and sufficient condition that a σ-additive μ-continuous function ψ from $\mathscr{M}$ to $\mathscr{X}$ be weakly derivable is that the function ψ be a Pettis integral; $\psi(M) = \int_M f\,d\mu$.

Proof. Suppose first that $\psi(M)$ is the Pettis integral of f over M. By 2.1, for any z' in $\mathscr{X}'$,

$$\langle z', \psi(M)\rangle = \int_M \langle z', f(x)\rangle\, d\mu(x)\,. \tag{2.2.1}$$

By virtue of the strong Vitali property, the scalar function $\langle z', \psi\rangle$ is derivable μ-a.e. in x to $\langle z', f(x)\rangle$, which proves that the weak derivative of ψ exists and equals $f(x)$ μ-a.e.

Next, suppose that ψ is weakly derivable μ-a.e. in x to $f(x)$. This means that for any $z' \in \mathscr{X}'$, the scalar function $\langle z', \psi\rangle$ is derivable μ-a.e. in x to the value $\langle z', f(x)\rangle$. The function $\langle z', \psi\rangle$, being σ-additive and μ-continuous, is the integral of the derivative;

$$\langle z', \psi(M)\rangle = \int_M \langle z', f(x)\rangle\, d\mu(x)\,, \tag{2.2.2}$$

which proves that ψ is the Pettis integral of f.

Remark. Prop. 2.2 shows that if ψ is σ-additive and μ-continuous, there is equivalence between the derivability of ψ and the existence of an integrand (i.e., a Radon-Nikodym integrand) f such that $\psi(M) = \int_M f\,d\mu$. If we assume only that f is σ-additive and μ-continuous, then the proposition fails to hold.

3. Counterexample. (J. A. CLARKSON, Uniformly Convex Spaces, Trans. Amer. Math. Soc., v. 40, 406 (1936)). $R = [0,1[$, $\mathscr{M}$ is the family of Borel sets of $[0,1[$, μ is the Borel measure on $[0,1[$. $\mathscr{X} = \mathfrak{L}_1$ = space of the μ-integrable functions on the interval $[0,1[$. To each point t, $0 \leqslant t < 1$, let correspond the element l_t of $\mathscr{X}$ defined as follows: $l_t(s) = 1$ if $s \leqslant t$; $l_t(s) = 0$ if $t < s$. The point function l_t satisfies the condition $\|l_{t_1} - l_{t_2}\| = |t_1 - t_2|$. For an interval $I = [a, b[$, we set $\psi(I) = l_b - l_a$. We extend ψ to the Boolean algebra of the figures F by additivity. We obtain a function ψ defined on a Boolean algebra, additive and satisfying the Lipschitz condition $\|\psi(F)\| \leqslant \mu(F)$. Consequently, $\mathscr{X}$ being weakly complete (DUNFORD-SCHWARTZ, loc. cit., p. 68 and p. 290) and ψ σ-additive, ψ can be extended as a σ-additive set function defined on $\mathscr{M}$ (PAUC, C. R., Prolongement d'une mesure vectorielle jordanienne en une mesure lebesguienne, Paris 223, p. 606 (1946)), and satisfying the Lipschitz condition $\|\psi(M)\| \leqslant \mu(M)$. For $J = [t', t''] \subset [0, 1[$, $t' \leqslant t_0 \leqslant t''$, $t' < t''$, we have $\psi(J)/\mu(J) = (l_{t''} - l_{t'})/(t'' - t')$. We assert that the function ψ is not weakly (*a fortiori* strongly) derivable at any point t_0. Since the dual of $\mathfrak{L}_1$ is $\mathfrak{L}_\infty$, we choose a function g bounded on $[0, 1[$ which is discontinuous at t_0, and let z' be the corresponding element of $\mathscr{X}'$. Then $\langle z', \psi\rangle$ is derivable at t_0 iff $\int_0^{t'} g(s)(l_{t''}(s) - l_{t'}(s)/(t''-t')ds$ has a limit when J contracts to t_0, which is not the case when g is discontinuous at t_0. Consequently, ψ is not weakly derivable at this point.

2°. Functional derivatives

1. Preliminaries. Concrete cases. (DUNFORD and SCHWARTZ, loc. cit., 218–222). Let us suppose that the function f is defined on $\mathbf{R}$, takes its values in a Banach space $\mathscr{X}$ and is Bochner integrable on $\mathbf{R}$. For $t \in \mathbf{R}$ we denote by $\varphi(t)$ the integral $\int_{-\infty}^{t} f(s)ds$, and by q_n^0 the function defined so that

$$q_n^0(t) = n \quad \text{if} \quad |t| \leqslant 1/2n, \qquad q_n^0(t) = 0 \quad \text{if} \quad |t| > 1/2n.$$

We have

$$\frac{f\left(t+\frac{1}{2n}\right) - f\left(t-\frac{1}{2n}\right)}{\frac{1}{n}} = \int_{-\infty}^{+\infty} q_n^0(t-s)f(s)ds. \tag{1.1}$$

The difference quotient on the left-hand side of (1) converges to $f(t)$ when $n \to \infty$ iff

$$\lim_{n\to\infty} \int_{-\infty}^{+\infty} q_n^0(t-s)f(s)ds \quad \text{exists and equals } f(t). \tag{1.2}$$

We term the latter property "convergence at t". According to the theorem on the derivability of Bochner's integral, it holds a.e. on $\mathbf{R}$ (cf. Complement 1⁰).

Definition. The positive functions q_n have the "Dirac behavior" for $t = 0$ iff

$$\int_{-\infty}^{+\infty} q_n(s)ds = 1 \quad \text{(normalization)}$$

and

$$\lim_{n\to\infty} \int_{-\varepsilon}^{\varepsilon} q_n(s)ds = 1, \qquad \varepsilon > 0.$$

As examples of sequences of functions exhibiting the same phenomenon we mention

$$q_n^1(t) = (ne^{-n^2t^2})/\sqrt{\pi}, \qquad q_n^2(t) = (1/\pi)(n+1)k_n(t))$$

(where k_n is Fejér's kernel) for $-\pi \leqslant t \leqslant \pi$, and $q_n^2(t) = 0$ for $|t| > \pi$; $q_n^3(t) = (1/\pi)d_n(t)$ (d_n is Dirichlet's kernel) for $-\pi \leqslant t \leqslant \pi$ and $q_n^3(t) = 0$ for $|t| > \pi$.

The question arises: does the convergence property hold a.e. or not? DUNFORD and SCHWARTZ prove that the answer is affirmative for a class of functions $(s,t) \to r_n(s,t)$ that are not assumed to be positive nor of the form $r_n(s,t) = q_n(t-s)$.

In consideration of (1.1) we can envisage the convergence of $\int_{-\infty}^{+\infty} q_n(t-s)f(s)\,ds$, (or, more generally, of $\int_{-\infty}^{+\infty} r_n(s,t)f(s)\,ds$), when $n \to \infty$, as a derivation process, the contracting sets of the ordinary derivation being replaced *at each point t* by the functions $q_n(t-s)$, (or, more generally, by the functions $r_n(s,t)$). We propose to call such derivatives „functional derivatives". They comprise as special cases the usual derivatives when the deriving functions are indicatrix (or characteristic) functions multiplied by constants (depending on n).

That there exists a close connection between the two types of derivation is shown, for example, by the following theorem of B. JESSEN, J. MARCINKIEWICZ, and A. ZYGMUND (Fund. Math. 25 (1935), 217–234):

If f is a real function defined on $\mathbf{R}^2$, then the integrability of $|f| \operatorname{Log}^+ |f|$ on $\mathbf{R}^2$ secures the derivability a.e. of $\int f\,d\mu$ on the interval basis and, on the other hand, the convergence property of the Fejér's means of the Fourier series generated by f. As another example, see an article by A. ZYGMUND (Fund. Math., 23 (1934) 143–149).

The question arose: can we find Vitali properties of functional type?

2. Abstract setting. Vitali properties. (R. DE POSSEL, C. R. Acad. Sci., Paris 224, 1137–1139 (1947); C. R. Acad. Sci., Paris 224, 1197–1198 (1947).) $(E, \mathscr{B}, \mu)$ is a measure space, the measure μ being σ-finite. The function to be derived is a measure v defined on $\mathscr{B}$, taking its values in a Banach space $\mathscr{X}$, the total variation of which is finite on each set B of finite measure. We assume that v admits a *Radon-Nikodym integrand* with respect to the measure μ. Q denotes the set of functions q defined on E, finite, nonnegative, vanishing outside a set of finite measure, and such that $\int q\,d\mu > 0$. To each point t there is associated a filter basis, more briefly a filter, $\mathscr{F}_t$ on Q. The functional derivate $Dv(t)$ is defined as

$$\lim_{\mathscr{F}_t} \frac{\int q\,dv}{\int q\,d\mu},$$

$\int q\,dv$ being defined as in DUNFORD and SCHWARTZ (loc. cit., 323), the limit referring to the norm, provided this limit exists. The functional derivation basis is denoted by $\mathfrak{F}$. Although it was just defined by means of the filters $\mathscr{F}_t$, it could as well have been defined by deriving (Moore-Smith) sequences q_F, indexed by the sets of $\mathscr{F}_t$ so that $q_F \in F$, the direction of the indices being prescribed by: $F_1 \succ F_2$ iff $F_1 \subset F_2$.

If v is an integral, $v(B) = \int_B g(s)\,d\mu(s)$, then the "difference quotient" $\int q\,dv = \int q g\,d\mu$. The deriving process of Dunford and Schwartz for integrals when $q \geqslant 0$ and $\int q\,d\mu > 0$ is a special case of de Possel's;

his deriving functions in t are $q_{n,t}: q_{n,t}(s) = q_n(t-s)$, the filter $\mathscr{F}_t$ consists of the sets $\{q_{n,t}, q_{n+1,t}, \ldots\}$, for $n = 1, 2, \ldots$.

We call *constituents* of $\mathfrak{F}$ those functions occurring in any set of $\mathscr{F}_t$ for any $t \in E$. An *$\mathfrak{F}$-fine covering V of E* is defined as a set of constituents including, for each t, a set V_t meeting all the sets of $\mathscr{F}_t$.

Theorem 1. *Let V be any $\mathfrak{F}$-fine covering of E, A any subset of E of positive finite outer measure. The following statements are equivalent:*

(1) *$\mathfrak{F}$ derives any $\mathscr{X}$-valued Lipschitzian integral to its integrand a.e.;*

(2) *$\mathfrak{F}$ derives any real-valued Lipschitzian integral to its integrand a.e.;*

(3) *For any V, A, and $\varepsilon > 0$ there exists a point t of A, a function ψ belonging to V_t, and a number $\lambda > 0$ such that*

$$\int \inf(c_{\bar{A}}, \lambda\psi)\, d\mu \geqslant (1-\varepsilon) \int \lambda\psi\, d\mu .$$

(4) *For any V, A, and $\varepsilon > 0$, there exists a finite set of points t_i of A, a finite set of functions ψ_i and numbers $\lambda_i > 0$ such that each $\psi_i \in F_{t_i}$ and*

$$\int \Big|c_{\bar{A}} - \sum_i \lambda_i \psi_i\Big|\, d\mu < \varepsilon .$$

Comments on Theorem 1. (1) Counterparts of the preceding properties in the setting of derivation by sets are given by R. DE POSSEL (Jour. Math. Pure et Appl. *15*, 400–405 (1936)). See also Part. I. II. 2.7 and Ch. III. 1. In this setting, (3) becomes the density property at a point and (4) the Vitali μ-property.

(2) We may notice that an $\mathfrak{F}$-fine covering covers the set E and that the V-functions ψ_i refer to the points t_i.

(3) The derivability of scalar Lipschitzian integrals implies the derivability of $\mathscr{X}$-valued Lipschitzian integrals. No special property of the value space is required.

Theorem 2. *In order that the basis $\mathfrak{F}$ derives any $\mathscr{X}$-valued integral to its integrand a.e., the following condition is sufficient:*

For any V and A one can find a subset A' of A of positive outer measure, and a number $\theta > 0$ such that for any $\varepsilon > 0$, there exists a finite set of points t_i from A', a finite set of functions ψ_i each belonging to V_{t_i} and λ_i satisfying the inequality $\int |c_{\bar{A}} - \sum_i \lambda_i \psi_i|\, d\mu < \varepsilon$ and additionally fulfilling the relation $\sum_i \lambda_i \psi_i(t) \leqslant \theta$ at almost all points t not belonging to $\bar{A}'$.

Theorem 3 gives a sufficient condition for the derivation of a $\mathscr{X}$-valued measure admitting a Radon-Nikodym integrand.

3°. Topologies generated by measures

1. (O. HAUPT and C. PAUC, La topologie approximative de Denjoy envisagée comme vraie topologie, C. R. Acad. Sci., Paris, 234, 390–392

(1952). Über die durch allgemeine Ableitungsbasen bestimmten Topologien. Ann. Mat. Pura Appl. Ser. IV, 36, 247–271 (1954)).

$(R, \mathscr{B}, \mu)$ is a measure space, the measure μ is complete and σ-finite, $\mathfrak{B}$ is a derivation basis in the sense of I. I. 1, with R as domain of definition. A point x is called "approximately interior" or "D-interior" to a set $X \subset R$ iff $x \in X$ and the lower $\mathfrak{B}$-density exists and equals 1. Explicitly, for any $\varepsilon > 0$ and any $\mathfrak{B}$-sequence (M_ι) converging to x, there exists an index ι' such that $\mu(X \cap M_\iota) > (1-\varepsilon)\mu(M_\iota)$ whenever $\iota > \iota'$. Denoting by $J(X)$ the set of D-interior points of X, we have $J(\emptyset) = \emptyset$; $J(X) \subset X$; $J(X) \subset J(Y)$ whenever $X \subset Y$; and $J(X \cap Y) = J(X) \cap J(Y)$. The *weak Vitali property* (I.II.2.7) is equivalent to: for each $X, J(X)$ is a μ-kernel of X. From now on, we assume that this property holds. Thus $J^2 = J$; J defines a topology called "D-topology". The D-closure of X is a μ-measure cover of X, the μ-measurable sets coincide with the Jordan measurable sets (i.e., their boundary is a μ-nullset). The D-open sets are μ-measurable. To each equivalence class $\mathscr{C}$ modulo $\mathscr{N}$ of μ-measurable sets, there exists a special representative O ($\mathscr{C}$), namely, the set of those points at which each set of $\mathscr{C}$ has density 1. If $\mathfrak{B}$ is the interval basis $\mathfrak{J}^n$, then a D-open set may be nowhere dense in the Euclidean sense; the space $\boldsymbol{R}^n$ is not D-locally compact.

See also: C. Ionescu Tulcea, On the lifting property and the disintegration of measures, Bull. Amer. Math. Soc. 71, § 4 (1965).

2. (R. J. Troyer and W. P. Ziemer, Topologies generated by outer measures, J. Math. Mech. 12, 485–494 (1963)). In previous papers by C. Goffman and D. Waterman (Proc. Amer. Math. Soc. 12, 116–121 (1961)), and C. Goffman, C. J. Neugebauer, and T. Nishimura (Duke Math J. 28, 497–506 (1961)), it was shown that the n-dimensional Lebesgue measure generates a topology on Euclidean n-space in the following manner: a set S is said to be *open* iff it is measurable and the classical metric density on S exists and is equal to 1 at every point of S. In the present paper a similar topology on $\mathbf{R}^n$ is defined in a similar manner by means of a regular outer Carathéodory measure φ finite on compact sets. For reasons of convenience, $\varphi(Q)$ is assumed to be positive for any cube Q. Q_x^r denotes the open cube with x as center and r as its diameter. For a φ-measurable set A, the φ-density $d(\varphi, A, x)$ is the limit, if it exists, of $\varphi(A \cdot Q_x^r)/\varphi(Q_x^r)$ when $r \to 0$. A set A is called "φ-open" iff it is φ-measurable and the φ-density is equal to 1 at all points of A. From a theorem of A. P. Morse (see (I. VI. 4)) the basic theorem follows; namely, the φ-density of A exists and is equal to 1 at φ-almost all points of A. Indications are given as to how the topologies corresponding to different outer measures compare. If S is a metric space and f is a φ-continuous mapping of $\mathbf{R}^n$ into S, then f is of Baire class 1. The φ-topology for $\mathbf{R}^n$ is not normal. The main theorem is an analogue of the Lusin-

Menchoff theorem: Let $E \subset \mathbf{R}^n$ be a φ-measurable set and $X \subset E$ be a closed set such that $d(\varphi, E, x) = 0$ for each $x \in X$. There is a closed set F such that (*i*) $X \subset F \subset E$, (ii) $d(\varphi, F, x) = 1$ for each $x \in X$. The proof specifically for $n = 2$ is euclidean since the construction of F involves weighted points and weighted horizontal and vertical segments. The complete regularity of the φ-topology follows from an easy generalization of a lemma due to Z. ZAHORSKY (Trans. Amer. Math. Soc. 69, 1–54 (1950)). This theorem implies that the φ-topology is the coarsest one for which the φ-approximately continuous functions are continuous. A condition is given for an open connected set to be φ-connected. We may observe that in the present paper only the measure induced by φ appears effectively.

4°. Vitali's theorem for invariant measures

(W. W. COMFORT and HUGH GORDON, Trans. Amer. Math. Soc. 99, 83–90 (1961)).

§ 2. X is a Hausdorff space; $\emptyset$ is the empty set; G is a group acting transitively on X; μ is a non-negative, countably additive measure such that gU is μ-measurable and $\mu g U = \mu U$ whenever U is a measurable subset of X and $g \in G$; $\bar{\mu}$ is the outer measure determined by μ. It is assumed that for each $x \in X$ and each $\varepsilon > 0$ there exists a μ-measurable neighborhood V of x for which $\mu V < \varepsilon$. Φ (the "origin") is an arbitrary but fixed point of X. For $B \subset X$ and $C \subset X$, $B^{-1}C = \cup \{gC : g \in G, F \in gB\}$. A sequence (S_n) of subsets of X is called a *sequence of quasispheres* iff, for some $\varepsilon > 0$ and each positive integer n, $S_{n+1}^{-1} S_{n+1} \subset S_n$ and $\bar{\mu} S_{n+1} > \varepsilon \bar{\mu} S_n$. A family $\mathscr{U}$ of closed μ-measurable subsets of X, each with positive μ-measure, is said to be a *regular Vitali cover* for $A \subset X$ iff there exists a sequence (S_n) of quasispheres and a real-valued function M ("parameter of regularity" in the classical theory) on A such that, for each $x \in A$ and each neighborhood V of x, there exists $U \in \mathscr{U}$, $g \in G$, and an integer n for which $x \in U \subset V$, $gU \subset S_n$, and $\bar{\mu} S_n \leqslant M(x) \mu U$.

§ 3. **Main theorem.** *Let $\mathscr{U}$ be a regular Vitali cover for $A \subset X$, and suppose that there is a sequence (V_n) of open subsets of X for which $A \subset \bigcup_k V_k$ and $\mu V_k < \infty$ for $k = 1, 2, , \ldots$. Then there is a (possibly finite) sequence (U_n) of pairwise disjoint elements of $\mathscr{U}$ for which $\bar{\mu}(A - A \cdot \left(\bigcup_n U_n\right)) = 0$.*

Sketch of the proof. It is first assumed that A is included in an open set K of finite outer measure and M is constant. The desired sequence (U_n) is defined in a manner similar to that used in the Banach proof of the Vitali theorem (I. IV. 2.2). $T_n = \cup \{U \in \mathscr{U} : \mu U < 2 \mu U_n, U \cdot U_n \neq \Phi\}$; n_0 denotes a positive integer such that $2 \mu U_n < 1$ whenever $n \geqslant n_0$; $C_n = \cup \{U^{-1} U : U \in \mathscr{U}, \mu U < (1/n)\}$. There exists P such that

$n\bar{\mu}(C_n C_n) < P$ for each n. If N denotes the integral part of $(1/2\mu U)$, then $\bar{\mu} T_n \leqslant \bar{\mu}(C_N C_N) < (P/N) \leqslant 4P\mu U_n$ whenever $n \geqslant n_0$.

§ 4. The existence of regular covers is asserted whenever G satisfies at some point a "local equi-Lipschitz condition" and X is locally euclidean.

The proof of the main theorem can be shortened using Morse's pattern (I. IV.) to prove the strong Vitali property.

It would be interesting to know if Boclé's results on global derivation in locally compact topological groups (see Complement 5⁰) can be extended to the preceding homogenous spaces.

5°. Global derivatives in locally compact topological groups

(J. Boclé, Sur la théorie ergodique. Ann. Inst. Fourier (Grenoble) 10, 1–45 (1960)).

The terminology is that of P. Halmos, Measure Theory. I denotes a locally compact topological group, e its unit element, and r, s, t are elements of I. If A is a subset of I, then A^{-1} represents the set of those elements a^{-1}, where $a \in A$. A is said to be *symmetric* iff $A = A^{-1}$. We use the notation sA to denote the set of elements of the form st, where $t \in A$; As is defined analogously, $\mathscr{C}$ denotes the class of compact subsets of I; $\mathscr{B}$ is the Boolean σ-ring generated by $\mathscr{C}$. The elements of $\mathscr{B}$ are the Borel subsets of I. A Borel measure is a measure defined on $\mathscr{B}$ and finite on each compact set. A Radon measure is a regular Borel measure. I admits a left Haar measure λ on $\mathscr{B}$; that is, a measure positive for each open subset in $\mathscr{B}$, and invariant under left translation. Thus, for any $A \in \mathscr{B}$ and $r \in I$, $rA \in \mathscr{B}$, and $\lambda(rA) = \lambda(A)$. λ is defined to within a positive multiplier.

Theorem 1. *Let φ be a (positive) Radon measure that is absolutely continuous with respect to λ; let f be the λ-integrand of φ.*

If the open and bounded Borel set U tends to e, then the function φ'_U defined by $\varphi'_U(t) = \varphi(Ut)/\lambda(Ut)$ converges in the mean to f on each bounded Borel set.

Theorem 2. *Let ρ be a Radon measure with Lebesgue decomposition $\rho = \varphi + \psi$, where φ is absolutely continuous with respect to λ and ψ is a singular function with respect to λ; f is the λ-integrand of φ.*

If the open, bounded, symmetric Borel subset W tends to e, then the function ρ'_W defined by $\rho'_W(t) = \rho(Wt)/\lambda(Wt)$ converges in measure to f with respect to λ on each bounded set.

The cases where φ and ρ are signed measures are reduced to the non-negative ones by the Jordan decomposition.

6°. Submartingales with decreasing stochastic bases

(K. KRICKEBERG, Absteigende Semi-martingale mit filtrierendem Parameterbereich, Abh. Math. Sem. Universität, Hamburg 24, 109–125 (1960)).

The setting is the measure space $(E,\mathscr{B},\mu)$, together with a decreasing stochastic basis $(\mathscr{B}_\tau)$; i.e., a filtering family (Moore-Smith sequence) of Boolean σ-algebras $\mathscr{B}_\tau$ of $\mathscr{B}$ with E as unit and such that $\mathscr{B}_\tau \subset \mathscr{B}_\rho$ whenever $\rho \ll \tau$. We denote by $\mathscr{B}_{\omega'}$ the intersection of the families $\mathscr{B}_\tau$, $\tau \in \Theta$, which is again a Boolean σ-algebra. We assume that $\mu(E) = 1$, although the results can easily be extended to a measure that is σ-finite over $\mathscr{B}_{\omega'}$. A stochastic process (f_τ) relative to $(\mathscr{B}_\tau)$ defines a submartingale or martingale, respectively, when the following are true: (1) $f_\tau, \tau \in \Theta$, is $\mathscr{B}_\tau$-measurable and integrable; (2) *if* $\rho \ll \tau$ and $A \in \mathscr{B}_\tau$, then $\int_A f_\tau \, d\mu \leqslant \int_A f_\rho \, d\mu$, with equality in the case of a martingale. If $0 \leqslant p \leqslant \infty$, then (f_τ) is said to be *L_p-bounded* when the set of numbers $\|f_\tau\|_p$, $\tau \in \Theta$, is bounded. The q-overlap of a subsequence $(L_1, L_2, \ldots, L_n)$ of $\mathscr{B}$-members is defined in II. III. 1.2, the Vitali property $(\mathbf{V}')$ in II. III. 5.2, and $(\mathbf{V}_q)$ in II. III. 2.

The main theorems. (1a) *Each submartingale for which* $\int_E f_\iota \, d\mu > -\infty$ *for some* $\iota \in \Theta$ *converges stochastically to a* $\mathscr{B}_{\omega'}$*-measurable and integrable function* $f_{\omega'}$ *such that* $\int_E f_{\omega'} \, d\mu < +\infty$.

(1b) *If* $\lim_\tau \int_E f_\tau \, d\mu > -\infty$, *then* (f_τ) *converges strongly in* L_1.

(2) *Let* (f_τ) *denote a submartingale in* L_p, $0 \leqslant p < \infty$. *If there exists a martingale* (g_τ) *in* L_p *such that* $g_\tau \leqslant f_\tau$ *holds for each* $\tau \in \Theta$, *then* (f_τ) *converges to* $f_{\omega'}$ *strongly in* L_p.

(3) *If* $(\mathscr{B}_\tau)$ *satisfies condition* $(\mathbf{V}')$, *then each submartingale* (f_τ) *such that* $\int_E f_\iota \, d\mu < +\infty$ *for some* $\iota \in \Theta$ *converges with respect to the order.*

(4) *If* $(\mathscr{B}_\tau)$ *satisfies condition* $(\mathbf{V}_q)$ *and* $q^{-1} + p^{-1} = 1$, *then every martingale is order convergent. If* $(\mathscr{B}_\tau)$ *satisfies the condition* $(\mathbf{V}_{+\infty})$ *(strong Vitali property), then every martingale is order convergent.*

Sketches of the proofs. The proofs of (1a) and (3) are quite similar to those of their counterparts Ths. II. II. 4.7, and II. III. 5.4 in the increasing case. In the indirect proof, because the separating element belongs to each $\mathscr{B}_\tau$, $\tau \in \Theta$, we do not need an approximating element C from $\mathscr{B}_\rho, \rho \in \Theta$. (1b) and (3) are inferred from (1a) by means of propositions on terminal uniform integrability (II. II. 3.1) of (f_τ). An essential tool in the proof of (4) is the Hölder inequality

$$\sum_{i=1}^{r} \int_{L_i} f \, d\mu - \int_L f \, d\mu \leqslant \omega_q(L) \cdot \|f\|_p$$

where $\mathscr{L} = (L_1, L_2, \ldots, L_r)$ and $L = \bigcup \mathscr{L}$. The final section contains examples of decreasing stochastic bases satisfying (**V**′), and an example that does not possess the Vitali property ($\mathbf{V}_1$).

7°. Vector-valued martingales and derivation

(M. MÉTIVIER, Martingales à valeurs vectorielles. Applications à la dérivation, Rennes, 1966.)

$(\Omega, \mathscr{F}, \mu)$ denotes a measure space with $\mu \geqslant 0$, bounded and complete, I an index set which is ordered by $\prec$, filtering to the right. A martingale basis is an increasing family $(\mathscr{F}_\alpha)_{\alpha\in I}$ of Boolean σ-subalgebras of $\mathscr{F}$. The σ-Boolean extension of the Boolean algebra $\bigcup\limits_{\alpha\in I} \mathscr{F}_\alpha$ is denoted by $\mathscr{F}_\infty$. **V** denotes a locally convex space in duality with **V**′ (N. BOURBAKI, Elements de Mathématique. Espaces Vectoriels Topologiques. Ch. IV).

Weak **V**-*valued martingale.* This is a family $(\mathscr{F}_\alpha, f_\alpha)$ satisfying: $(M1)$ For each $\alpha \in I$, f_α is defined on Ω, takes its values in **V**, and is scalarly $\mathscr{F}_\alpha$-measurable. $(M2)$ For each α and each $F \in \mathscr{F}_\alpha$, the weak integral $\int\limits_F f_\alpha d\mu$ is an element of **V**. $(M3)$ For any α and β with $\alpha < \beta$ and each $F \in \mathscr{F}_\alpha$, $\int\limits_F f_\alpha d\mu = \int\limits_F f_\beta d\mu$.

Strong **V**-*valued martingale.* **V** is a Banach space. A **V**-valued martingale is a strong martingale if it satisfies $(M'1)$, $(M'2)$, and $(M'3)$, obtained by replacing scalar measurability by strong measurability, and weak integrability by strong integrability (i.e., integrability in Bochner's sense), in $(M1)$, $(M2)$, and $(M3)$.

The following problems arise under the assumption $(\mathscr{F}_\alpha, f_\alpha)$ is a weak (resp. strong) martingale.

Problem 1. If I has no largest element, we add one, namely ∞, to I, the enlarged set being $\hat{I}$. Does there exist an application f_∞ from Ω into **V** such that $(\mathscr{F}_\alpha, f_\alpha)_{\alpha\in\hat{I}}$ is a weak (resp. strong) martingale?

Problem 2. If a solution f_∞ to Problem 1 exists, does the family $(f_\alpha)_{\alpha\in I}$ converge to f_∞ according to a suitable mode of convergence?

Problem 3. Independently of the existence of a solution to Problem 1, does the family (f_α) converge for a suitable choice of convergence to an application f of Ω into **V**?

Convergence theorems are given. The second half of the paper is devoted to derivability.

Here we can only mention the following solution to Problem 2 given by A. and C. IONESCU TULCEA and J. NEVEU: If a solution f_∞ exists for Problem 1 in the strong case, then f_α converges to f_∞ in $\mathrm{L}^1(\Omega, \mathscr{F}, \mathbf{V}, \mu)$. If $I = \mathbf{N}$, then also strong $\lim\limits_{n\to\infty} f_n(\omega) = f_\infty(\omega)$ μ-almost everywhere.

8°. A theorem of Ward for cell functions. A martingale convergence theorem of Ward's type

(1) (D. Rutovitz and C. Y. Pauc, Theory of Ward for cell functions, Ann. Mat. Pura App. Ser. IV, 47, 1–58 (1959)). In Part I of the paper (by D. Rutovitz), Ward's theory of the derivation of (finitely) additive euclidean interval functions F as expounded in S. Saks' Theory of the Integral (Warsaw, 1937), is developed in an abstract setting. A measure space $(\mu, \mathscr{M}, R)$ replaces the euclidean space, a family $\mathscr{I}$ of $\mathscr{M}$-sets of finite positive measure called *cells*, is postulated and endowed with some properties of the family of euclidean intervals sufficient to ensure the validity of Ward's theorems. Compared with the Burkill theory for cells expounded in Part II of the present report, the novel features of the present setting are the *bordering property* and a *regularity property* with respect to an enumerable monotone graduated subnet $\mathscr{A}$ of $\mathscr{I}$, whose cells take over the part of euclidean cubes. The derivation theorems obtained assert the existence almost everywhere of derivates corresponding to the euclidean strong and ordinary derivates. The weak Vitali property is assumed.

Part II of the paper (by C. Y. Pauc) is concerned only with the strong derivability of F on the set of points at which both strong (extreme) derivates are finite. The weak Vitali axiom is discarded. Instead of a pointwise derivate DF, a so-called *total integrand* is defined by *fusion of Radon-Nikodym integrands*, linked with F by an integral relation. The total integrand coincides almost everywhere on E with DF when the weak Vitali property holds.

(2) (Y. S. Chow, A martingale convergence theorem of Ward's type, Illinois J. Math. 9, 569–576 (1965)). In this paper, by following Doob's approach (Notes on martingale theory, Proc. Fourth Berkeley Symp. Math. Stat. and Prob. 95–102 (1960)), the author obtains a convergence theorem (Theorem I) which includes some martingale convergence theorems and extends a theorem of Rutovitz to a non-atomic basis. Theorem IV puts the Ward's theorem, cited above, into a martingale setting.

9°. Derivation of measures

(D. Kölzow, Lecture Notes in Mathematics, no. 65 (1968), Springer-Verlag).

Setting. $\underline{M} = E, \mathfrak{m}, \mu)$ denotes a measure space, $\mathfrak{m}$ coincides with its Carathéodory extension, $\mathfrak{n}_1$ is the family of the local φ-nullsets. A measure ψ defined on $\mathfrak{m}$ is called φ-continuous iff any φ-nullset is a ψ-nullset. The theorem of Radon-Nikodym holds for $\underline{M}$ iff for each φ-continuous function ψ there exists a non-negative and measurable function

$f \in \bar{\mathbf{R}}^E$ (the derivative of ψ) such that for any φ-summable set S, fc_s is φ-summable iff S is ψ-summable, and thus $\psi(S) = \int_S f\, d\varphi$. An M-partition $\mathfrak{z}$ is a family of pairwise disjoint φ-summable sets of positive φ-measure such that to each φ-summable set with $\varphi(S) > 0$ there exists $Z \in \mathfrak{z}$ for which $\varphi(S \cap Z) > 0$. A linear lifting of $\underline{M}$ according to VON NEUMANN (Algebraische Repräsentanten der Funktionen „bis auf eine Menge von Masse Null", J. Crelle 165, 109–115 (1931)) is a map L of $\mathfrak{m}$ into itself such that $M \Delta L(M) \in \mathfrak{n}_1$ (symmetric difference), for $M \in \mathfrak{m}$, $L(M_1) \subset L(M_2)$ when $M_1 - M_2 \in \mathfrak{n}_1$, $L(M_1 \cap M_2) = L(M_1) \cap L(M_2)$, and $L(M_1 \cup M_2) = L(M_1) \cup L(M_2)$ for M_1 and M_2 belonging to $\mathfrak{m}$.

Selected results. The existence of an $\underline{M}$-partition implies the validity of the Radon-Nikodym theorem. The following propositions are equivalent: (1) $\underline{M}$ admits an $\underline{M}$-partition. (2) $\underline{M}$ admits a linear lifting. (3) There exists on $\underline{M}$ a strong Vitali basis of derivation. (4) There exists on $\underline{M}$ a weak Vitali basis of derivation.

Index

Absolute continuity 148
Atoms 138
Axiom, de Possel's heredity 6
– (**E**) 7
–, Fréchet's convergence 6

Basis, Busemann-Feller 9
–, D- 8
–, derivation 6
–, Dieudonné 9
–, interval 83
–, metrical 8
–, Morse, generalized 45
–, –, special 44
–, parallelepipedon 104
–, pre- 6
–, strip 8
–, sub- 9
–, topological 8
–, ultrafilter 9
–, weak 21
Blanket, Morse 8, 110
–, star 114
Bohr's construction 98

Cell basis 138
– functions 139
– – of bounded variation (from above, below) 140
– –, additive, subadditive, and superadditive 140
– –, μ-absolutely continuous 165
– –, μ-singular 165
$\mathscr{C}$-measurable function 144
Complex 138
Components 138
Concepts at finite measure 163
Condition (C) 79
–, weak halo ($W H$) 62
– ($W H$)′ 64
– ($W H$)″ 73, 68
–, weak halo evanescence 60
– (**B**) 140
– ($\mathbf{C}^0$) 171

Condition (**L**) 176
– (**M**) 176
– (**P**) 142
– (**R**) 180
– (**U**) 180
– (**V′**) 170
– ($\mathbf{V}_q$), $1 \leqslant q \leqslant +\infty$ 169
– ($\mathbf{V}_Y$) 168
– ($\mathbf{W}_Y$) 169
– (**Z**) 179
Conditional expectation 149
Constituents 6, 138
Covering, $\mathfrak{B}$-fine 9
–, deficiency of 16
–, ε-, 0- 16
–, excess of 16
–, full $\mathfrak{B}$-fine 10
– in measure 16
–, overlap of 16
–, redundancy of 16
– relation 112
– –, Δ-restrained 112
– –, diametrically restrained 113

Deficiency of σ-additivity 128
Density, mean 59
Derivable (L_X) 178
Derivates in the mean 166
–, lower and upper 6
Derivative 6
Deriving family 181
Domain 9

Essential fine covering 174
– infimum, supremum 173
– union 59
Excess function 168

Fine cell covering 171
– covering 168, 175
Frequency, $\mathscr{E}$- 16
Function, bounded 14
–, disentanglement 41
–, $\mathscr{E}$-excess 16

Function, halo 94
–, minimum halo 97
–, $\mathfrak{L}^q$- 24
–, $\mathcal{M}$- 14
–, $\mu^{(q)}$- 23
–, μ-finite 14

G-pruning principle 12

Halo 41
–, dilation 41
–, η-weak 59
–, η-weak partial 59
–, weak 59
–, weak partial 59
–, Morse 41
–, ψ- 65
Haupt's adaptation property 17
Hive 111
H_p-property 36
Hub 114
– radius 114

Indicatrix 145
Inferior envelope 175
Infimum of a family of additive setfunctions 125
– – – family of martingales 130
Integral, μ- 14
–, μ-finite μ- 14
–, Radon μ- 14
– of a cell function 141
– – – premartingale 131
– – – submartingale 131
– representation 151
Integrand representation 151, 164
Internal radius 114

Martingale 129
–, induced 137, 152
–, in L_X 164
–, μ-absolutely continuous 150
–, μ-singular 150
–, of bounded variation 129
–, purely simply additive 136
–, σ-additive 136
Mean convergence 161
Measure cover 5
– kernel 5
–, $\mathcal{M}$- 14
–, μ-finite 14
–, Radon 14

Measure, signed $\mathcal{M}$- 14
–, signed Radon 14
M-family 15
Morse halo properties 41, 45

Natural extension 152
Nets 110
Nullsequence 60
–, μ- 60

Order convergence 167
– density theorem 169
– derivatives 171
– limits 167
Overflow 16
Overlap 16, 24
– of order Y 168

Partition 126
– fineness 138
Pointwise limits 175
Premartingale 128
–, norm of 129
– of bounded variation (from above, from below) 129
Pretopology; T- and P-concepts 9, 12
Property (L) 56
Purely additive set function 128

Reduced strong Vitali property 27
– Vitali ψ-property 16
Redundancy 16

Saks' counterexample 98
– "rarity" theorem 107
Sequence, contracting 6
–, convergent 6
–, deriving 6
Sets, D-closed, D-open 11
–, family of 5
–, measurable 5
–, μ^*-entangled 5
–, μ_S-measurable 12
Sharp reduced Vitali property 42
– Vitali property 42
Singularity 150
Spread 9
Stochastic basis 129, 167
– convergence 153
– derivates 165
– limits 153, 167
– process 151, 168

Strong Vitali property 27
Submartingale, supermartingale 129
Superior envelope 175
Supremum of a family of additive functions 126

Terminally uniformly absolutely continuous premartingale 152
— — bounded premartingale 129, 152
— — integrable premartingale 152

Uniform σ-additivity 146
Universal lower and upper approximation property 18

Upper μ-approximation property 17

Variations (positive, negative, total) of a cell function 140
— (—, —, —) of a martingale 131
Vitali property of order Y 168
— —, strong 169
— —, weak 169
— — for Radon integrals, $\mathscr{M}$-measures 23
— ψ-property 16

Weak halo properties $(W\,H)$, $(W\,H)'$, $(W\,H)''$ 62, 64, 68, 73

Ergebnisse der Mathematik und ihrer Grenzgebiete

---

1. Bachmann: Transfinite Zahlen. DM 38,—; US $ 10.50
2. Miranda: Partial Differential Equations of Elliptic Type. DM 58,—; US $ 16.00
4. Samuel: Méthodes d'Algèbre Abstraite en Géométrie Algébrique. DM 26,—; US $ 7.20
5. Dieudonné: La Géométrie des Groupes Classiques. DM 38,—; US $ 10.50
6. Roth: Algebraic Threefolds with Special Regard to Problems of Rationality. DM 19,80; US $ 5.50
7. Ostmann: Additive Zahlentheorie. 1. Teil: Allgemeine Untersuchungen. DM 38,—; US $ 10.50
8. Wittich: Neuere Untersuchungen über eindeutige analytische Funktionen. DM 28,—; US $ 7.70
11. Ostmann: Additive Zahlentheorie. 2. Teil: Spezielle Zahlenmengen. DM 28,—; US $ 7.70
14. Coxeter/Moser: Generators and Relations for Discrete Groups. DM 32,—; US $ 8.80
16. Cesari: Asymptotic Behavior and Stability Problems in Ordinary Differential Equations. DM 36,—; US $ 9.90
17. Severi: Il teorema di Riemann-Roch per curve, superficie e varietà questioni collegate. DM 23,60; US $ 6.50
18. Jenkins: Univalent Functions and Conformal Mapping. DM 34,—; US $ 9.40
19. Boas/Buck: Polynomial Expansions of Analytic Functions. DM 16,—; US $ 4.40
20. Bruck: A Survey of Binary Systems. DM 36,—; US $ 9.90
21. Day: Normed Linear Spaces. DM 17,80; US $ 4.90
25. Sikorski: Boolean Algebras. DM 38,—; US $ 9.50
26. Künzi: Quasikonforme Abbildungen. DM 39,—; US $ 10.80
27. Schatten: Norm Ideals of Completely Continuous Operators. DM 26,—; US $ 7.20
28. Noshiro: Cluster Sets. DM 36,—; US $ 9.90
29. Jacobs: Neuere Methoden und Ergebnisse der Ergodentheorie. DM 49,80; US $ 13.70
30. Beckenbach/Bellman: Inequalities. DM 30,—; US $ 8.30
31. Wolfowitz: Coding Theorems of Information Theory. DM 27,—; US $ 7.50
32. Constantinescu/Cornea: Ideale Ränder Riemannscher Flächen. DM 68,—; US $ 18.70
33. Conner/Floyd: Differentiable Periodic Maps. DM 26,—; US $ 7.20
34. Mumford: Geometric Invariant Theory. DM 22,—; US $ 6.10
35. Gabriel/Zisman: Calculus of Fractions and Homotopy Theory. DM 38,—; US $ 9.50
36. Putnam: Commutation Properties of Hilbert Space Operators and Related Topics. DM 28,—; US $ 7.70
37. Neumann: Varieties of Groups. DM 46,—; US $ 12.60
38. Boas: Integrability Theorems for Trigonometric Transforms. DM 18,—; US $ 5.00
39. Sz.-Nagy: Spektraldarstellung linearer Transformationen des Hilbertschen Raumes. DM 18,—; US $ 5.00
40. Seligman: Modular Lie Algebras. DM 39,—; US $ 9.75
41. Deuring: Algebren. DM 24,—; US $ 6.60
42. Schütte: Vollständige Systeme modaler und intuitionistischer Logik. DM 24,—; US $ 6.60
43. Smullyan: First-Order Logic. DM 36,—; US $ 9.90
44. Dembowski: Finite Geometries. DM 68,—; US $ 17.00
45. Linnik: Ergodic Properties of Algebraic Fields. DM 44,—; US $ 12.10
46. Krull: Idealtheorie. DM 28,—; US $ 7.70
47. Nachbin: Topology on Spaces of Holomorphic Mappings. DM 18,—; US $ 5.00
48. Ionescu Tulcea/Ionescu Tulcea: Topics in the Theory of Lifting. DM 36,—; US $ 9.90
49. Hayes/Pauc: Derivation and Martingales. DM 48,—; US $ 13.20
50. Kahane: Series de Fourier Absolument Convergentes. DM 44,—; US $ 12.10
51. Behnke/Thullen: Theorie der Funktionen mehrerer komplexer Veränderlichen. DM 48,–; US $ 13.20
52. Wilf: Finite Sections of the Classical Inequalities. DM 28,–; US $ 7.70